The Touchstone of Life

The Touchstone of Life

**Molecular Information,
Cell Communication,
and the Foundations of Life**

Werner R. Loewenstein

New York   Oxford   Oxford University Press   1999

Oxford University Press

Oxford   New York

Athens   Auckland   Bangkok   Bogotá   Buenos Aires   Calcutta
Cape Town   Chennai   Dar es Salaam   Delhi   Florence   Hong Kong   Istanbul
Karachi   Kuala Lumpur   Madrid   Melbourne   Mexico City   Mumbaí
Nairobi   Paris   São Paulo   Singapore   Taipei   Tokyo   Toronto   Warsaw

and associated companies in
Berlin   Ibadan

Cataloging-in-Publication Data
Loewenstein, Werner R.
The touchstone of life : molecular information,
cell communication, and the foundations
of life / Werner R. Loewenstein.
p.   cm.   Includes bibliographical references and index.
ISBN 0–19–511828–6
1. Information theory in biology.
2. Cell interaction.
3. Cellular signal transduction.
I. Title.
QH507.L64   1998
571.6—dc21   97–43408

1 3 5 7 9 8 6 4 2

Printed in the United States of America
on acid-free paper.

To Birgit

Contents

Acknowledgments

In the entrance hall of an old Cape Cod house on the bank across the pond where I live, one is greeted with the saying on a tapestry, "Be blessed with friends." I have been, and I am grateful for that. It was through the stimulating and comfortable discussions with friends that the ideas about biological information presented in this book took shape and that I plucked up courage to bell the cat.

But I owe a special debt to my colleague Wolfgang Nonner for help with how to present the mathematical kinship between information and entropy. He also gave me valuable advice concerning ion channels, which was crucial in the formulation of the concept of cross-correlator Maxwell demons, and on more than one occasion he put the damper on me when I tried flying untried balloons.

William Jencks introduced me to his work on the thermodynamics of cyclic coupled vectorial processes and to the peculiar energy economics of macromolecular binding. This was an eye-opener for me. Indeed, it was during a delightful discussion with him on a flight across the Atlantic (a chance meeting) that I first saw how advantageous it is to treat such binding in terms of information theory rather than in the traditional way of Gibbsian thermodynamics.

Peter Läuger shared with me his deep insights into the transitional states of membrane proteins. We bandied about many an idea concerning the workings of such proteins and their information transfer while sailing the Cape Cod and Florida waters or hiking in the Swiss Alps. I was lucky to draw from the depth and breadth of his knowledge over more than a decade, before his life was so sadly cut short.

I am thankful to my colleague Birgit Rose for wading through many chapters of an early draft, and for reading me the riot act whenever she found a show-off passage or a problem I had glibly skated over.

I thank Kirk Jensen, Executive Editor of Oxford University Press, for sound counsel. He urged me as early as 1990 (or was it already in the 1980s?) to write a book, before I was persuaded, and stayed with me through all these years. And when I finally sent him my first draft, he offered trenchant comments, and he would not stop bugging me until I had weeded it and polished it up. That some parts still are not easy to read is my fault, not his.

I have been fortunate to have James Levine as my agent and Helen Mules of Oxford University Press as my Production Editor. All in all, Oxford University Press and I turned out to be a good match. Oxford's dictionary is the only one I know of that gives the nod to the splitting of infinitives, and I am thankful for that license. This got all sorts of pen-wielding gadflies off my back—I am as adept at splitting infinitives as the leading characters in my book are at splitting covalent bonds.

Ronald Bagdadi, Robert Golder, Miguel Luciano, Paul Oberlander, and Erik Petersen turned my sketches into professional figures. I thank them for drawing and cheerfully re-drawing so many of them. My assistants Linda Howard and Cecilia Schmid labored to decipher my handwriting to produce page after page, version after version, and did much of the library search. My special thanks to both of them.

I am indebted to many individuals and institutions in the United States and abroad for organizing and hosting the lectures I gave over the years. My hosts were too many to be named here, but the opportunities they gave me for airing my views were invaluable. One speaks to learn, and I learned much from the reactions of so wide an audience.

Perhaps, also an apology should be added here. I chose to put the references in a section at the end of the book because long lists of authors in the text make one stumble. To save space, I cited books or reviews wherever I could, though I strived to quote original papers that were seminal. If, despite my best efforts, I missed some essential references, I apologize in advance for that.

It was a long and exciting adventure to see this project through—at times, it seemed an interminable one. It was finally a relief not to have to hear my family and friends asking me, "When will the book be finished?" And I am grateful to my wife Birgit for bearing with my constant obsession with its writing . . . and for not putting me up for adoption.

Introduction

Mark Twain once remarked that when he penned his *Pudd'nhead Wilson* "he was unable to follow his original plot because several of the characters started to develop wills of their own and did things quite different from what he had planned for them." I found myself here in the same fix. No sooner had I settled on a plot than my main character, Information, acted up and began to commandeer *all* roles.

I originally intended to write a book about intercellular communication — about how the cells in an organism exchange information among each other — a question I had worked on for the past thirty years. But as I got to the heart of it, which required me to track the spoor of intercellular information to its source, a picture materialized seemingly out of the blue: a continuous intra- and intercellular communication network where, with DNA at the core, molecular information flowed in gracefully interlaced circles. That apparition had an allure I could not resist, and so this became a book about information.

I was twice fortunate. First, the spectacular advances in molecular biology of the past three decades had paved the way for my information tracking; they had laid bare the cells' macromolecular information core and were just beginning to break the seal of an information shell where molecular carriers, large and small, swarm all over the organismic map. Second, information theory, a thriving branch of mathematics and physics, allowed me to view this information flow in a new light.

I shall argue that this information flow, not energy per se, is the prime mover of life — that molecular information flowing in circles brings forth the organization we call "organism" and maintains it against the ever-present disorganizing pressures in the physics universe. So viewed, the information circle becomes the unit of life.

There are other self-organizations in nature, of course. The world of inorganic crystals is full of beautiful examples of self-organization; and research in areas as diverse as atmospheric and water turbulence, and chemical catalysis has brought to light that quite ordinary matter, when raised enough above thermodynamic equilibrium, has a surprising propensity toward self-organization. But all this pales against the self-organizing power of living beings, and we will see why. Through the strong loupe of information theory, we will be able to watch how such beings do what nonliving systems cannot do: extract information from their surrounds, store it in a stable molecular form, and eventually parcel it out for their creative endeavors. It is this three-ply capability that allows organisms to defy the relentless disordering pressures in and around them—the forces of formless darkness—to use Yeats's richly evocative words.

This book is written for the scientist *and* for the reader with an interest in science—no specialized knowledge of biology or physics is assumed in advance. With the general reader in mind, I have dispensed with mathematical apparatus. Thus, there will necessarily be some loss of precision, but the main points, I hope, will get across. Richard Feynman, a physicist renowned for his probings of the quantum world, showed that even the most abstract, and often counterintuitive, concepts in his field could be explained in simple terms. For him the freshman lecture, not precision, was the yardstick of mental grasp—"If we can't reduce something to the freshman level," he said, "that means we really don't understand it."

Well, eventually, I had to make a small compromise in order to tidy up the concept of information itself. *Information* means something different in science from what it does in everyday language—something deeper. It connotes a cosmic principle of organization and order, and it provides an exact measure of that. So I use three equations to help define these things. All three are simple and meet the Feynman test. But the reader who wishes to skip them will not lose the thread of the book; the concepts they embody are explained in plain language along the way.

The book falls into four parts. The first part introduces the entity Information, traces its cosmic origin, and follows its vagaries over some 15 billion years as it pulls a molecular organization out of the erratic quantum world and engenders the primal organizations of matter in the universe. We then pick up its trail on the cooling Earth and fix on a fateful eddy that occurred there some four billion years ago, a weensy whirl in the cosmic stream where information, instead of flowing straight as everywhere else, flows in circles. This was the pregnant twist, I will argue, that gave birth to life. We will view with the mind's eye the primeval molecules that may have made this happen and get acquainted with their modern-day successors, the RNAs, DNA, and proteins. Through our information loupe we will see how these heavyweights hand on immense amounts of information to one another, laying circle upon information circle; and we are let in on their trick—the most dazzling of magicians' acts—whereby they conjure up form from formless darkness.

The second part takes up the full-grown weft of circles, the *intra*cellular communication network, as found in every living being today. It charts the network—a two-tier organization which has no equal in our technological world—and then zeroes in on what makes cells tick: the intracellular information flow. It lets us into the secret of the cellular codes and ciphers and tells us what is behind the amazing fidelity of the cellular information transmissions. And with this knowledge in the bag, we will take a journey on the cellular information stream and see—though always at a safe distance from the dizzying circles—how information from the DNA is meted out in three dimensions and whirls the molecular wheels of life.

The third part deals with the *inter*cellular communication network, the web that ties all cells of an organism together, coordinating and subordinating their activities. This is the information flow that makes an organism out of the mass of cells. We will meet the leading characters in this age-old play and eavesdrop on their communications. They are kin of the characters ruling the intracellular stage—the whole intercellular network, I will show, is but a far-flung extension of the intracellular one and uses the same transmission and encoding strategies to overcome the background noise. An information theorem will give us the rules that govern the encodings in both networks, from the paltry steroid hormone encryption to the mighty genetic code. And as we explore the subtler registers of the theorem, we perceive the full meaning of molecular coding and its central role in the development of life.

This part winds up with a young offshoot of the intercellular network, the neuronal web, which owes its existence to a few phrases in the genetic information script. Here, an information aspect sui generis comes to the fore, consciousness, and the argument turns from the molecular to the quantum information world.

Through all three parts runs the question of how the biomolecular information system evolved. I shall argue that it advanced by an accretion of information circles along a path of least information cost. And this, the *principle of information economy* of self-developing systems, I submit, is the guiding principle of biological evolution.

The fourth part is a short philosophical foray where we see the principle through to its heuristic conclusion, an Occam's Razor for biology, and ponder over the prospects of a unifying theory of that strange and bewilderingly time-dependent biological world.

The book ends with a brief history of Information, tracing the concept to its physics and surprisingly old biology roots.

W. R. L.
Woods Hole
February 1998

Part One

Information and the Origins of Life

Information and Organisms

Maxwell's Demon

No one can escape a sense of wonder when looking at a living being from within. From the humblest unicellular organism to man, from the smallest cell organelle to the stupendous human brain, living things present us with example after example of highly ordered matter, precisely shaped and organized to perform coordinated functions. Where does this order spring from? How does a living organism manage to do what nonliving matter cannot do—bring forth and maintain order against the ever-present disordering pressures in the universe?

The order prevailing in living beings strikes the eye at every level of their organization, even at the level of their molecular building blocks. These macro-molecules are made of the same kind of atoms the molecules of nonliving things are made of and are held together by the same kind of forces, but they are more complex. Take a hemoglobin molecule, for example, the major protein of our blood, and set it against a sodium chloride molecule, the common salt from the inanimate mineral world. Both are orderly structures. The salt is an array of sodium and chloride atoms—always the same two atoms in an endless crystal repeat. The hemoglobin is an array of 574 amino acids constituting four chains that twist about each other, forming an intricate coil; the overall structure is essentially the same in zillions of hemoglobin molecules, and there normally is not an amino acid out of place in the linear sequence of the chains. Such a structure clearly embodies a higher level of organization than the monotonous salt repeat. It is also less likely to form spontaneously. While the salt crystal readily assembles itself from its atoms in solution—we can watch this day in and day out when seawater evaporates—it is highly improbable that a hemoglobin molecule would ever do so spontaneously.

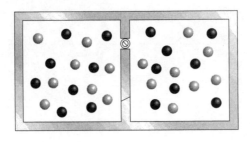

"*. . . If we conceive of a being whose faculties are so sharpened that he can follow every molecule in its course, such a being, whose attributes are still as essentially finite as our own, would be able to do what is at present impossible to us. . . . Let us suppose that a vessel is divided into two portions A and B by a division in which there is a small hole, and that a being who can see the individual molecules opens and closes this hole, so as to allow only the swifter molecules to pass from A to B, and only the slower ones to pass from B to A. He will, thus, without expenditure of work raise the temperature of B and lower that of A . . .*" James Clerk Maxwell, "Theory of Heat" (1871) p. 309.

Figure 1.1   A free rendering of Maxwell's paradox, representing the system at equilibrium (*top*) and after intervention of the demon (*bottom*). The dark spheres represent the swift (hot) molecules and the lighter spheres, the slow ones.

Is there some special organizing principle in living beings, some entity that manages to create order in the liquid pandemonium of atoms and molecules? About 150 years ago, the physicist Maxwell, while musing about the laws that rule the tumultuous molecular world, came up with an intriguing idea. He conceived an experiment, a thought experiment, in which a little demon who picks out molecules seemed to escape what was thought to be the supreme law, the second law of thermodynamics. This demon comes as close as anything to the organizing biological principle we are angling for. So, let's take a look at that gedankenexperiment.

Maxwell chose a system made of just two sorts of molecules, a mix of hot and cold gas molecules moving randomly in two chambers, *A* and *B*, connected by a friction-free trap door (Fig. *1.1*). Suppose, said Maxwell, we have a demon at that door, who can see individual molecules and distinguish a hot molecule from a cold one. When he sees a hot one approaching from *B* to *A*, he opens the door and lets it through. As he repeats that operation over and over, we would wind up with an organization in which the two kinds of molecules are neatly sorted out into a hot and a cold compartment—an organization which would cost nothing because the demon only selects molecules but does no work.

One literally seemed to get something for nothing here. The assumption of a friction-free operation is valid in a thought experiment, because there is no intrinsic lower limit to performance. So, to all appearances, this cunning organizing entity of Maxwell's seemed to flout the Second Law. And adding insult to injury, it would be able to drive a machine without doing work—the old dream of the perpetual motion machine come true.

This paradox kept physicists on tenterhooks for half a century. Two generations of scientists tried their hands at solving it, though to no avail. The light at the end of the tunnel finally came when Leo Szilard showed that the demon's stunt really isn't free of charge. Though he spends no energy, he puts up a precious commodity called *information*. The unit of this commodity is the *bit*, and each time the demon chooses between a hot and a cold molecule, he shells out one bit of information for this cognitive act, precisely balancing the thermodynamics accounts.

Thus, the hocus-pocus here has an exact information price. As has happened often in physics, the encounter with a deep paradox led to a higher level of understanding. Szilard's introduction of the concept of information as a counterweight to thermodynamics disorder (entropy), opened a whole new perspective of the organization of energy and matter. It marked the beginning of information theory. This theory already has shown its pizzazz in a number of fields in physics and engineering; the present-day communication and computer revolution provides ample proof of that. But perhaps its greatest power lies in biology, for the organizing entities in living beings—the proteins and certain RNAs—are but Maxwell demons. They are the most cunning ones, as we shall see; their abilities of juggling information are so superlative that it boggles the mind.

In this and some of the following chapters, we will try to get to the bottom of their tricks. We will see how they finagle information from their molecular neighbors—how they extract it, transmute it, and use it to organize their own world. But first of all we need a clear picture of what Information is. I capitalize the word here to signal that it has a somewhat different meaning in science than

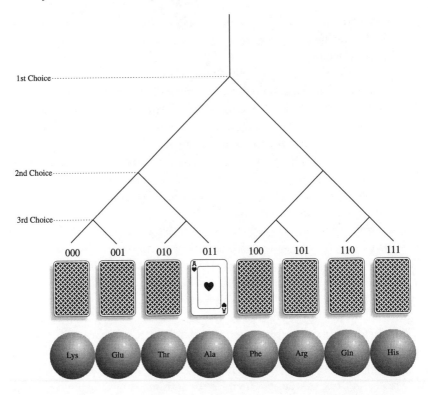

Figure 1.2 To arrive at state 011 in a system of equally probable states, one needs to make three correct consecutive choices. Thus, the (nonredundant) message that would guide us along the bifurcating paths to position 011, the message identifying the Ace of Hearts in a deck of eight cards in that position, and the message specifying alanine (Ala) in the chain of eight amino-acid residues below, each contain 3 bits of information (assuming equiprobability of asociations for the residues, to make it simple).

in daily usage. This meaning is defined by two formulas, the only occasion we resort to mathematics in this book. The concepts embodied in these formulas are straightforward and will be explained as we go along.

What Is Information?

Information, in its connotation in physics, is a measure of order—a universal measure applicable to any structure, any system. It quantifies the instructions that are needed to produce a certain organization. This sense of the word is not too far from the one it once had in old Latin. *Informare* meant to "form," to "shape," to "organize." There are several ways one could quantify here, but a specially convenient one is in terms of binary choices. So, in the case of Maxwell's demon we would ask, how many "yes" and "no" choices he makes to achieve a particular molecular organization, and we would get the answer here directly in *bits*. In general, then, we compute the information inherent in any given arrangement of matter (or energy) from the number of choices we must make to arrive at that particular arrangement among all equally possible ones. We see intuitively that the more arrangements are possible, the more information is needed to get to that particular arrangement.

Consider a simple linear array, like the deck of eight cards or the chain of eight amino acids in Figure *1.2*. How much information would be needed to produce such an array with a perfectly shuffled set? With the cards, the problem is like finding out in a binary guessing game their order when they are turned face down. Let's start, for example, with the Ace of Hearts. We are allowed in such a binary game to ask someone who knows where the Ace is a series of questions that can be answered "yes" or "no"—a game where the question "Is it here?" is repeated as often as needed. Guessing randomly would eventually get us there; there are eight possibilities for drawing a first card. But, on the average, we would do better if we asked the question for successive subdivisions of the deck—first for subsets of four cards, then of two, and finally of one. This way, we would hit upon the Ace in three steps, regardless where it happens to be.

Thus, 3 is the minimum number of correct binary choices or, by our definition above, the amount of information needed to locate a card in this particular arrangement. In effect, what we have been doing here is taking the binary logarithm of the number of possibilities (N); $\log_2 8 = 3$. In other words, the information required to determine the location in a deck of 8 cards is 3 bits.[1] If instead of 8 cards our deck contained 16, the information would be 4 bits; or if it contained 32 cards, it would be 5 bits; and so on.

In general, for any number of possibilities (N), the information (I) for specifying a member in such a linear array, thus, is given by

$$I = \log_2 \frac{1}{N} \,.$$

I here has a negative sign for any N larger than 1, denoting that it is information that has to be acquired in order to make the correct choice.

Leaving now this simple example, we consider the case where the N choices are subdivided into subsets (i) of uniform size (n_i), like the four suits in a complete deck of cards. Then, the information needed to specify the membership of a subset again is given by the foregoing equation, where N now is replaced by N/n_i, the number of subsets

$$I = \log_2 \frac{n_i}{N} \,.$$

1. The reader used to computer language will arrive at this readily by designating the states binary fashion, 000, 001, 010, 011, 100, 101, 110, 111. To specify any one state, one needs three binary digits.

If we now identify n_i / N with the proportion of the subsets (p_i), we have

$$I_i = \log_2 p_i.$$

As we take this one step further, to the general case where the subsets are nonuniform in size, that information will no longer be the same for all subsets. But we can specify a mean information which is given by

$$I = \sum_i p_i \log_2 p_i. \qquad (1)$$

This is the equation that Claude Shannon set forth in a theorem in the 1940s, a classic in information theory.

Thus defined, information is a universal measure that can be applied equally well to a row of cards, a sequence of amino acids, a score of music, an arrangement of flowers, a cluster of cells, or a configuration of stars.

Information is a dimensionless entity. There is, however, a partner entity in physics which has a dimension: *entropy*. The entropy concept pertains to phenomena that have a developmental history, events that unfold and cannot be undone, like the turning of red hot cinders into cold ashes, the crumbling of a mountain into dust, or the decay of an organism. Such events always have one direction, due to the spontaneous dispersal of the total energy, and tell us which way time's arrow is pointing. They belong to the world of atoms and molecules, the world whose statistical effects we experience through our senses and perceive as never turning back in time—our everyday world of which we say "time goes on."

Such an arrow of time is absent *inside* the atoms, the sphere of the elementary particles. In that world entropy has no place; there is complete symmetry of time (at least in ordinary matter)—the constraints prevailing at the level of atomic interactions are lacking, and the time course of the particle interactions can be reversed. Of this world we have no sensory experience and no intuitive feeling. We will concern ourselves with it only in passing in this book.

Entropy, in its deep sense brought to light by Ludwig Boltzmann, is a measure of the dispersal of energy—in a sense, a measure of disorder, just as information is a measure of order. Any thing that is so thoroughly defined that it can be put together only in one or a few ways is perceived by our minds as orderly. Any thing that can be reproduced in thousands or millions of different but entirely equivalent ways is perceived as disorderly. The degree of disorder of a given system can be gauged by the number of equivalent ways it can be constructed, and the entropy of the system is proportional to the logarithm of this number. When the number is 1, the entropy is zero ($\log 1 = 0$). This doesn't happen very often; it is the case of a perfect crystal at absolute zero temperature. At that temperature there is then only one way of assembling the lattice structure with molecules that are all indistinguishable from one another. As the crystal is warmed, its entropy rises above zero. Its molecules vibrate in various ways about their equilibrium positions, and there are then several ways of assembling what is perceived as one and the same structure. In liquids, where we can have atoms of very many different kinds in mixture and in huge amounts, the number of equivalent ways for assembly gets to be astronomical. Enormous changes of entropy, therefore, are necessary to organize the large protein and nucleic acid molecules that are characteristic of living beings. The probabilities for spontaneous assembly of such molecules are extremely low; probability numbers of the order of 10^{-50} are not uncommon. Thus, it is practical to express entropy quantities (like information in equation *1*) logarithmically.

Boltzmann's entropy concept has the same mathematical roots as the information concept: the computing of the probabilities of sorting objects into bins—a set of N into subsets of sizes n_i. By computing how many ways there are to assemble a particular arrangement of matter and energy in a physical system, he arrived at the expression of entropy (S), the statistical mechanical expression of the thermodynamic concept

$$S = -k \sum_i p_i \ln p_i \qquad (2)$$

where k is Boltzmann's constant (3.2983×10^{-24} calories/°C).

Shannon's and Boltzmann's equations are formally similar. S and I have opposite signs, but otherwise differ only by their scaling factors; they convert to one another by the simple formula $S = - (k \ln 2) I$. Thus, an entropy unit equals $-k \ln 2$ bit.

The Information-Entropy Trade-Off

Information thus becomes a concept equivalent to entropy, and any system can be described in terms of one or the other.[2] An increase of entropy implies a

---

2.   In Boltzmann's equation, the probabilities (p_i) refer to internal energy levels, the thermodynamicist's way of fully describing his system. In Shannon's equation, the p_is are not a priori assigned such specific roles; thus, by using the proper p_i, the expression can be applied to any physical system and provide a measure of order. Indeed, only in terms of information (or entropy) does "order" acquire scientific meaning. We have some intuitive notion of orderliness—we can often see at a glance whether a structure is orderly or disorderly—but this doesn't go beyond rather simple architectural or functional periodicities. That intuition breaks down as we deal with macromolecules, like DNA, RNA, or proteins. Considered individually (out of species context), these are largely aperiodic structures, yet their specifications require immense amounts of information because of their low p_is. In paradoxical contrast, a periodic structure—say, a synthetic stretch of DNA encoding nothing but prolines—may immediately strike us as orderly, though it demands a relatively modest amount of information (and is described by a simple algorithm). The word "order," in the sense used here, has its roots in the weaver's experience; it stems from the latin *ordiri*, to lay the warp. Order refers to the *totality* of arrangements in a system, something immediately grasped in the case of a warp but not in the case of a macromolecular structure like DNA.

decrease of information, and vice versa. In this sense, we may regard the two entities as related by a simple conservation law: *the sum of (macroscopic) information change and entropy change in a given system is zero.* This is the law which every system in the universe, including the most cunning biological Maxwell demons, must obey. So, when a system's state is fully determinate, as in our two eight-state examples in Figure *1.2*, the entropy or uncertainty associated with the description is evidently zero. At the other extreme, the completely naïve situation where each of the eight states is equally probable (when all cards are face down), the information is zero and the thermodynamic entropy is equivalent to 3 bits. In general, for a system with 2^n possible states, the maximum attainable information and its equivalent maximum attainable thermodynamic entropy equal n bits.

With this we now can go beyond our earlier illustrative statement of information in terms of choices made by entities endowed with brains (the appropriate statement for human signals and communication) to a definition in the deeper sense of the two equations, in terms of the probabilities of the system's component states, that is, in terms of how far away the system is from total disorder or maximum entropy—how *improbable* it is. This is the sense of information that will be immediately useful when we concern ourselves with the *self*-organizing molecular systems of living beings, which pull themselves up from disorder.

To illustrate the information-entropy liaison, let us take a look at the evolution of a puff of smoke in a closed room, one of the simplest molecular systems one might think of. Early on, when the puff has just been delivered, most of the smoke molecules are close to the source. At that instant, the system has significant order and information. With time, the smoke molecules spread through the room, distributing themselves more and more evenly; the system evolves toward more probable states, states of less order and less information (Fig. *1.3:1–3*). Finally, the distribution of the molecules is completely even in the statistical sense (Fig. *1.3:4*). The system then lacks a definable structure—its molecules merely move about at random; order and information have decayed to zero. This is the state of *thermodynamic equilibrium*, the condition of highest probability, which all systems, abandoned to themselves, sooner or later will run down to—once a system reaches that state, it is highly unlikely that it will go back to a state with more order.

Thus, the molecular system loses its information with time; or, to use the other side of the coin, its entropy increases—which puts this in the traditional terms of thermodynamics, its Second Law. The information escapes, as it were, and no wall on earth could stop that; even the thickest wall won't stop the transfer of heat or shield off the gravitational force. Because we can see and feel this loss of information with the passage of time in everything around us, no one will feel like a stranger to the Second Law—it's time's arrow inside us. Indeed, the basic notions of that law are deeply ingrained in our folklore in phrases like

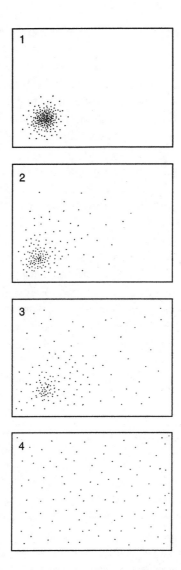

Figure 1.3 Dissipation of order and information. When smoke is puffed into a closed chamber (one puff), the molecules are initially concentrated near the source (*1*). This structure decays as the molecules spread through the room (*1–4*) until all macroscopic structure is lost (*4*, thermodynamic equilibrium). Stage *4* is the most probable condition, with information content zero; stage *1* is the least probable of the four stages. The information content at stage *1* is the displacement of the probability in respect to *4*.

"that's the way the cookie crumbles," "it does no good to cry over spilled milk," "you can't have your cake and eat it too." But, perhaps, nothing says it better than that old nursery rhyme at the end of the chapter (Figure *1.4*).

The time's arrow of thermodynamic events holds for nonliving as well as living matter. Living beings continuously lose information and would sink to thermodynamic equilibrium just as surely as nonliving systems do. There is only one way to keep a system from sinking to equilibrium: to infuse new information. Organisms manage to stay the sinking a little by linking their molecular building blocks with bonds that don't come easily apart. However, this only delays their decay, but doesn't stop it. No chemical bond can resist the continuous thermal jostling of molecules forever. Thus, to maintain its high order, an organism must continuously pump in information.

Now, this is precisely what the protein demons do inside an organism. They take information from the environment and funnel it into the organism. By virtue of the conservation law, this means that the environment must undergo an equivalent increase in thermodynamic entropy; for every bit of information the organism gains, the entropy in the demon's environment must rise by a certain amount. There is thus a trade-off here, an information-for-entropy barter; and it is this curious trade which the protein demons ply. Indeed, they know it from the ground up and have honed it to perfection. Bartering nonstop, they draw in huge information amounts, and so manage to maintain the organism above equilibrium and locally to turn the thermodynamic arrow around.

The Act of a Molecular Demon

The information-for-entropy barter brings us a little nearer to what makes living beings tick. So, it will pay to keep an eye on those demons as they juggle the bits and entropies. But let me say from the start that the juggling is entirely aboveboard; protein demons always obey the laws of thermodynamics. Though, to get to the bottom of their tricks, we may have to crane our necks, for they bend and twist, going through quite a few contortions. But we will see that much of this is only sleight of hand. What these proteins really do is shift their stance—they switch between two positions, two stable molecular configurations. They are like digital computers in this regard. Indeed, in a sense, they are minicomputers and, at bottom, their switching is cast in the same information mold as the switching of the logic gates of a computer memory register.

Let me explain. Every protein is endowed with a certain amount of information. That information is inherent in the orderly structure of the molecule, just as the information in our deck of cards was inherent in its orderly sequence. The information arises as the protein molecule is assembled from its components in a unique configuration in the organism—the molecule then gets charged with information, as it were.

With this information, its native intelligence, the protein now shows its demon side: it recognizes other molecules and selects them out of many choices. Though it may not *see* the molecules as the demon in Maxwell's thought experiment does (it needn't sense photons), it recognizes them just as surely at short range by direct molecule-to-molecule interaction; it fingers them, feeling their shape, so to speak. If a molecule fits the protein's own configuration, it will stick to it by dint of the intermolecular attractive forces. The protein, thus, can pick a particular amino acid molecule or a particular peptide molecule out of a myriad of choices.

Such are the first cognitive steps from which biological order emerges, but the act still isn't over. Next, upon interaction with its chosen prey, it undergoes a change in configuration. This change doesn't necessarily involve the entire bulk of the protein molecule, but it is enough for it to disengage. It is then available for another round of cognition, but not before it is switched back to its original state of information; its configuration must be reset for each cognitive cycle.

The kinship with the computer memory now begins to shine through. In a computer each of the possible logical states is represented by a certain configuration in the hardware—each of n logical states has its physical representation in the hardware. And when the computer memory of n bits is cleared, the number of 2^n possible states is reduced to 1—an increase in information which must be balanced by an increase in entropy in the computer's environment, in compliance with the conservation law.

It is thus the forgetting that is thermodynamically costly. And that is why a computer, indeed, any cognitive entity, cannot skirt the Second Law. This solution to Maxwell's perennial riddle finally came in recent years through a penetrating inquiry into the energy requirements of digital computers by the mathematicians Rolf Landauer and Charles Bennett. It is in the resetting of the memory register where the unavoidable thermodynamic price is paid.

The same holds for the biological demon. For each cognition cycle, the demon's memory must be reset; after each molecular selection, he must forget the results of that selection in order to select a molecule again. There may be, in addition, some thermodynamic cost owing to the thermal jitter in the biological demon's molecular structure (as there may be in the case of Maxwell's original demon owing to such jitter in the trap door), but *it is in the forgetting where the irreducible energy cost of the demon's cognitive work is paid*—where the thermodynamics accounts are settled by an increase in entropy exactly equal to the information gain.

How the Molecular Demons in Organisms Gather Information

This tells us more about the act of a protein or RNA demon than a thousand chemical analyses. But where do the entropies come from in the memory resetting, and how do they flow? The entropy-information to-and-fro can get quite tangled in demons that do their thing in conjunction with other molecules, as most protein and RNA demons of our body do. So to see through their machinations, to see how they balance the thermodynamic accounts, we return to Maxwell's original thought experiment where the complexities of the entropy flow of the cognitive act are reduced to the minimum, because the demon has no molecular partner. There, the demon *sees* his cognitive target—he relies on photons bouncing off the target—and so he can do the balancing with entropy derived directly from photons. This keeps things relatively simple.

Recall that Maxwell's demon starts with a molecular system in total disarray, a system at thermodynamic equilibrium. Everything is pitch-black then by definition. Thus, to see, the demon needs a source of light, say, a flashlight (in principle, any source of photons that is not at equilibrium will do). A source like that allows us to readily follow how the entropy comes and goes in the demon's world. So let us take advantage of this to pry into his well-guarded secret.

Consider the flashlight first. That consists of a battery and a light bulb with a filament. The battery contains the energy but produces no entropy; the filament radiates the energy and produces entropy. Now, one of two things can happen, depending on what the demon does. When he does nothing, the radiated energy is absorbed by the molecules in the system, producing an overall increase in entropy—energy which goes to waste. When he intervenes, there is an additional increase of entropy as a quantum of light is scattered by a molecule and strikes his eye. But, in this case, not all of the energy is wasted; part of it is used as information.

And this takes us to the essence of the cognitive act: *the demon uses part of the entropy and converts it to information*; he extracts from the flashlight system negative entropy, so to speak, to organize a system of his own.

There is, then, nothing otherworldly in the demon's act. His operation is perfectly legal; the sum total of the entropies is decreased as he selects the molecule, decreasing the probability of the system (p_i, equation *1*), exactly satisfying the second law of thermodynamics. All the demon does is redeem part of the entropy as information to organize his system. His stunt is to reduce the degradation of energy infused into the system from without.

This analytically simple system affords us a glimpse into the demon's secret. The demon's act here is not very different from that of some of the molecular demons populating our body. Drawing negative entropy from a battery or a light source for molecular organization is not just a figment of a thought experiment, but is precisely what certain proteins do. In our own organism, only few proteins (like those in the receptors of our eyes) can use a light source; most of our protein demons get their thermodynamic pound of flesh from high-energy phosphates in their surrounds, and some get it from the 0.1-volt battery of the cell membrane, as we shall see. But in plants and bacteria there are proteins that pump in information directly from the sunlight and, in a wider sense, this pumping is what keeps all organisms on earth alive.

It may seem odd at first to think of energy as an information source, but energy, indeed, can hold much information; it can be as ordered as matter. Just as a gas will spread into a volume, energy will spread over the available modes as it interacts with matter. This simply reflects the tendency of the energy quantum to partition into various modes; the probability is higher for the entire quantum to exist in several modes than in only one particular one. Thus, in the sense of equation 2, the unpartitioned quantum, the less probable state, embodies more information.

Consider now what happens when a photon of sunlight gets absorbed by matter. It ends up in many modes of smaller quantum—a state with less information. By degrading the energy of one photon into smaller quanta of thermal energy, the information content thus gets reduced by a certain amount. This way, with one kilocalorie of energy at ordinary room temperature, there are about 10^{23} bits of information to be had. Now, a plant protein demon, who specializes in photons, manages to recoup a good part of that degraded energy in information. Thus, he can build quite a nest egg over a lifetime.

The amounts that animal protein demons gather by extracting information from phosphate and other molecules are no less impressive. Every interatomic bond in a molecule represents a certain energy; a covalent bond, for instance, amounts to the order of 100 kilocalories per mole (a mole contains 6.022×10^{23} molecules). The average protein demon in our organism can extract nearly one-half of that in information—about 5×10^{24} bits per mole (the rest heats the environment). A few moles of covalent bonds in the hands of a competent demon, thus, go a long way informationwise.

Organisms Are Densely Packed with Information

Organisms have whole legions of protein demons at their beckon, hauling in information from their surrounds. Virtually all of this information ends up being compressed into the microscopic spaces of the cells (spaces often no more than 10^{-15} cubic meters). Order also exists elsewhere, of course. The beautiful world of mineral crystals, the splendid patterns of nonequilibrium hydrodynamic and chemical structures, the constellations in the night sky, all provide examples of information in the universe. But there nowhere seems to be so much information for a given outlay of space and time as there is in organisms—and if we came across an unfamiliar thing from outer space as cramfull with information, we would take it for a live one.

The large molecules in living beings, the *macromolecules*, offer the most striking example of information density. Take, for instance, our DNA. Here on a 1-nanometer-thick strand are strung together a billion molecular units in an overall nonrecurring, yet perfectly determined sequence; all the DNA molecules of an individual display that same sequence. We are not yet up to the task of precisely calculating the information inherent in this structure, but we need no precision to see that its potential for storing information is huge. The possible positions in the linear order of units, the four types of DNA bases, represent the elements of stored information. So, with 10^9 bases and four possible choices for each base, there are 4^{10^9} possible base sequences—4^{10^9} states that are in principle possible, yet they occur always in the same sequence. Even expressed in the logarithmic units of information, this is an impressive number. The number of possibilities is greater than the estimated number of particles in the universe!

This gives us a feel for the immense potential for storing information in macromolecules and prepares us to be at home with the notion—as far as one can ever be with immense numbers—that the basic information for constructing an organism would fit on such a molecular string, which in humans, taking all 46 chromosomes together, is over a meter long.

There is less storage capacity in the shorter DNA strings of lower organisms, though it is still quite impressive. For some single-celled organisms whose DNA nucleotide strings have only a fraction of the length of those in our own cells, we can make a crude estimate of the information content. The DNA of an amoeba (a nonsocial one), for example, holds on the order of 10^9 bits. In other words, one billion yes/no instructions are written down in that four-letter script—enough to make another amoeba. This script contains everything an amoeba ever needs to know—how to make its enzymes, how to make its cell membrane, how to slink about, how to digest the foodstuffs, how to react when it gets too dry or too hot, how to reproduce itself. And all that information is enscrolled into a space so small you would need a good microscope to make it out. If you wanted to give all these instructions in the English language, they would fill some 300 volumes of the size of this book (the information content of an average printed page in English is roughly 10,000 bits).

As for the script of our own body cells, that would fill a whole library—a library with more volumes than in any library now in our civilized world. This immense information amount, too, is compressed into a microscopic space—it all fits into the nucleus of a cell. And nowhere else on this Planet is there such a high concentration of information.

The Law that Rules It All

Immense as they are, those amounts of information in cells are always finite. This also is true for the amounts in any conglomerate of cells, even a conglomerate as large as our own organism, for it can be shown from first principles that, however large, the number of possible states in any physical system cannot be infinite. This is a good thing to know from the start, as we will shortly embark on a journey along the organismic information stream where the amounts and circularities of the flow might easily deceive one to think that the information is boundless.[3]

It is also good to know that organisms come into possession of all that information by legal means. The molecular demons who enable them to gather the information may be crafty, but they are not crooked; they always scrupulously settle their entropy accounts. So, what the living system gains in order comes at the expense of its surrounds, exactly in compliance with the second law of thermodynamics. And when the time comes and the system can no longer haul in enough information to balance the accounts, the game is up. The system falls apart, and all the king's horses and all the king's men couldn't put it together again.

Thus, the living system gets no free rides to organization. Like anything else in the universe, it is subject to the laws of thermodynamics. There have been claims to the contrary—such claims crop up in one disguise or another from time to time. But it has invariably proven a safe bet—and Maxwell's stubborn demon turned out to be no exception—not to fall for explanations that violated these laws. Of all our mental constructs of the world, those of thermodynamics are the most solid ones. Albert Einstein, once musing over the rankings of scientific theories, expressed his conviction that classic thermodynamics is "the only physical theory of universal content which . . . , within the framework of its basic notions, will never be toppled."

3. "Immense" is a subjective term, but "infinite" is not. An infinite number means a quantity greater than any number, no matter how large. If you stand between two (perfect) mirrors, you get an endless number of images of yourself, each the reflection of another—this is an infinite number. None of the numbers we will come across in this book—numbers of bits, cells, molecules, atoms, particles—are of that sort. We may call them immense, for that is what they are from our habitual point of view. Mathematicians, accustomed to speculate with extremely large numbers, may find those 4^{10^9} possible DNA states I mentioned earlier to be rather on the short side of the market. They trade in googols or googolplexes, units that even an astronomer would think large. Those names have a charming history. The mathematician Edward Kasner once asked a nine-year-old nephew of his to make up a name for something he was working on at the time, a number of the order of 10^{100}—one followed with a hundred zeros. The boy called it a *googol*. And that name stuck. A googol is much larger than that DNA number, yet it is not infinite: it is exactly as far from infinity as is the number 1.

Humptp Dumptp sat on a wall,
Humptp Dumptp had a great fall;
All the King's horses
And all the King's men

Couldn't put Humptp together again.

Figure 1.4 The Second Law in an older version.

The Cosmic Origins of Information

The Bigger Picture of Organizations

The information in the biological world is deployed in organizations of matter that span twelve orders of magnitude on a length scale, from 10^{-8} to 10^4 cm (Fig. 2.1). These biological organizations constitute several hierarchies: the building units of the molecules, the biomolecules themselves, their aggregates (supramolecular structures), the cells, the cell associations (tissues), and the organisms.

The rest of the organizations in the universe span 39 orders of magnitude. These form hierarchies stretching from the elementary particles (10^{-13} cm) to the groupings of stars (10^{26} cm): quarks combine to form the protons and neutrons; protons and neutrons associate to form the nuclei; nuclei and electrons form atoms; atoms assemble to molecules; molecules form supramolecular structures; and so, continuing upward in scale, we get to the organizations of moons, planets, and stars (10^5–10^{11} cm), to the star galaxies, the clusters of galaxies, and, finally, to the clusters of galaxy clusters (10^{26} cm).

That "rest" of information, thus, is not something to be sneezed at. If we used *immense* before in describing the amounts of information in the living world, we quickly run out of vocabulary in the global sphere. The sheer number and girth of the organizations in the universe boggle the mind. Take just the galaxies: some hundred billion of these huge structures bound by gravity populate the universe, and each of them, in turn, contains on the average one hundred billion star masses.

The Universal Quartet of Forces and Its Soloist in Molecular Organization

All organizations in the universe, regardless of their kind and scale, are ultimately composed of elementary particles. Thus, the secret of how they hold together, a

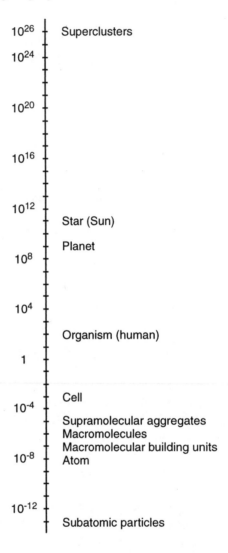

Figure 2.1   Organizations of matter and their dimensions. [The organization of stars has three levels: clusters of stars or galaxies, galaxy clusters, and clusters of galaxy clusters (superclusters)].

stability stretching over distances of 39 orders of magnitude, must ultimately lie with these particles. At that level and depending on the set of particles one is dealing with and the distance of interaction, one can distinguish four categories of force: the gravitational force, the electromagnetic force, the weak nuclear force, and the strong nuclear force.

The gravitational force acts on all particles, drawing them together according to their mass or energy. It is the weakest of the four forces, but it acts over long distances. The gravitational force between the nucleus and electron of a hydrogen atom amounts only to 10^{-39} of the electromagnetic force. But those tiny long-range forces between individual particles add up to a substantial attraction in large chunks of matter, like the earth or the moon, enough to keep the moon in orbit and to pull the water in the ocean up in every tide. Each particle here may be envisaged as the source of a gravitational field where, in the quantum view of things, the force is carried by massless particles traveling at the speed of light, called *gravitons*; two particles are drawn together because they exchange gravitons between them.

The electromagnetic force acts on electrically charged particles like electrons and quarks. It causes the electrons to go around the nucleus, much as the gravitational force causes the moon to go around the earth. This force is carried by another massless particle, the *photon*. This photon is a kin of the photon we are accustomed to in our visible world, but it's not quite the same. It is unobservable (*virtual*); it moves over extremely short distances as it gets exchanged between electrons and protons (the product of its momentum and the distance satisfies Heisenberg's uncertainty condition). When an electron going around a nucleus is switched to an orbit nearer to the nucleus, the virtual photon converts to a real one which is given off to the environment. Conversely, a real photon coming in from the environment is absorbed when it strikes an atom and its energy is used up in switching an electron to a farther orbit. *This transit from real to virtual photon, we shall see, is the bridge between the nonbiological and biological organizations in the universe; it is the window through which living beings let the cosmic information in.*

The weak nuclear force acts on various types of nuclear particles, driving their transmutations, like those occurring in radioactive atoms. Its range is extremely short: about 10^{-16}cm. The strong nuclear force acts on quarks, gluing them together in the protons and neutrons and, in turn, holding these particles together in the atoms. Its range is about 10^{-13}cm.

Because their range is so short, the two nuclear forces have no hand in macroscopic organizations. As for the gravitational force, this is too weak to assert itself between individual atoms or molecules, though the gravity of the earth is felt by these particles and their aggregates and plays a role in supramolecular organizations.

Of the four forces, the electromagnetic one concerns us most in dealing with biological organization, at least at a first go. This is the force that rules the roost in the society of atoms; it governs their endless bouncings off each other, as well as their rarer interpenetrations (the chemical bondings) that engender molecules. So, we shall come across this force quite often as we journey along the biological information stream. But let us keep in mind that these molecular constructions have an infrastructure, and that it takes the entire quartet of forces to stabilize (and understand) the whole.

Origins

By virtue of equation *1*, we can wean ourselves from force and distance in evaluating organizations. Information is information, regardless of what force stabilizes an organization or what distances it spans. All organizations in the universe are open systems as far as their information is concerned. Thus, under the conservation law, information can be exchanged between systems, including between biological and nonbiological ones, and this exchange, we have seen, sustains all life.

This is all very well and good, but where did the information that is being exchanged come from in the first place? Where did the ordered organization of energy and matter spring from? Not so long ago such questions about the origins were best left to philosophers and theologians, but a discovery in 1929 by the astronomer Edwin Hubble brought the problem of the beginnings into the realm of science. Observing the many galaxies scattered throughout the universe, Hubble found that they were rushing apart with a speed approaching that of light. On the scale of distances of the universe, this rushing presently amounts to a doubling of the distance between any pair of typical galaxies about every 10 billion years—an expansion of the universe somewhere between 5 and 10 percent every billion years.

One can mathematically trace this expansion back to a time about 10 to 20 billion years ago when all the galaxies would have been at the same place, when things were so close together that neither stars nor even atomic nuclei could have had separate existences. This moment of infinite density is, by general consensus, the origin of the universe. This singularity is generally called the "Big Bang." It will be our zero mark of time in this chapter.

In the past two decades our understanding of physics has become deep enough to reach into the distant past of the universe—not quite as far as the Bang, but pretty close to it. We cannot bodily move back in time, but fortunately a good telescope will do this for us. We can, in a sense, look back into the past with it, because light travels at only a finite speed. When we blink at the sun, we see it not as it is at that instant, but as it was 8 minutes before, when that flash of light left that star. When we gaze through a modern telescope into the far reaches of the universe, we see emissions from matter of billions of years ago. By analyzing the emission spectra, one can learn about events that took place a few million years after the Big Bang. With radio telescopes, one can get even closer. These instruments can detect the glow left over from the time when the universe was still young and hot. This glow originated when the universe was only a few thousand years old, and eventually it got red-shifted to microwave length by the expansion of the universe. From studies of this sort and from experimental work in high-energy physics, the following picture has emerged of how an organization of energy and matter is likely to have evolved in the early universe.

In the beginning there was the Big Bang. The bang filled all space, with every particle rushing apart from every other particle. At about one-hundredth of a second after the Bang, the earliest time about which one can speak with any confidence, the universe was filled with an extremely hot (10^{11} degrees Kelvin) and dense undifferentiated brew of matter and radiation—a mix of photons, electrons, positrons, and neutrinos (and their antiparticles), which were continually springing from pure energy, and a small amount of protons and neutrons—everything flying apart. The universe then was exclusively populated with fundamental particles; it was still too hot for molecules or atoms, or even the nuclei of atoms, to hold together. The organizations of matter and energy in the universe then were simpler than they would ever be again.

After about three minutes, the temperature had dropped to 10^9 degrees Kelvin (about a tenth of the peak temperature in an H-bomb explosion). At this temperature, protons and neutrons no longer had enough energy to escape the force of their attraction (the strong nuclear force). They combined to form nuclei, first the simple nucleus of one proton and one neutron of deuterium (heavy hydrogen), then the stable nucleus of two protons and two neutrons of helium. So, we have the first smatterings of organization—nuclear organization. Apart from these nuclear structures, the universe then mainly contained photons (whose glow the radio telescopes still can detect today), neutrinos, and some electrons. Most of the electrons and positrons, which dominated the earlier picture, had disappeared when the temperature fell below the energy threshold for these particles.

The universe, now populated with elementary particles and nuclei, kept expanding, but not much happened for almost a million years. The temperature had to drop to where even nuclei no longer had enough energy (10^3 ºK) to overcome the electromagnetic attraction between them, before new organizations, namely, stable atoms, could assemble. This happened about 700,000 years after the Big Bang.

From this basic level, the organization of matter could go on to the next tiers in suitable regions. While the universe as a whole kept expanding, the regions which had higher-than-average density were slowed down as the gravitational force pulled their matter together. As some of these regions got denser, they began to spin under the gravitational force of outside matter and formed disk-shaped galaxies, like the Milky Way where we were to come on the stage some ten billion years later.

So far, there were only the simplest elements around: hydrogen and helium. Higher nuclear organizations had to wait for further condensations of matter to take place in the hydrogen-and-helium clouds of galaxies. With time, these broke up into smaller clouds that contracted under their own gravity, their local temperature thereby rising enough to start nuclear fusion reactions that converted the hydrogen into more helium. Thus, stars were born.

The matter in these first generations of stars still was not dense enough to permit the organization of larger nuclei. Heavier nuclei cannot be built by simply adding neutrons or protons to helium nuclei, or by fusing two helium nuclei, because nuclei with five or eight particles are not stable. Hydrogen fusion is a dead-end process, and that is why even today 99.9 percent of the atomic matter in the universe is hydrogen and helium. To go further, higher densities (or higher neutron fluxes) are needed than those prevailing in common stars. Now, a star is ordinarily kept from contracting beyond a certain point by the pressure produced by the heat of its nuclear reactions. However, in the long run it must run out of nuclear fuel. Only then, as it starts to contract, may it reach a density where the helium nuclei collide frequently enough to generate the next stable nucleus, a nucleus of twelve particles.

And that is a nucleus we are specially interested in: the nucleus of carbon, the cornerstone of organic matter. That nucleus (C^{12}) forms from three helium nuclei (He^4). Now, from carbon, oxygen (O^{16}) can be made by adding a helium nucleus; and with carbon, oxygen, and helium, heavier elements can be built up under a bombardment of neutrons. And such a bombardment occurs when a collapsing star explodes, as massive stars tend to do, blowing off into space all but their inner cores.

Such stellar explosions are not uncommon in the universe. Three have been observed in our galaxy in recorded history. One occurred in 1054 and was noted by the astronomers of the Chinese Imperial Court. This explosion must have been of colossal proportions. The remains can still be seen in the constellation Taurus; the fulgent debris are still flying apart and, at its center, the remnant of the star, now a tiny body of only a few thousand miles in diameter, sends out pulses of light, X rays and radio waves at the rate of 30 per second.

Colossal as they are, these explosions—or *supernova*, to use the astronomers' term—are but humdrum in the universe. The chances are that right now there is one happening somewhere out there. Those explosive events are pregnant with information. Under the bombardment of the vast numbers of neutrons liberated from the collapsing star core, heavier elements may form that are blasted into space by the enormous pulse of energy released from the imploding core. One such explosion may have occurred about 5 billion years ago in our province of the Milky Way, giving rise to all the elements heavier than hydrogen we now find here. Our sun, a second-generation star, is thought to have formed out of the condensing debris of that explosion and to have spewed out the heavier elements which eventually condensed to solid bodies. One of them became the planet Earth.

The Elusive Wellspring of Information

The reader may wonder how confident one can be about things that happened so long ago. Digging into ancient history is always risky business, but the risks are tolerable when good fossils are on hand. And here the background radiowave radiation, this microwave leftover of the heat of the Big Bang, offers an excellently preserved fossil. The assurance comes from accurate measurements of this microwave leftover and from experiments in particle accelerators simulating the heat of the early universe as far back as one ten-billionth of a second after the Big Bang. The confidence in the picture of the early phase of cosmic evolution—from one-hundredth of a second to 700,000 years after the Big Bang—is high, indeed; some cosmologists go as far as to call it "the Standard Model."

As for the origin of information, the fountainhead, this must lie somewhere in the territory close to the bang. By virtue of the conservation law, everything must have been in that primordial grain as it began to expand—and what a grain it must have been, laden with so much! Alas, time zero, the moment of the Big Bang itself, is impenetrable to scientific inquiry. At that instant when all energy and matter in the present universe were squeezed into a single point[1]—which for want of a better word we will call *Unum*—the density and the curvature of space-time, as Stephen Hawking and Roger Penrose have shown, are infinite. Here we reach the edge of scientific understanding. And a little further off the start, the first 10^{-43} seconds are off limits, too. This is the *Planck time*, the shortest intelligible instant, the irreducible minimum below which the quantum effects overwhelm everything and the concept of space-time breaks down.

Thus, it is after that 10^{-43} seconds time mark, during the first instants after it, where the fountainhead should be located. Here things are still shrouded in mist. The earliest moment one may presently speak about with any confidence is 10^{-2} seconds (the time at which the preceding chronicle began). So, information's wellspring gives us the slip, as all things concerning the origin have had the habit to do, ever since human beings have tried to grab hold of them. But there is no denying that to get within one-hundredth of a second of that effluvium is pretty good by human standards.

Debut of the Cosmic Quartet

Although we cannot pinpoint the wellspring, we may try to estimate its boundaries by paying heed to the four forces of nature, the entities that move everything and hold together all organizations in the universe. So, before we leave the knotty question of origins, we turn once more to this foursome, to view its origin in the light of the current unification theories in physics. When did these entities come on the cosmic scene? The latest theories, we shall see, give a surprising answer: the forces burst forth one by one during the first instants of the universe.

1.   In some cosmological theories, our present universe is but one in a succession of universes; the Big Bang would mark then the instant when the information about the preceding universe was destroyed.

These unification theories, which bring three of the forces under one roof, constitute the most exciting advances in theoretical physics in the past twenty-five years. They are built on the precept which every physicist holds dear: that the universe is symmetric. The symmetry of things is often quite obvious, as in the parity of systems whose components are rotated, translated, or reflected in a mirror. But it also can be more subtle, hidden in the mathematics, as it were. To perceive it requires the envisaging of more dimensions than those we ordinarily see and feel—the recent unification theories work with seven extra dimensions, and some with more. The symmetry creed has been a genial guide, at times the only one, in the human probings of nature. Newton, Maxwell, Bohr, and Einstein were all devotees of this creed, and their crowning achievements eloquently affirm its validity.

The current unification theories predict an astounding property of the two nuclear forces: they get effectively weaker as the energy or temperature increases. The electromagnetic force, in contrast, grows stronger, as one might intuitively expect. This opposite behavior holds the key to our question, for if one goes to environments of higher and higher energy, one should eventually get to the point where the three forces have equal effective strength—the point of their unification. This point calculates out to be at about 10^{23} electron volts and corresponds to a temperature of about 10^{27} degrees Kelvin. Before that point, there is one of partial unification where the weak nuclear and the electromagnetic force get to be of equal strength. And this happens at 10^{15} degrees Kelvin.

Such enormous energy levels don't occur ordinarily on earth, and even the most powerful particle accelerators now available can only go up to 10^{15} degrees Kelvin (which was just enough to verify the theory concerning the merging of the weak nuclear and electromagnetic forces). But those levels are precisely of the sort that must have prevailed in the extreme environment of the early universe. A temperature of 10^{27} degrees Kelvin may be estimated for the 10^{-35}-second instant, and a temperature of 10^{15} degrees for the 10^{-10}-second one. So, with these guideposts one may begin to reconstruct the history of the forces.

Let us start as close to the Big Bang as we are permitted to: 10^{-43} seconds— the Planck time. At this instant the thermometer reads 10^{32} degrees Kelvin. We are well above the triple unification point; there is then only one force. An instant later, at about 10^{-35} seconds, the temperature has dropped below that unification point. The strong nuclear force makes its debut. And at about 10^{-10} seconds—the thermometer now marks 10^{15} degrees Kelvin—the weak nuclear and the electromagnetic forces follow suit (Fig. *2.2*).

This stakes out the boundaries of time for the earliest organizations of energy and matter, however fleeting they may have been, produced by these three forces. The organizations produced by the electromagnetic force are the laggards here. That force is the last to come on the cosmic stage, as seems befitting to a majestic entity that eventually will dominate the living world. Not included

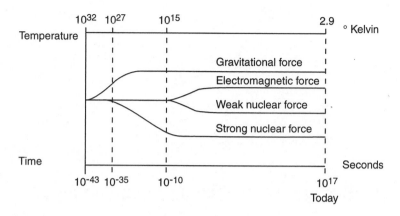

Figure 2.2   The forces of nature and their evolution, according to the unification theories. The four forces burst out from a single one, as the temperature in the universe falls during the first instants after the Big Bang. Their development is represented on a time scale in seconds after the bang; the corresponding temperatures in degrees Kelvin are indicated on top. The gravitational force is not comprehended by the current unification theories; it is presumed here to emerge first. (After J.D. Barrow and J. Silk, *The Left Hand of Creation,* Basic Books, 1983, and H. Sato, *Symposium on Relativistic Physics,* Ann. N.Y. Acad. Sci. 375:44, 1981.)

in this chronicle is the gravitational force; it is not comprehended by the current unification theories. Presumably it is the first force to emerge, perhaps right after the Planck instant.

These unification theories are splendid pieces in a long stream of efforts of the human mind to render the complex simple. They continued what Maxwell began a century ago when he unified the electric and magnetic forces. In this recent quest the first successful step was taken by Abdus Salam and Steven Weinberg who, in 1967, amalgamated the electromagnetic and the weak nuclear forces into the *electroweak force.* Next, in 1973, Sheldon Glashow and Howard Georgi merged that force with the strong nuclear one. The gravitational force is still the odd man out. But the unification enterprise is booming, and the day when the muses sing perhaps may not be too far off.

three

The Growth of the Cosmic Inheritance on Earth

From Symmetric to Asymmetric Molecules

At this juncture we take up again our chronicle of the first organizations of energy and matter in the cosmos. We had left the account at a stage when our solar system emerged out of the remnants of a star explosion which happened some 5 billion years ago in the Milky Way, a supernova whose heavier atoms were hurled out into space and eventually condensed to form the planets. From this point on we taper our view to planet Earth.

There, the workshop of organizations idled for a while, until things cooled down. Most everything on Earth then was prefabricated, so to speak. What elements there were, came from somewhere else; the carbon and oxygen nuclei came from the ancestral supernova, and the hydrogen and helium were leftovers of the syntheses during the first three minutes of the universe. There would be little chance for further development until the temperature dropped to a degree where atoms could interact electrostatically and stick together to form molecules. A pair of hydrogen atoms, for example, will interact at a short distance: the electron of one atom is attracted by the nucleus of the other, and vice versa; at a distance of 0.74 angstroms, there is an energy trough where the two atoms can build a stable organization in which their electrons are shared. Such covalent bonding requires no energy outside that contributed by the two atoms themselves. In fact, energy is given off. Thus, as the right temperature was reached, many species of molecules would form by simple downhill reactions, and the surface of the earth would become settled with a host of molecules: carbon, nitrogen, oxygen, silicon, metals, and the like.

It is easy to see where the information for this simple first generation of earthly molecules came from: it was inherent in their prefabricated atoms. Thus, for hydrogen the information would date back to Minute Three and for the rest, to the five-to-fifteen-billion-year later vintage of nuclear organizations in the ancestral star.

The stage then was set for further molecular growth by input of extrinsic information. This information came from photons—photons from sunlight and eventually, as an atmosphere formed, photons from electric lightning. Everything on Earth was then being continuously bombarded by these energy particles, and in this crucible, the first-generation molecules gained enough information to enter into combinations. The photons colliding with those molecules were captured and converted to virtual photons, and the corresponding gain of information was cashed in to build molecules of somewhat higher complexity. In turn, some of these new molecular structures, including some unstable ones of short life time (the so-called intermediaries), supplied the information for yet another molecular generation.

So, molecule would beget molecule, and our planet, its surface and atmosphere, became populated with all sorts of small molecular structures, including small organic ones. The heaviest structures sank to the center of the earth to form a core 3,500 kilometers in diameter (mainly of iron and nickel), around which various metal silicates (mainly iron-magnesium silicate) layered according to density. The least dense of the silicates, those made of silicon and oxygen, like clay, floated to the surface, forming a crust about 5–33 kilometers thick, where here and there some rarer stuff, made of carbon, nitrogen, phosphorus, sodium, calcium, magnesium, and so on, was wedged in.

This was still a world populated by structures of little information content, by and large a world of symmetric structures. Inorganic molecules are rather symmetric, and so are the simplest organic ones, like propane gas or alcohol; these structures look the same if we turn them about an axis or view them in a mirror. Some inorganic molecules can constitute large crystals, and those structures are showpieces of self-organization. Take, for example, the crystals that form in a water cloud. As the cloud cools down, large numbers of water molecules get joined together because the oxygen and hydrogen atoms of neighboring molecules attract one another ever so slightly. And that same interaction, then, sees to it that the groups are frozen down to crystals—a perfectly ordered grid of molecules—which eventually, in turn, order themselves to become a soft and brilliant snowflake. But, as large and beautiful as these symmetric organizations are, they hold only relatively modest amounts of information because of their monotone regularity.

However, things became ripe for higher informational structuring as soon as the first asymmetric organic molecules were forged. A certain degree of asymmetry is needed if you want to construct something with round balls which themselves are all fairly symmetric, as atoms are, and have it come out containing more than a trifle of information. Equation 2 here tells us that the number of

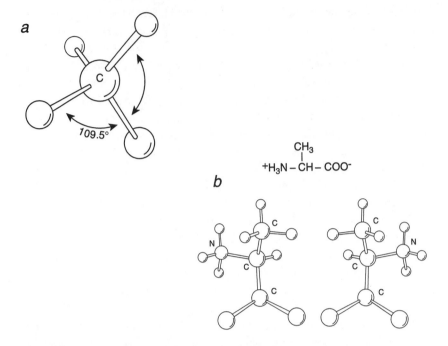

$$CH_3$$
$$+H_3N - CH - COO^-$$

Figure 3.1 *a,* When a carbon atom (C) shares its four electrons with other atoms (H), the four electron pairs repel one another equally; at equilibrium the four bonds point tetrahedrally away from the central carbon atom.

b, Any carbon atom that has four atoms bonded to it will have two possible configurations that are mirror images of one another. The two mirror-symmetric forms of the amino acid alanine are illustrated here. The one on the left is L-alanine and that on the right, D-alanine. Only the L-form is found in the proteins of living organisms. The carboxyl group (COO^-) and the amino group (NH_3^+) are represented in ionized state.

equivalent ways of putting the structure together (p_i) must be kept low. So, the atoms, aside from occupying determinate positions, must be distributed nonuniformly in space. In other words, the atoms in the modular units not only must be strongly bonded but also must be of a different kind.

Among the elements on hand, carbon filled this bill to perfection. It has exactly half as many electrons as are needed to fill its outer shell; it neither loses nor gains electrons in chemical reactions, and so it makes no ionic bondings. Ionic bonds are fuzzy, hence unsuited for precise and directional positioning of atoms. Instead, carbon shares its four electrons with other electron-sharing atoms—and it can do so with atoms and groups of atoms of many different kinds—forming covalently bonded structures where the electrostatic fields constrain the atoms to exactly occupy four points in space around itself, the corners of a tetrahedron (Fig. *3.1*). Moreover, its quantum energy levels are few and wide apart, and so it is good at storing information. Thus, in combinations with hydrogen, oxygen, and nitrogen, carbon can build up a variety of asymmetric modular structures: chains, branched chains, sheets, and composites of these. In fact, it is so uncannily good at this that I doubt we could have come up with a better thing even in our most inspired dreams.

With such a talented atom in the lead role of molecular productions, a variety of asymmetric molecules eventually found themselves on Earth. Experiments simulating the prebiotic environment and its photon crucible suggest that many kinds of such molecules, some even with appreciable complexity, could have formed spontaneously in a water medium: glycoaldehydes, sugars, purines (like adenine and guanine), pyrimidines, amino acids (the building blocks for protein), and even nucleosides and nucleotides (the building blocks for DNA and RNA).

The Advantages of Molecular Complementarity

These molecules and some of the smaller ones from before could exchange their wares when they bumped into one another. They traded electrons, energy, matter, and information. The water pools on Earth in which the concentration of the molecules was high must have been busy marketplaces. It was a market free for all and the trading was strictly quid pro quo—we shall see the rules in detail later. The rare commodity, however, was not electrons, or energy, or matter, but information. That was the really precious stuff, and the name of the game was to bag as much of it for one's own molecular self as was legally possible.

It takes time for a molecule to accumulate information under the law of conservation. But, over some 100 million years of trading under an unchanging law, it is possible to feather one's nest. To judge by their successors now in us, some of these early molecules did all right by themselves, indeed.

But what is it that makes some perform better than others in this game—a game played under the stern rules of the Second Law that will not tolerate cheaters? We cannot track the molecular interactions that took place so far back—there are no nice fossils here, like the primeval microwave leftover—but we can see the property that made the winners. This property has to do with the forces whereby molecular information is transmitted.

When we speak about the transmission of information from one molecule to another, we mean a transfer of information inherent in the molecular configuration—in the linear sequence of the unit structure or in the three-dimensional disposition of the atoms. Since molecules cannot talk or engage in other human forms of communication, *their method of transmitting information is straightforward: the emitter molecule makes the atoms of the receiver deploy themselves in an analogue spatial pattern*. Now, there is only one force that can make this happen: the electromagnetic force; the gravitational force is much too weak and the nuclear forces are of much too short range (10^{-16} and 10^{-13} cm).

It is then in the electrostatic field around the molecules, the field of electro-magnetic attraction, where we must look for the answer to our question. This field is effectively limited to distances of 3 to 4 x 10^{-8} cm in organic molecules. *Thus, to transfer information from one organic molecule to another, the participating atomic components of the two molecules have to be situated within that distance. In other words, the two molecules must fit together like glove and hand or like nut and bolt, and this is the sine qua non for all molecular transfers of information.*

Thus, it takes matching contours for molecules to operate at this information marketplace. There may have been few, if any, such molecules to start with. But, as more and more molecular specimens turned up on the earthly scene, we may imagine that here and there a pair by chance may have had some complementary features. Not many at first; one would hardly expect a precision-fit between molecular surfaces from happenstance. It may have been only a sloppy fit, not better than that of a mitten or a wobbly nut, but so long as it was enough to transfer a little information, it served the purpose. At the first go the information exchange was not necessarily restricted to organic molecules, but may well have had inorganic participants. Clays, for instance, exhibit a rough-hewn complementarity for some organic molecules, sufficient to catalyze some syntheses. So, these abundant silicon oxides may have been the information donors that orientated atoms in a rudimentary production of organic molecules, and time would ensure that, bit by bit, the organization of these molecules would grow in the photon crucible, giving rise to ever fresh and larger structures.

The Animated Moment: The Begetting of Information Circularity

Among those organic structures, and largely generated by the wheel of chance, a few eventually would come out having complementary features—say, a couple of molecules fitting the contours of a larger one. That would start a whole different ball game. The larger molecule now would serve as a matrix—a sort of template—for making composites from the smaller ones. Such a template would have a lot more to offer for catalyzing molecular syntheses than clay had. Its asymmetric organic structure would hold vastly more information for positioning and linking molecular building blocks, and its tight fit would permit a more faithful information transfer. Our molecular workshop now would gather speed and precision. Good molecular replicas would come off the assembly line and the place, we may imagine, would soon teem with products of some complexity, all alike and of the same ancestry—the first *biomolecular species.*

However, it seems unlikely that the inventory of the workshop would be limited to just one species for very long. On a stage showered with photons, change is always waiting in the wings. Inescapably, under the bombardment of high-energy photons in the ultraviolet- and X-radiation in the environment, the atoms of the organic template would undergo some translocations. Not all of those changes

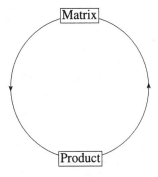

Figure 3.2 A self-promoting molecular information loop.

would necessarily rub off on the products, but those with complementary features appropriate for information transfer would. Thus, a wider and wider variety of organic molecular species would make the scene —the first *biomolecular mutants.*

If the scene so far was rather placid—the genesis of molecules, though sped up, still proceeded at a stately pace—that serenity one day would be shattered by an unprecedented event: among the mutants, one turned up—one in a myriad—who was able to catalyze the synthesis of its own matrix. The product here donated the information to construct its template! So, information flowed in circles: from matrix to product to matrix— *an information loop where the product promoted its own production* (Fig. 3.2).

In later chapters we will find out who the two molecular characters might have been who managed this feat; there are a few fossils still around to lend us a helping hand. Here we only draw the outline of the information loop, to show its puissance: the product makes more template makes more product makes more template . . . —a self-reproducing system whelping molecules in ever-increasing amounts.

Such a system would outreproduce anything then on the scene. And if another system looped like that would turn up, the competition was on. The two would vie for the information at hand, each trying to snatch the biggest chunk from the surrounds. Competition, it is said, brings out the best in individuals, but it also brings out its share of greed. The circumstances attending the conservation law kept a lid on the information lust, while the molecules were still trading on the market separately on their own. Alas, with the advent of information loops, the greed became rampant, and the most rapacious systems inherited the earth.

A Synoptic View of the Beginnings

It would still take a few billion years more until the winners in that game—the molecular systems best at grasping information—sorted out and filled the niches of what we call the living world. We will follow the systems in their long trek in the next sections as well as we can, guided by those which endured. The time chart of the information flow on the opposite page shows the landmarks that may help us keep our bearings (Fig. *3.3*).

Let us take a bird's eye look at that flow: 10 to 20 billion years ago the universe starts with a bang. Information welling from an uncomprehended source and transferred by forces unfolding from an ever-increasing space, engenders organizations: the elementary particles and the hydrogen and helium nuclei during the first three minutes, their atoms some 700,000 years later, and all the other atoms and the stellar structures over the next 10 billion years. On the cooling Earth the evolution continues from atoms to molecules; first the simplest molecules, then, nursed by the same cosmic information now flowing with photons, a little more complex molecules, including the first asymmetric organic ones. Eventually, the formerly always straight information flow makes loops which, by the power of their feedback, whip it to a more lively pace; higher-tiered molecular systems are then engendered. With time the flow gets more and more convoluted and the systems engendered, more and more complex. After some 4 billion years, they stand up and wonder about it all.

The Lively Twig of the Cosmic Tree

It may perhaps have seemed odd in the preceding lines to see information put before the evolution cart. I used the word "engender" (for lack of a better term) in the sense of *immanent cause*, so that it has an etiological connotation but not the ordinary lexical one of "generate," an action which cannot beg force. In terms of equations *1* and *2* it is as correct to say that order is inherent in information as is the more familiar reverse. And in such things who is to say what comes first, the chicken or the egg? We have seen already how, by virtue of equation *2*, information takes on a deeper meaning, changing from a quantitation of organization with engineering overtones to something immanent. By the time the next four chapters are wound up, I trust, this metamorphosis and we will be on more intimate terms.

But the purpose of this synoptic view was not to unfurl the flag; it was to show the continuity between the cosmic and the biological evolution and to see this in perspective. It is hard to spot where one evolution ends and the other begins. This is in the nature of the beast. However, we may agree perhaps to look upon the emergence of the information looping—that very special twist— as the beginning of the biological part. Others may prefer to make their count of biological time from the emergence of asymmetric molecules. On the scale of the adjacent chart it hardly matters from which of these two points we start. Either way we get about 4 billion years to the present time. And onward, if all goes well, another 4 billion years are in the offing. By then our sun will have

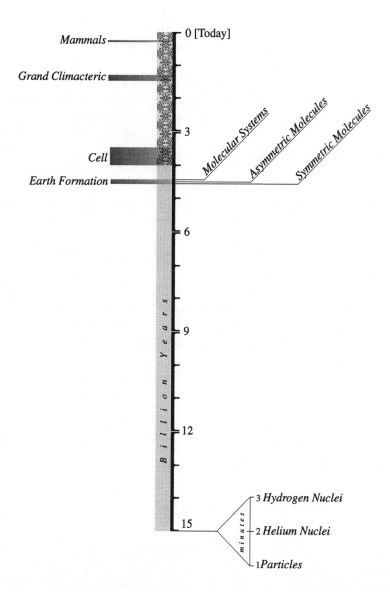

Figure 3.3 The cosmic time chart. The Big Bang is thought to have occurred 10 to 20 billion years ago (mark 15 on this scale). Life on earth started 3.5–4 billion years ago (the earliest fossils of prokaryote cells are 3.4 billion years old). The span occupied by life is shown in zebra pattern; the "Grand Climacteric" denotes the transition from single- to multicelled organisms (eukaryote cells appear just below that mark). The grey zones indicate spans of uncertainty. The first humans appear about a hundred thousand years ago, nearly 0 on this scale.

burned out, and when that photon source dries up, there is no help for it, the biological information stream will come to a halt, and all its information content will return to the large cosmic conservatory.

So, biological evolution is an ephemeral branch of the cosmic evolution tree. We don't know whether such branches exist elsewhere in the depths of space. Contrary to popular science fiction, it is by no means certain that anything like this has sprouted outside of Earth; and if it has, we can't say where or what its organizations might be like. But if there are more branches around, I would wager that they have one thing in common: circular information flow.

In the tree itself the information flow is straight: it goes from one region in space to another; it may reverse its direction but it rarely, if ever, goes in circles. By contrast, in the earthly branch the flow is constantly twisting and circling back onto itself, and these feedback loops and convolutions add to the languorous flowing of cosmic information a pulse of their own. This is what gives the branch that wondrous strange quality we know no equal of.

As for quantity, the information amount in the branch is huge. But what is enormous by everyday standards has a habit of shrinking on a cosmic scale. If we could add up the information gathered by biological systems from their evolutionary start to this day, all that prodigious hoard would amount to only a minuscule part of the information existing in the universe. And even if we added to that the caches that might pile up between now and the time all goes back to the conservatory, it's a safe bet that the ratio wouldn't change by much.

So, biological evolution is more like a twiglet than a branch, and having settled these proportions, we drop down the curtain on the cosmic pageant to focus on things biological and their special brand.

The Missing Hand: A Phylogenetic Mystery

We begin with the development of carbon chains. We have seen that with the formation of asymmetric molecules the channeling of information into molecular structures gathered way. The prime movers of that trend were molecules with a spine of carbon atoms. These molecules were somewhat lopsided from the start; other atoms, at first probably only a few, were positioned unevenly about that spine. So, these molecules were either left- or right-handed. With time, and under the influx of information, the carbon spine got longer and the number of attached atoms, larger, but the original hand remained, perpetuated by the intermolecular transfer of information. So, today every biological molecule worth its name has a handedness that is the same in every organism, from microbe to man—glucose and amino acids are left-handed, the sugars in DNA and RNA are of one hand, proteins corkscrew to the right, and so on, to the highest tiers of molecular organization.

Such a molecular asymmetry is not something one would expect from the rules of chemistry or physics. Indeed, when one synthesizes a relatively simple organic molecule like glucose in the laboratory, one gets glucoses that are mirror-images of each other, a mix with about half as many left-handed glucose molecules as right-handed ones. And this is precisely what we would expect from the laws of physics. Nevertheless, in organisms we find only the left-handed type. Why?

What wakes the detective here in us is the thermodynamic clue that in equilibrium conditions one type has as good a chance to be as the other. Then, why is the other type missing? This question reaches deep into the sap of life; it pertains to all biological molecules and their systems. We find symmetry, it is true, at higher structural levels—we only have to look at ourselves in the mirror. But this sort of symmetry never reaches down to the molecular floor. The biomolecules of our left body half have the same hand as those of our right half and, as much as we may try, we'll never meet our mirror-image partners!

One might wonder whether the mirror-image molecules would be functional. But, in fact, there is no physical reason why they should not be. They match their existing biological partners in all functionally significant respects: the distance between every pair of atoms is the same and the energy is the same, as would be true for any reflection-symmetric system. Take, for instance, a pair of mirror-symmetric clocks. Such clocks match each other like a pair of gloves, left and right. Their springs are wound the opposite direction, their gears move the opposite way, their screws have opposite threads, but the two clocks will work alike and give exactly the same time. It is immaterial whether they are run by electromagnetic, gravitational, or nuclear force; reflection-symmetric machines always behave exactly alike under the laws of physics.

So, what happened to the mirror-image types? There are only two possibilities if we keep our hands clean of metaphysics: either the opposite partner was destroyed as it was made, nipped in the bud, so to speak, or there was no information to make it. The latter turns out to be the case, and this has to do with the way molecular information is transmitted.

The agent for transmission of information between molecules, the sole agent, as we have seen, is the electromagnetic force, and such transmission works only when the transmitter and receiver molecules have complementary shapes. Now, a molecule can either be complementary or mirror-symmetric to another one, but it cannot be both. Just try to fit the glove of your right hand on your left hand. Thus—and here is the catch—the mirror-symmetric counterpart of a transmitter molecule cannot transmit information to the receptor of that transmitter molecule. Hence, a biomolecular evolution, based as it is on intermolecular transfer of information, can progress only with components that have the same hand. Let a single molecule with the other hand in, and the works are jammed.

A simple analogy, the evolution of a modular structure from nuts and bolts, will illustrate the point. The bolts (of the sort without a head) are fastened

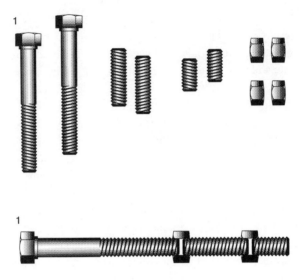

Figure 3.4   "Evolution" of a nut and bolt structure.
The hand of the structure is fated by the first happenstance selection (*1*) of the left-handed thread.

together one by one at their ends by the nuts, and in this monotone way that basic structure grows. Suppose, we started the erection job with an abundant supply of bolts, but with only one nut. Suppose further that nuts, in scarce supply, can be reordered only one by one as they are added to the structure. Then, the instant we fasten the first two bolts with the available nut, the die is cast: henceforth, the structure can grow only with units which have the thread of the original nut. If that thread happens to be a left one, the structure is fated to become left-handed throughout; however long it grows, it can never change its hand (Fig. *3.4*).

This analogy highlights the element of chance in the evolutionary growth scheme. Growth is set into motion only when a molecule chances upon a partner capable of information transfer. Such encounters must have been rare at the dawn of biological evolution, for molecules with complementary shapes must have been rare. But when such an encounter did occur, the system's handedness was fixed for life; its future development became as tied to the hand of that original match as the development was tied to the hand of the original nut in our nuts-and-bolts game. The opposite hand is absent, not because, God forbid, biological evolution violates the laws of physics, but simply because the hand was unfit for growth. It could as well have been the other way around if the opposite hand had been the first to make the fitting match.

When we talk about growth here, we mean future growth, as distinct from present growth. The opposite hand was potentially perfectly fit for one-time growth but not for iterative growth or, to call the baby by its name, phylogenetic growth. So, the mystery of the missing partner turns out to be a case of nipping in the bud, after all—the phylogenetic bud.

The Hobson's Choice of Molecular Evolution

We don't seriously expect a villain in this case, but there certainly must be a force behind a one-handedness that lasted 4 billion years. We couldn't lay the blame on information itself. Information sits on the throne; it can guide, it can give the coordinates for the molecular movement, but it doesn't do the actual work of moving the molecular pieces. So, if we want to get to the bottom of this case, we must go after the power behind the throne, the electromagnetic force.

That force is bipolar; it can attract or repulse. In interactions between two molecules, both energy components are important. When the molecules get very close, namely, to within a few tenths of an angstrom where their electron clouds overlap, the repulsive energy dominates and the molecules repel one another. When they are farther apart, at separations of a few angstroms where the interacting atoms of the two molecules get electrically polarized through synchronized fluctuations in their electron densities, the attractive energy dominates and the molecules are pulled together. A balance is struck someplace in between—for the atoms in organic molecules, between 3 and 4 angstroms—where the two molecules are held in place stabilized by their energy fields. That marks out a zone, a fine-drawn intermolecular domain, where one molecule can move another one or pieces of that other one by its magnetic influence to determinate positions.

And this is the zone that interests us, the domain where the information in one molecule can be passed on to another. However, no single pair of interacting atoms can hold two molecules in place for long; their attractive energy is too weak to stand fast against the constant thermal jostling of molecules. To bond two molecules stably by this electrostatic influence, several atom pairs must combine their energies, and for that all pairs must be aligned along the critical 3–4 Å zone. When atoms form linear monotone arrays, as they do in many inorganic molecules, such alignment is achieved by simply sliding one molecule alongside another. *But when different atoms or groups of them are unevenly spaced in three dimensions, as they are in all biological molecules, the alignment condition can be met only if the molecules have closely complementary shapes. No amount of sliding or turning will do here. Nor will any amount of pushing; there is not much give and take in molecules.*

We now see clearly why molecular complementarity is all-important in biology, and why anything conflicting with it like molecular asymmetry (the missing hand) would be forbidden. For Evolution it was a Hobson's choice. It was, "either complementarity or the game is over!" So, pursuing this necessitarian view, we might say that the instant the electromagnetic force came into the cosmos, it was ordained that we should never meet our mirror-image partners.

Chance and Determinism

In tune with the molecular theme and covering the enormous spans of time in the preceding topics, we emphasized the deterministic in evolution. However, like everything else in nature, the evolutionary enterprise is based on elementary particles and, at their level, chance plays first fiddle. One does not notice the underlying chance fluctuations when one looks at the macroscopic movement of things; the individual quantum whim gets lost in the bulk, and when there is a force behind, that bulk flows in a statistically predictable way. It is a little like watching a herd of cattle stampeding by. A lot of individual bucking, seesawing, and zigzagging is going on in there, but in our eyes, all this brouhaha blurs into one streaming mass, blinding us to the singular incidents.

Are the microscopic chance fluctuations always overshadowed by the mass in the evolutionary flow? This question is meant, of course, for the flow away from the cosmic origin, after the flow gathered way in a rapidly expanding space; near the origin, at the Planck time, such randomness of motion probably was all that was. Did the randomness stay snowed under the flowing mass forever, or did it come out into the open sometimes? We are asking, thus, in information terms the age-old question: Whether evolution is ruled deterministically or by chance?

Here first of all, we need a definition of determinism and chance. Casual use of these words will not do. What we really need to define, if we want to learn something new, is whether the flow is straight or convoluted. If it is straight from A to Z, the random fluctuations could never surface—no manner of means could pull macroscopic chance fluctuations out of quantum physics. If it is convoluted, one could think of several ways of channeling the information flow to boost the microscopic fluctuations to macroscopic magnitude. The question, determinism or chance, thus distills to informational linearity or nonlinearity.

The concept of determinism implies that an event, say the motion of a thing, can be predicted. Motion is governed by Newton's law, $F = ma$. This algorithm is a prescription for predicting the future. The formula is more powerful than any crystal ball. It says that if the force F acting on a given mass m is known, its acceleration a is predictable. So, if the position and velocity of a thing can be measured at a given time (t), all we have to do is apply the rules of calculus to compute the position for a subsequent time, t + dt; each of the locations is completely determined by the previous one—the locations are specified for all times. *Thus, we can know what will happen before it actually happens; the future is completely determined by the past.*

The definition of chance may be couched just in opposite terms: the future is not completely determined by the past. This definition admits shades of unpredictability. When we deal with events in the particle microcosm, the unpredictability is complete. Heisenberg's principle states the fundamental uncertainty, the limitation to the accuracy with which the position and velocity of a particle can be measured. Here the indeterminacy is absolute. It is not just that we don't have enough information to make the prediction, but the information is impossible to obtain because it is not there. Indeed, if we put Heisenberg's principle in terms of information, we get its full epistemological tang: *the information about the location of a particle and the information about its motion do not exist at one and the same time.*

This absolute uncertainty at the individual particle level gives place to predictability when we deal with particles in bulk. The degree of certainty depends then on the amount of information at hand. Thus, when the significant variables are known, the set of quantities that characterize the state of a system—like position and speed, temperature and pressure, chemical constitution and reaction rate—are pinned down with reasonable certainty by the Newtonian dynamics algorithm and its offsprings. We then only need to compute the probability of the event, which is simply the square of a measured quantity (the *probability amplitude*) or the sum of such squares when the event can occur in several alternative ways. *So, the future here can be predicted in terms of probabilities. And when we say that something is determinate, we mean it in this probabilistic sense: usually we mean that the probability is high.*

This sort of predictability also applies to systems with some complexity. Equipped with a good dynamic algorithm, which generally takes the form of a second-order differential equation, one can predict the whole sequence of events in a system. The instantaneous state of each piece in the system is described as a point and defined by the piece's position and the first and second derivatives of that position, velocity, and acceleration. The succession of states then unfolds upon integration. This succession of states is called *trajectory*; it contains all the information that is relevant to the system's dynamics.

Those mathematical maneuvers are not without pitfalls. They can easily create a system which, for all its mathematical self-consistency, does not exist. Such airy nothings can be pretty and as tempting as mermaids. There is only one recipe to stay out of harm's way, and that is to put the algorithm to this acid test: Does it encompass all states of the system, does it have a solution that is unique, does it itself behave independently of the system states? If the algorithm does all three things, we can be reasonably sure that it is the real McCoy whose logic exactly reflects the deterministic links between the physical states. And with these safeguards, dynamics becomes a powerful tool of understanding and predicting the workings of systems.

What all those nicely predictable systems have in common is linearity. Cause and effect are, to a good approximation, proportional here; when the cause doubles in intensity, the effect doubles too—a relationship that plots as a straight line on a graph. This simplicity makes it possible for one to come to grips with most systems based on periodic motion—sound waves, electromagnetic waves, and, as of late, even gravitational waves—so long as their amplitude is not too high. These motions are specified by simple periodic mathematical functions. Also the bulk flow of many sorts of particles—the diffusion of heat, gases, and liquids—behaves linearly, hence, predictably.

However, not all systems are so open-and-shut. Many possess nonlinearities that amplify the innate randomness of the movement of individual particles enough to make even their bulk manifestations random. Such behavior does not lend itself to scientific forecasting. Just consider a game of billiards. Could we predict the trajectories of the balls resulting from a perfectly controlled strike? Never, not even if we could make things friction-free. The tiniest variation in the gravitational field caused by a faraway planet would be enough to muck up our crystal ball because of amplification of outside perturbations or internal fluctuations. The balls are round and as they collide with one another, small variations at the points of impact give rise to disproportionately large effects. This amplification is compounded with each collision, and in this exponential way the microscopic variations rapidly grow to macroscopic size.

The Retreat of Scientific Determinism: Deterministic Chaos

Much of the world around us behaves a little like those billiard balls. The systems there may exhibit both linear and nonlinear behavior. A classic example is the movement of water. Water can flow smoothly when its molecules slide evenly alongside one another—a flow pattern called *laminar*. It then behaves linearly; position and motion in the systems are nicely predictable. But, with no special warning, it can shift gears and go over to nonlinear behavior and swirl and seethe in unpredictable ways. This transition to turbulence can come about quite abruptly, as anyone knows who has canoed down a mountain stream. Similarly, the movement of air can switch from laminar to turbulent, something we are made woefully aware of when the plane we travel in flies into a thunderstorm.

Systems that switch from linear to nonlinear behavior are by no means uncommon in nature, though they are not easy to make friends with. Understandably physicists have traditionally stayed away from such schizoid demeanor. But that is changing. In recent years a number of penetrating attacks have been made at those nonplusses and baffles, and a science is emerging that encompasses the complexities of nonlinear systems, including those inside living organisms.

At this juncture, what interests us are systems that shift from deterministic to random behavior. This randomness is called *deterministic chaos* to denote its origin. Such behavior is found all over the dynamic map: in air and water turbulences; in rhythmic motions as diverse as the swinging of a pendulum, the reverberations of sound or the heartbeat; in abnormal electrical activity of muscles, heart, and brain; in the growth of populations; in monetary economics; in video feedback; in the blending of plastics; and . . . well, add your own.

The randomness in this kind of chaos is not a state of total disorder, but one that is above thermodynamic equilibrium. It is a randomness that contains information. Nevertheless, it is unpredictable. One might suppose that if one knew all the variables in a chaotic system, one could predict it. Well, one cannot. It has now been proven that this randomness transcends the limits of knowledge.

An old clung-to notion thus was stung to the quick. It had long been assumed that the laws of physics held unlimited determinative power, and this presumption permeated all scientific thought. This started in the seventeenth century. We even can allocate the date: April 28, 1686. On that day Newton presented his "Principia," which contained the basic laws of motion, at the Royal Society of London. Never before in the history of mankind had there been anything that held so much predictive power—and Newton wrote with undisguised pride in the "Principia," "I now demonstrate the frame of the System of the World."

That power was further enhanced in the next 250 years by William Hamilton's and Henri Poincaré's elegant formulations of the laws. These and a number of other scientific triumphs nursed the notion that, if all the factors were known, one could predict or retrodict everything. The eighteenth-century mathematician Pierre Simon de Laplace stated this most eloquently in his "World System": "if we conceive of an intelligence which at a given instance comprehends all the relations of the entities of this universe, it could state the respective positions, motions and general effects of all these entities at any time in the past or future."

This proud view held sway for two centuries, pervading all sciences. Laplace's omniscient demon became the talk of fashion in the salons of Europe, and so from the natural sciences the view spilled over into sociology, economy, and politics. Its influence was all-pervading. Metternich carried Laplace's World System in his baggage when he traveled through Europe to shape his balance of power, and the fathers of the American Constitution used it as a source book. In physics, the Laplacian view eventually developed into the "Hidden-Variable Theory": deterministic prediction was limited in practice, but it was possible in principle; if some things escaped one's comprehension, this was only because one did not know all the variables involved—a human shortcoming that eventually would be overcome.

That theory, the cornerstone of scientific determinism, was dealt two humbling blows in the present century. First, Heisenberg's uncertainty principle made the theory untenable at the quantum level; then, von Neumann showed its fallacy at the information level. Thus, the Hidden-Variable Theory already was as good as dead. But scientists like to make sure. A few years ago the mathematical proof was brought forward that deterministic chaotic systems are *fundamentally* unpredictable. Their motions are so complex and their trajectories so varied, that no observation, whatever its precision, can give us the exact initial condition in the system from which we could derive its dynamic behavior. We can at best determine an ensemble of trajectories from the probability function of the initial condition. Quantum theory tells us that initial measurements are inherently uncertain, and the analysis of deterministic chaos went on to show that the dynamics of chaos amplify the uncertainties at an exponential rate, always putting the system beyond the reach of prediction. In contrast to linear dynamic systems where the information processing keeps pace with the action, the uncertainties grow faster here than the capacity of information processing.

Thus, the randomness in deterministic chaos is fundamental; gathering more information won't help.

The Randomness that Contains Information

The randomness in deterministic chaos, however, is not simply the result of amplification of the microscopic fluctuations. Those fluctuations do get amplified, but chaotic systems are more than just amplifiers of noise. These systems continuously shuffle the information they contain, and so bring forth new organization independently of noise or, better said, above noise. The degrees of freedom of the system's motion can be represented as the coordinates of an abstract space called *state space* (in the case of mechanical systems the coordinates are position and velocity). The dynamics of the system, its temporal evolution then becomes the motion of a trajectory through a sequence of points in the state space. This space undergoes stretching and folding that repeats itself over and over, creating folds within folds, within folds . . . ad infinitum.

The operation is a little like kneading bread dough: the dough is spread out by rolling and then it is folded over itself, and so on. If we sprinkled some red dye and followed that colored blob as it is being kneaded in with the rest, we would see it stretching and folding many times, forming layers of alternating red and white. After only some twenty kneadings, the blob would be stretched out more than a million times and would be thinned to molecular size. Throughout this mixing operation, the stretching and folding produces all sorts of patterns and shapes—a motion picture of ever-changing forms.

The continuous stretching and folding of the state space in a deterministic chaotic system will eventually thin it down in spots to the bare microscopic fluctuations. These fluctuations are amplified locally to macroscopic proportions by the nonlinearities in the system. Since the stretching and folding is repetitive, such local amplifier effects eventually will dominate the behavior of the system and its dynamics become independent of the microscopic noise level. Once launched on the way to chaos, the system is not easily stopped. Lowering the temperature will quiet down the microscopic noise, but the system keeps going and will generate macroscopic randomness on its own. Under those whirling wheels of change, the system becomes a kaleidoscope of form for as long as it stays above thermodynamic equilibrium.

A Farewell to Arms

Let us summarize then what transpired about randomness from our discussion about chaos and information. Above thermodynamic equilibrium there are two kinds of randomness: one for the microscopic state at the quantum level, as embodied by the Heisenberg principle, and one for the macroscopic state at the level of deterministic chaos. Either kind of randomness is fundamental; it admits neither prediction nor retrodiction or, put in terms of Laplace's metaphor (though hardly the way he intended it to be), his demon is as helpless as we are.

There is another type of fundamental randomness, of course, the one prevailing at thermodynamic equilibrium. But this type is utterly run-down and, in a book on information, we are entitled to snub something that has sunk so low.

Now, what of the old polemic about determinism and chance, the dispute that has been going on in philosophical circles since the time of the ancient Greeks to the present. In the light of the new knowledge about randomness, that issue changes entirely the cut of its jib. For two thousand years it has been a polemic vehemently fought over two seemingly mutually exclusive alternatives. With the benefit of hindsight, we see now that there has been a crucial piece of information missing: the knowledge that there can be a fundamental randomness that is above thermodynamic equilibrium, a randomness encompassing a spectrum of information-containing states. Rather than two stark extreme alternatives, there is a whole gamut running from the random to the determinate, that is, from the absolutely unpredictable to the probabilistically predictable.

So, the old polemic goes poof. It reduces to a quantitative problem of information that, for any given system or set of systems, can be couched in terms of equation *2* or equation *1*: How high above thermodynamic equilibrium is the system or how much information is in it? What matters then for evolution is how much of this information can be used for organization of energy and matter. And this takes us to the question of how to pull an organization out of randomness.

Molecular Organization from Chaos

Few will have difficulties envisioning a deterministic system as a source of organization. The advances made in biology over the past thirty years have enlightened us in the mechanisms of replication and synthesis of biomolecules, and it has become cliché to say that DNA supplies the information. However, one may not feel so comfortable with the thought of randomness as a source of information. Yet, in principle, the conditions are the same: if there is information in a system and if there are ways to transfer it, there is organizational potential.

A few examples from random systems may set us at ease. In these cases one can literally see the organization arising. The systems here are kept high above thermodynamic equilibrium by input of energy, and they become destabilized by dint of intrinsic nonlinearities. Form then suddenly appears out of nowhere, as it were.

The jumbling of a TV picture by computer action affords a most vivid demonstration of such creation of form. One starts by digitizing the information in the picture, say, a drawing of Snoopy, and then lets the computer do the stretching and folding, simulating the operations of deterministic chaos. As the information for Snoopy gets scrambled, his image dissolves into an undifferentiated gray. However, as the computer keeps stretching and folding a billion times over, we get glimpses of varying shapes. Snoopy himself never returns; the points in his picture don't come back to their original locations. The chance for such a recurrence—a "Poincaré recurrence," as it is called in statistical mechanics—is extremely small in the presence of background fluctuations. But many different shapes will turn up. Like apparitions, they heave into sight and melt away, and for as long as we hold the machine above equilibrium, we are regaled with an amusing, sometimes eerie, display.

That combination of video and computer technology affords a fine tool for demonstrating the morphogenetic capacity of chaos and for analyzing its dynamics. But we really don't need such fancy gimmickry if we want to see organization coming out of chaos, especially molecular organization. We can have this for free; nature is generous in offering spectacles of this sort. Anyone who has stood by a racing stream and has watched the eddies and swirls forming has witnessed this firsthand: in places where the water moves slowly, the flow is smooth because the water molecules slide evenly alongside one another, but where it is fast, the flow abruptly becomes turbulent, chaotic; there, the seemingly formless flow abruptly breaks up into complex whirling shapes.

These shapes are dynamic molecular organizations where billions of molecules move coherently. They are states of matter quite different from the organizations of crystals and biomolecules. Those are molecular organizations, static equilibrium structures in which the bonding forces have ranges of the order of 1 angstrom (10^{-8} cm) and the time scales can be as fast as 10^{-15} seconds, the period of vibration of the individual molecules; whereas the dynamic structures here are supramolecular organizations with macroscopic parameters reflecting the global situation of nonequilibrium that produces them—distances of the order of centimeters and times of the order of seconds, minutes, or hours. Such supramolecular organizations occur in all sorts of flows, in liquids or gases, sufficiently away from equilibrium.

A thermodynamics has been developed for these organizations by the physicist Ilya Prigogine. Unlike classical thermodynamics, this is a nonlinear one. It shows that, far from equilibrium, the classical time-dependent downward slide to disorder is incomplete (Boltzmann's order principle no longer holds), and some of the original information can reemerge in new dynamic forms called *dissipative structures*. Prigogine chose that name to underscore that the struc-

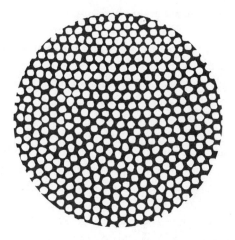

Figure 3.5   Self-organization in water. This honeycomb pattern, a far-from-equilibrium dynamic organization along a uniform temperature gradient (Bénard phenomenon), ensued as the water was gently heated from below. The water (in a round beaker) is shown here in a top view.

tures stem from what in classical thermodynamics would be considered dissipation, energy that goes to waste. The point is that not all of the energy here is wasted; a part is recovered as information in new organization. Since we already made friends with Maxwell's demon, this point will strike a familiar chord.

Some of these organizations are quite spectacular. In the Bénard phenomenon, for instance, water gets organized into a plainly visible honeycomb pattern of macroscopic dimensions (Fig. *3.5*). The phenomenon is easily produced in a home kitchen. All it takes is a pot of water and a device that can heat it uniformly from below and maintain a uniform temperature gradient. The patterning begins when the upthrust of the bottom layer of water overcomes the viscosity of the layers above, producing an instability. Then the system no longer obeys the theorem of minimum entropy production of classical thermodynamics: the driving temperature difference remains unchanged, yet for that same constraint the entropy increases faster, as the heat transfer is accelerated by coherent convection. Part of the information from the heater, which otherwise would have ended up as entropy, emerges as information directing large ensembles of molecules to move in the same direction and to shape a dynamic spatial organization of hexagonal cells.

Like all dissipative structures, the hexagonal water organization arises spontaneously. There are no molds in the water pot to shape the honeycombs, just as there are no templates in the river bed to form the whirls. Chaotic systems are self-organizing. The structural molecules move coordinately as if each molecule knew the system's global state.

How do they know? We are asking here not where the information comes from—the information, as we defined above, is present in the system at the moment of the instability or before—but how it is transferred. The mystery is that the transfer occurs over macroscopic distances. In the two hydraulic examples the dissipative structures have dimensions of centimeters, vast distances for transmission of molecular information. Such transmission, we have seen, is limited to distances of 10^{-8} centimeters, the range of electromagnetic molecular interactions. So, to produce a macroscopic organization, the information can only be transferred serially from molecule to molecule. Each molecule transmits information about the direction of its movement to its immediate neighbors, that is, to whichever molecule happens to be within its 10^{-8} centimeter range, and through such sequential short-range transfer, large molecular ensembles are instructed to move coherently. The system is governed by short-range electromagnetic interactions but, as a whole, it behaves as if it were governed by a long-range force.

This principle of serial information transfer is at the root of many supramolecular organizations. In our next example, such serial information transfer brings forth rather complex patterns in a simple blend of organic and inorganic molecules—malonic acid, potassium bromate, and cerium sulfate—all quite small. Just a little local heating will set things in motion here: with the cerium metal as the catalyst, the malonic acid gets oxidized by the potassium salt, giving rise to a dazzling display of spiral chemical waves (Fig. 3.6).

Well, this makes quite a contrast with the dull reactions of chemicals in homogeneous solution we are accustomed to from high school, which are brightened at best by a uniform change in color. What pulls this striking organization out of chaos is the amplification provided by an information loop, a positive feedback loop between catalyst and reaction product. This loop closes as that product binds to the catalyst, stabilizing it in a conformation appropriate for catalysis. Thus, the product promotes its own production. The loop amplifies the quantum chemical fluctuations in the system, propelling it to instability; at a threshold above equilibrium the formerly random movement of molecules becomes coherent, generating the macroscopic patterns shown.

These patterns get even more spectacular when the flow and mixing of the reagents is appropriately controlled. Then, the organizations become time-dependent: the chemical patterns alternate rhythmically like a clock; as if on cue, the system pulsates with different colors!

The New Alchemy and the First Stirrings in the Twiglet of the Cosmic Tree

Those natural motion pictures give us an idea of the informational potential of deterministic chaotic systems. Evidently, a little bubbling, whirling, and seething

Figure 3.6 Chemical wave patterns. *1–4* are snapshots taken over a period of 6.5 seconds of a chemical reaction in a dish containing a mix of malonic acid, potassium bromate, and the catalyst ceric sulfate ("Belousov-Zhabotinsky reaction"). The macroscopic order here—a coherent movement of diffusing molecules forming spiral patterns—originates from an instability due to a positive (autocatalytic) feedback loop throwing the system off equilibrium. (Reproduced with permission from Arthur T. Winfree, *Scientific American,* 230:82–94, 1974 by copyright permission of Freeman Press.)

goes a long way in organizing matter. This fact did not escape those who once upon a time peered into alchemist retorts and witches' brews. However, only now are we beginning to understand the reasons. That understanding has led to the birth of a new alchemy, an undertaking which is after something more precious than gold: the extraction of order from chaos. The dynamics used here is more complex than the one Newton started, but it might well have appealed to him. Though hardly renowned for that, Newton was an active practitioner of the old alchemy; he did many experiments trying to synthesize gold and wrote a number of papers on that pursuit.[1]

Nature uses many tongues to speak to us. Through the phenomena of turbulent chemistry and physics she tells us clearly that deterministic chaos can be a source of form. Is she also trying to tell us here something about life? It is hard to put one's fingers on this, but there is something about these dissipative structures that summons up the image of life; and if one has been privy to the metamorphoses in a growing embryo, it may be hard to keep down an ah on seeing those forms in Figure 3.6 whirling in front of one.

Such feelings may well be caused by superficial likeness, but our question is prompted by a cooler reason than that: the very elements we have seen to be conducive to organization in chaotic systems are present also in living beings. Much of what goes on inside an organism is above equilibrium and kinetically nonlinear. Indeed, as much as we may yearn for linearity, we find little in the behavior of living beings that, even to a first approximation, could be called that.

Information is found circling at all levels of biological organization, and many of these loops operate in positive feedback mode, like the self-promoting information loop in Figure 3.2. Take, for example, the loops in the machinery that produces our proteins, the machinery that translates the information in DNA. A cell contains as many loops like that as there are proteins, and each loop sees to it that the corresponding DNA information gets amplified a zillion times. The punch of those information circles would be more than enough to propel the cell onto chaotic paths if the underlying quantum fluctuations were as prone to surface as they are in the liquid dissipative structures.

However, the fact is, they are not as prone, because any oscillations that do come up are damped, actively suppressed, by information loops operating in negative feedback mode. Thus, the elements of chaos are kept under a lid, and this is why organisms, at least the highly evolved ones on Earth today, show little of the erratic behavior and none of the fleetingness of chaotic systems. In fact, they are paragons of reproducibility and conservatism, and their core information is held in rather stable, solid structures, as we shall see.

---

1. Newton's papers on alchemy were rescued from oblivion after two and a half centuries, largely through the efforts of the economist Lord John Maynard Keynes. The collection includes Newton's descriptions of the works of ancient alchemists, as well as his own laboratory experiments attempting the synthesis of gold. In the preface of the collection, Keynes gives the context with this penetrating statement: "Newton was not the first of the age of reason. He was the last of the Babylonians and Sumerians, the last great mind which looked out on the visible and intellectual world with the same eyes as those who began to build our intellectual inheritance rather less than 10,000 years ago." *See* Dobbs, B. J. *The Foundations of Newton's Alchemy.* Cambridge University Press, Cambridge, 1975.

But what of the beginnings—to come back to the question—the first stirrings in the biological twiglet of evolution, when that splendid solid-state apparatus had not yet developed and a few smallish organic molecules were all that was? Was then the capture of information and the emergence of order governed by the dynamics of deterministic chaos? There is no clear answer at hand, but this would seem to be within the realm of possibility. In the twiglet's bud randomness above equilibrium may well have contributed its share to the first asymmetric organic molecular forms and even to the closure of the first molecular information loops. Such a use of randomness would have merely continued a standing custom in the main trunk.

That custom of fashioning spatial differentiations of energy and matter out of random fluctuations is as old as the universe. It started at the Planck time when the randomness in the prevailing quantum fluctuations gave rise to the first energy density differentiations (Chapter 2). Then, during the next instants, as the enormous number of particles were created, the fluctuations got drowned out in the gross. The fluctuations never leave the microcosmic floor, but they would not surface again, except when they are sufficiently magnified by nonlinearities. Then, above thermodynamic equilibrium, they would give rise to spatial differentiations of energy and matter again and again—randomness being the source of form.

Now, in the biological twiglet, the elfs eventually begin to dance to a different tune. As information here accumulates in large hard molecular structure and is propelled as well as controlled by information loops, the genesis of form becomes more and more deterministic. However, randomness never entirely gets the sack. From time to time the evolutionary information flow loses its necessitarian impulse and is in the hold of chance. The molecular systems then search for complementary partners who satisfy the electrostatic boundary condition for the continuation of the flow. The continuation is decided by the flip of the coin. That decision behind, the flow gathers way again and follows a predictable course until the next continuation stage, and so on. Thus, all along the biological twig deterministic behavior alternates with happenstance—a curious concatenation giving biological evolution a special stamp.

A Question of Time

And so the twig got layered with information loops—molecular loop upon molecular loop—circles within circles and between circles. But was there enough time for all that growth? This is by no means an idle question if we mean growth from scratch, for the available time on Earth was not so long by cosmic standards. It must have taken an astronomic number of evolutionary trials of chance to produce even one macromolecular loop, let alone the thousands we find in our cells today. So, let's estimate the odds and weigh them against the available time on our planet.

That time, we can read off our cosmic chart (Fig. *3.3*). The earth came into existence 4.5 billion years ago, and the fossil record suggests that cellular beings resembling modern bacteria existed by about 3.4 billion years. Thus, allowing for some cooling time of the planet, there were, at most, some 0.9 billion years (4.5×10^{14} minutes) available for the development of the basic macromolecular hardware. And this is a generous estimate—if anything, it stretches the window of evolutionary opportunity.

Now, to weigh those odds, we need to translate this window of time, which amounts to 4.5×10^{14} minutes, into an evolutionary trial number. For this we will use a path tread some time ago by the astronomer Fred Hoyle and the biochemist Robert Shapiro, and make a generous allowance about the trial space: we assume that the primeval evolutionary laboratory occupied the entire surface of the earth—a surface covered by an ocean 10 kilometers deep, with chemical building blocks in abundance everywhere. If we divide that maximized space then into compartments of cell size, 10 cubic micrometers, and grant a trial speed of 1 per minute—a super speed—we obtain a number of possible trials of the order of 10^{48} [$4.5 \times 10^{14} \times 5 \times 10^{33} = 2.25 \times 10^{48}$].

This is an astronomic number of trials, indeed. But is it enough to beat the odds against macromolecular genesis? Consider a macromolecule of modest size, a protein of 200 amino acids. To figure out the odds for its random assembly, we proceed as we did in the case of the random assembly of DNA from its units (p. 16). So, with 20 different amino-acid species in the soup—20 unit choices— the probability would come out to 1 in 10^{120}. However, things are not quite as bleak here, because there is much more leeway in the linear arrangement of the units. The arrangement need only be catalytically functional, not unique, and a large number of different unit arrangements will yield catalytically functional protein. This vastly improves the odds; for a polypeptide chain of 200 units, the probability for functional assembly works out to be about 1 in 10^{25}.

So far, so good. Our number of evolutionary trials would be more than enough to accommodate those odds. However, there are several thousand proteins to be accounted for—at least 2,000, the number of protein species we find in even the most primitive bacterium today. This pulls the probability for random assembly down to 1 in $10^{25 \times 2,000}$ or $10^{50,000}$.

Such abysmal odds may, at first sight, make one throw one's arms up in despair or spin creationist thoughts. But on cooler reflection, one sees that the very improbability here bears a momentous message: that the primeval macromolecular geneses were not isolated events, but must have been linked to one another. And this is not a weak hint, but an argument born of reasoning by exclusion. Indeed, those odds banish any thought one might have cherished about macromolecules arising one by one in the primordial soup, independently from one another—a piecemeal mode of evolution that had been widely taken for granted by biomolecular evolutionists. Those odds leave open only one possible mode, a mode of multiple macromolecular genesis with an information connection, namely, an information-conserving connection—this is the only possible way that the odds against macromolecular genesis could have been brought down enough from those extreme heights.

Now, this is precisely the coign of vantage the molecular information loop in Figure 3.2 has to offer. Recall that this is a positive loop where the products of molecular genesis promote their own production—a self-catalyzing system. Such a loop is inherently information-conserving, and a large set of loops, properly connected, can squirrel away immense information amounts. What matters, then, is the probability with which such loops get closed. And that probability, we have seen, depends on the complementarity between the molecular participants, the three-dimensional fit between the molecular surfaces required for catalytic information transfer. Thus, the more kindred macromolecules are present in the soup, the higher is the chance for molecular fits and, hence, the higher the probability for catalytic loop closure.

This much is seen intuitively. But, intuition is not enough; we need to quantitate that probability, or at least get a rough estimate of it. Alas, with so many macromolecules in the soup, the number game gets complex. Fortunately, the mathematical ways and means for analysis of such complex systems are now at hand, and we may lean on the results that Stuart Kauffman has obtained with systems of linear autocatalytic polymers. In such systems, the catalytic probability rises nonlinearly with the number of polymer species in solution; and when the number of species crosses a threshold, that probability rises so dramatically, Kauffman showed, that a whole self-sustaining autocatalytic network suddenly materializes in the solution.

Thus, given a large enough family of macromolecules of the sort we have considered, the closure of catalytic information loops becomes virtually a certainty. And so our question of time receives a heartening answer, after all. It was just a matter of directing the question to the right evolutionary unit, the information loop. So, assured on that account, we now take aim at this circular unit and inquire into its information and energy requirements.

From Energy to Bonds to Molecular Information: The Information/Entropy Balancing Act

How to Pull the Information Rabbit from the Cosmic Hat

The solid-state gear which embodies the deterministic information core in living beings is made of DNA, RNA, and protein. Vast amounts of information go into the making of such macromolecules, and all that information ultimately comes from photon energy, as we have seen. But how does this energy get to the molecular construction sites, and in what form?

Let us first take a bird's eye view of the energy flux through a forming macromolecule. We need to consider here the macromolecule and the immediate environment with which the macromolecule exchanges energy. So, we have then: an energy source—the sunlight; two connected systems where the energy flows through—the macromolecule and its environment; and an energy sink—the outer space that the environment gives off the unused energy to (as infrared radiation). The situation in the environment is readily defined. Because the source and sink are for all practical purposes constant, the energy flux is constant too, and this has been so for a long time. Thus, the environment is in a time-independent balanced state above thermodynamic equilibrium, or what is called *steady state*.

As for the nascent macromolecule, the situation is largely ruled by the energy flux in the environment. While this flux produces the work necessary to hold the environment in steady state, it also keeps the connected system—the amalgamating building blocks of the nascent macromolecule—above equilibrium. And this is the energy portion that interests us most, for it drives the work of making the macromolecule and nurses its organization.

Energy is not in scarce supply. There is a flux of 260,000 calories at every square centimeter of the earth—10^{24} calories per year—coming from the sun in the form of photons. So, the extent of macromolecular organization, the amounts of macromolecular information that can be generated, are limited only by the entropy produced in our two systems under the conservation law. In principle, then, the production of macromolecular information can be as large as you wish, so long as the gain of molecular information (ΔI_m) and the sum of the production of entropy in the macromolecule (ΔS_m) and of the entropy in the environment (ΔS_e) balance out:

$$\Delta I_m = \Delta S_e + \Delta S_m. \qquad (3)$$

This, then, is the formula to pull the information rabbit from the cosmic hat. It is a formula that all living beings have down pat and use each time they add an informational piece to themselves. We shall see more of how they do it further on — it is easily the most dazzling of all magicians' tricks — but it is good to know in advance that it passes muster with any thermodynamics accountant.

A Simple Organic Genesis

With this formula we are almost ready to go to the mat with the genesis of macromolecules. But it will be useful to take a look first at a creation of a simpler ilk: the formation of a small organic molecule — an aldehyde, for example.

Aldehydes, like HCO and $HCHO$, form from CO and H radicals when a mixture of water and carbon dioxide is irradiated with ultraviolet (or alpha particles). There is more organization in the emerging hydrocarbon chains than there was originally in the CO_2 and H_2O mixture, and the ultraviolet energy flux supplies the information for this. Overall:

$$CO_2 + H_2O \underset{\leftarrow}{\overset{h\nu}{\rightarrow}} HCO + HCHO + O_2.$$

Entropy (ΔS_e) here is produced in two steps, as the parent molecules are broken by the radiant energy, $h\nu$:

$$CO_2 + h\nu \rightarrow CO + O$$
$$H_2O + h\nu \rightarrow OH + H$$

and the information (ΔI_m) generated in the reactions

$$H + CO \rightarrow HCO$$
$$2\,HCO \rightarrow HCHO + CO$$
$$O + O \rightarrow O_2$$

can be shown to exactly balance out according to equation *3*.

The amount of information gained in this case is a pittance by macromolecular standards. But it is not impossible to get somewhat larger amounts by this simple principle; for example, amino acids will form in mixtures of CH_4, NH_3, H_2O, and N_2 when these are exposed to ultraviolet, ionizing radiation, or electrical discharge. In fact, a broad variety of amino acids then come out. The purine adenine (a constituent of DNA and RNA and, in combination with phosphate, a central actor in energy utilization) can form from HCN and H_2O in ultraviolet. Sugars and even complex nucleosides, like cytidine, assemble in appropriate test-tube conditions.

Thus, when enough energy is passed through equilibrium systems of appropriate atomic composition, many of the basic building blocks of macromolecules organize themselves. This is how one imagines that the first building blocks formed under the ultraviolet and lightning that prevailed on Earth before the advent of cellular life.

The Energy for Organizing Molecules with High-Information Content Must Be Supplied in Small Doses

So far, so good. But there is a world of difference between a building block and a macromolecule, and as we try to apply the foregoing principle to the generation of such huge amounts of information, we come up against the vexing problem of how to put so much energy into a molecular system without blasting it to smithereens.

When energy is pumped into a chemical system, the energy partitions into thermal and electronic components. The thermal component makes the molecules move faster, and the electronic component increases the number of their high-energy electronic states. Both energy components will foster molecular organization: the faster the molecules vibrate, rotate, and translate, and the more of them that are in electronic states above ground level, the higher is the probability that the molecules will interact and the more work can be done in organizing them. However, there is a limit to that. At very high energy levels, all chemical bonds become inherently unstable. Thus, as the energy flux increases, the system, having gone through a maximum of organization, comes apart. All its molecular structures eventually fall to pieces, the system entering into the state of stormy ions and free electrons, the *plasma state*.

This property of matter draws the line to the energy input, and this is why no one ever made a macromolecule in one run from scratch. However, nothing, in principle, stands in the way of making such a molecule piecemeal by supplying the required energy little by little. Imagine, for example, a synthetic program, where first a few atoms are joined to form small ordered aggregates, after the model of the synthetic aldehydes and amino acids, and then, in turn, the aggregates are joined, a pair at a time, to form larger ordered aggregates, and so on.

This way each step of molecular synthesis could be driven by a separate and tolerable energy input.

That is precisely the strategy used by living beings. They build up their macromolecules with modular pieces, and pay the energy price under an installment plan. The energy supplying the information for the synthesis doesn't flow through the nascent molecular structure all at once; rather, the energy stream is broken up into many parallel pathways, each supplying its dose to an individual synthesis site. Installment plans are never cheap—this one costs the living system a king's ransom, as we shall see—but there is simply no other way to accumulate large amounts of molecular information.

The Energy Flux Nursing Biomolecular Organization

To pay their installments, organisms generally use chemical coin, rather than photons. This way they can meet more readily the heavy energy needs of macromolecular synthesis; covalent bonds can pack a good deal of energy. With photons only a limited amount of molecular organization could be achieved. Organic molecules will not grow longer spontaneously; the bonds in systems made of C, H, O, and N atoms are not efficiently rearranged at equilibrium for chain elongation. They tend to be dominated by C=O, H-O, and N=N bonds, none of which can add atoms to their ends. Thus, only low grade CO_2, N_2, and H_2O structures will be on an equilibrium scene. Now, as we pump in energy, a dozen or so new covalent bonding possibilities arise, but most of them are short-lived configurations and the yield of molecular elongations is low. The C-O bond forming under ultraviolet in our example of aldehyde synthesis is a case in point. Only two of the bonding possibilities—C-C and C-N, which are stable under ordinary conditions—could substantially contribute to molecular growth. Repeated pumping in of photons would lead to modest chain elongation, but such an operation would yield progressively diminishing returns—hardly the sort of thing that would be useful to make macromolecules.

Living beings favor a more roundabout strategy to that end. They store photon energy in chemical form, and then trickle it down molecular chains to the individual molecular bonding sites. The energy flux that organizes all living matter on our planet—to start with the wholesale picture—is so channeled as to first pump CO_2 and H in the atmosphere and water up to the level of carbohydrate, namely glucose, and then to drop the level gradually from that reservoir back to ground again.[1] This gradient drives nearly all work in the biomass, not just the making of macromolecules.

1. The pumping is on a gigantic scale. Among all the organisms on land and in the sea, some 10^{14} kilograms of CO_2 get fixed in a year and converted into carbohydrate. If all this carbohydrate were in the form of sugar cane, it would make up a mile-high heap covering some 85 square miles.

The flow starts with the capture of photons by certain molecules, such as the chlorophyll of plants and similar pigments of microorganisms. This stage is not very different from that in some solar-energy devices of modern technology, which use a selenium mineral as light absorber: the photon energy raises electrons in the absorbing atoms to a higher energy level, enough to hurl the electrons out of their orbits so that they can be led off. But there the analogy ends. Instead of flowing freely in a wire, the electrons in biomatter jump from molecule to molecule—a whole series of jumps in higher organisms. Such acrobatics are rarely seen outside the living world, certainly not so many leaps again and again. It took Evolution aeons to prepare the field for such vaulting: a chain of molecules—many of them proteins—with exactly the right electrical potential. Besides, the photon energy, instead of being stored in an electrical battery, is stored in the covalent bonds of glucose—about 8 quanta of light in one glucose molecule. From this reservoir, energy then flows along various pathways, nursing everything, all organization and all work.

There can be a dozen or more molecules in series in such a pathway, yet the energy flows continuously and without undue losses. A physical chemist unaccustomed to the wonders of biochemistry may well shake his head at this point in disbelief. The situation is a little like that of the Sultan in an old story, who fretted about another sort of continuity. The Sultan woke up in a cold sweat in the middle of the night and summoned his wizard. "Wizard," he said, "my sleep is troubled; tell me, what is holding up the world?" "Great Sultan," replied the Wizard, "the earth rests on the back of a giant elephant . . . and you can stop worrying right there; it's elephants all the way!"

The elephants here are proteins with the right redox potential. Some are in the cell water, others in the fatty matrix of the cell membrane. They are the links of chains where molecules transfer electrons or chemical potential to one another. The overall efficiency in such chains—the percentage of the energy that is not lost as heat—is reasonably high. For example, the efficiency in the segment photon-to-glucose is between 30 and 40 percent. This is better than the efficiency of a standard gasoline engine, but by no means spectacular. But, we should keep in mind that it's not energy that counts in biological systems, but information; and when it comes to amassing that commodity, there is nothing on Earth that could hold a candle to those systems.

The chemical energy chains that nurse macromolecular organization commonly use *ATP* as their final link. This little phosphate molecule—an adenosine triphosphate—transfers the energy to the sites of molecular synthesis. It has served the task of transferring chemical potential between molecules since the early times of biological evolution (the adenine nucleotide NAD plays an equivalent coupling role in the transfer of electrons between molecules in the membrane matrix). The partial double bonds between its phosphate tetrahedra represent an energy package of 7.3 Kilocalories per mole, which is set free when one of the tetrahedra is split away by water. This package is given off at the sites

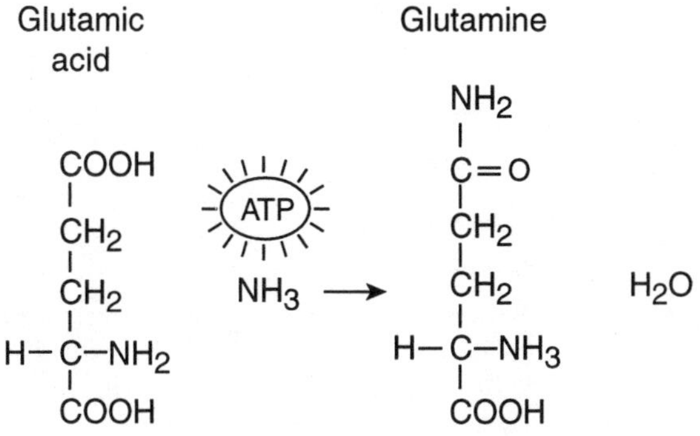

Glutamic
acid

Glutamine

$$NH_2$$
$$|$$
$$COOH \qquad C=O$$
$$| \qquad\qquad |$$
$$CH_2 \quad (ATP) \quad CH_2$$
$$| \qquad\qquad |$$
$$CH_2 \quad NH_3 \longrightarrow CH_2 \qquad H_2O$$
$$| \qquad\qquad |$$
$$H-C-NH_2 \qquad H-C-NH_3$$
$$| \qquad\qquad |$$
$$COOH \qquad\qquad COOH$$

(a)　　　(b)　　　　(c)

ATP + glutamic acid --> glutamyl phosphate + ADP

(c)　　　　　　(d)

glutamyl phosphate + NH$_3$ --> glutamine + phosphate

Figure 4.1 The ammonium (NH$_3$) ion is linked to the glutamic acid as the hydrolysis of ATP delivers the necessary energy package. The phosphate of ATP couples transiently to glutamic acid, supplying the energy for linking the NH$_2$ group to the carbon chain. The two-step reaction, with glutamyl phosphate as the intermediate, is shown below. The entire reaction is catalyzed by an enzyme, glutamine synthetase, which we will see in action later on (the letters *a, b, c, d* stand for the complementary sites on this enzyme).

of amalgamation of the molecular building blocks—one package for each site. And this is the payment for the "installment plan."

The advantages of the roundabout biological strategy of converting photon energy into chemical energy are thus clear. This way, energy can be delivered in sufficiently large packages; and not only that, the energy can be meted out with spatial precision to where it is needed. For example, in the simple case of molecular synthesis illustrated in Figure *4.1*, the 7.3 Kilocalorie package of ATP is targeted onto the spot of synthesis of a molecule of glutamine from its amino acid and ammonium. Here the ATP energy is fed into the site of elongation, the last member of the carbon chain, while the phosphate transiently couples to the amino acid and, by way of this phosphate intermediate (glutamyl phosphate), the energized molecular system stably links a NH$_2$ group to the chain.

Light into Bonds into Information

Given its abundance, it is not surprising that Evolution chose light as an energy source to drive living systems (Fig. *4.2*). But why visible light and not ultraviolet? Ultraviolet must have been as abundant as the visible spectrum early on on Earth; the layer of oxygen and ozone, which now shields the ultraviolet off,

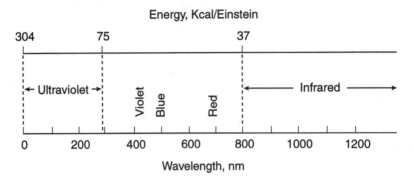

Figure 4.2   Light spectrum and energy.

arose only later as a product of life. Besides, ultraviolet is well absorbed by organic molecules and, like the longer visible wavelengths, specifically so. Ultraviolet not only heeds overall atomic composition but also the finer grain of molecular structure (the electronic transitions are sharply sensitive to wavelength here). Such a specificity is a good thing for a chemical system that transduces radiant energy. Its lack must have weighed heavily in natural selection against the wavelengths shorter than ultraviolet, whose energies spread rather randomly in the absorbing molecules. Infrared and radio waves were hardly ever contenders; they are randomly absorbed and convert right away to thermal energy modes.

So, ultraviolet may once have been important in fashioning prebiotic molecules after the aldehyde model. But what probably eventually decided the choice in favor of visible light is the good match between its energy quanta and the energy level of the covalent bond; the quanta of visible light are in the range where these bonds are stable, whereas the higher quanta of the ultraviolet break them apart.

Whatever the reason, the fact is that light became the energy input and life's evolution became inextricably geared to it. If there is something like a motto of evolution, in that early phase it must have been, "light into bonds." First and foremost this meant covalent bonds. These are the forces that hold atoms together by a sharing of electrons. Among the various types of bonds between atoms, these are the most durable ones, and evolution entrusted them with the function of photon energy conversion and conduction.

Julius Mayer, the discoverer of the law of conservation of energy, wrote about this more than a hundred years ago, long before our knowledge of the nature of chemical bonds:

> Life has set itself the task to catch the light on the wing to earth and to store the most elusive of all powers in rigid form.

One could hardly think of a more elegant phrasing, except perhaps for the terse dispatch in the Bible coming before the account of animation, "And there was light."

The Three-Dimensional Carriers of Biological Information

Slam-Bang Is No Substitute for Information

The energy transmissions in the pathways nursing biomolecular synthesis usually wind up, as we have seen, with a 7.3 Kilocalorie puff from ATP. This energy package serves to weld together the amino-acid units of proteins and the nucleotide units of DNA and RNA, as well as the units in sugar and phospholipid molecules that abound in cells.

However, this hardly fills the bill. Though the ATP package represents a significant amount of information, a lot more is needed to build those molecules up from their units. The units all must be in the right sequence and must have the right orientation before they can be welded together. The probability for these arrangements to arise spontaneously is extremely low; we would have to wait a very, very long time for any significant number of molecules to form. Biological molecules are not altogether unique in this regard. Most molecules on the earth's surface have fairly stable electronic configurations—a sign of our Earth's old age. Things just don't go sizzling or exploding around us; most molecules are not very reactive. It is not that reactions cannot occur, but their probability is low. A potentially explosive mixture of H_2 and O_2, for example, can stay for years at room temperature without appreciable reaction.

But these things can change quickly if the molecules are offered something to hold on to. Just sprinkle a little platinum black into that room with H_2 and O_2, and you have an instant explosion! The platinum black itself does not enter in the reaction here, nor does it affect the overall energy of the reaction. It merely offers a surface where the gas molecules can attach to—a matrix; and, as they attach, they are forced into a spatial orientation where their diatomic bonds get weakened, an orientation more favorable for reaction. Such matrices are called *catalysts.* They are the keystones of synthetic chemistry, both biological and non-biological.

Let's look at this a little more closely. First, what is a reaction? When chemists talk about a reaction, they mean an affair between molecules where their structures are broken apart and bonded together in a new atomic combination. Such breaking of atomic bonds and forming of new ones involves a good deal of pulling and pushing of electrons. The pertinent atoms or groups of atoms, therefore, must get within range of their electronic orbitals. However, it is not always easy for the atoms to get that close to each other, especially in the case of the organic molecules. Their motley atoms are strung out in various directions, and many of them may not be readily accessible for interaction; they may be surrounded by other atoms and by forbidding electrostatic fields. It is a little like trying to reach an important someone at a crowded cocktail party, without letting go of one's own retinue. Slamming through might help but, in the molecular case, such a brute-force approach is limited by the temperature around. An approach using intelligence is better, and here is where the catalyst comes in.

The Molecular Jigs

A little slam-bang actually is not a bad thing. Some kinetic energy is needed here or else the electron clouds won't interpenetrate, and instead of reacting, the colliding molecules simply bounce off each other again. A minimum of energy is required even when the molecules are optimally oriented for interpenetrating collisions. Such fortunate positioning happens rarely, though; the fields of potential energy of molecules with heterogeneous atoms are quite uneven, and there are not many paths by which two such molecules could approach each other. For molecules with just a few atoms, there is often only one good plane of approach, and in large molecules, only a narrow, tortuous path. All other approaches involve more hindrances and need more kinetic energy. Therefore, when billions of such molecules are moving about randomly, they need, on the average, a higher energy than the minimal one to interact. This statistical average is the *activation energy* of the reaction.

The activation energy requirement can be lowered by providing the molecules with information on how to orient themselves more favorably for interaction. It is like giving our participant in the cocktail party the spatial coordinates for homing in on his target. The information for the molecules must be raised above the background noise and the occasional outside perturbation, just as our message to the partygoer must be above the background din and the occasional boom of laughter.

But how does one guide molecules along such tortuous paths of approach? If we were dealing with strongly magnetic molecules, we could think of several dynamic ways of how to steer them. But organic molecules are rarely so obliging. Thus, the way to go is to make use of the thermal energy of the two reacting molecules to get them from one place to another and to guide them to the proper final position with a jig. We shall map out this strategy in the following pages. But, in short, the recipe is this: *let the reacting molecules jiggle and trap them in the right orientation on the electrostatic field of the jig.*

In living systems such jigs are proteins. Indeed, proteins are magnificently suited for that. They are made of modular units, the amino acids, offering a huge number of combinational possibilities. There are twenty different kinds of amino acids, and typically proteins contain from fifteen to a few hundred of them. Thus, in a protein of, say, 200 amino acids, 20^{200} variants are possible in principle, if all amino acids occurred with the same probability.

The strings of amino acid form three-dimensional structures, coils and pleats, which are stabilized largely by noncovalent interactions. The backbone is covalently bonded, but the coils and pleats are mostly held by weaker forces—hydrogen bonds, van der Waals, hydrophobicity, ionic pairs— with an occasional coordinated bond to a metal ion or a sulfur atom. All of this adds richness of shape to richness of unit sequence—a virtually inexhaustible repository for inscribing information. There is nothing in the inorganic world that could carry that much information and, not surprisingly, the platinum blacks, nickels, aluminums, and other catalyzing surfaces of that world are catalytic dwarfs by comparison. Such metals can guide in two dimensions only and, hence, are good solely for catalyzing reactions between very small molecules or molecules with straight and regular shapes. Proteins can guide in three dimensions and, beyond that, are moldable and nimble, adding speed.

Proteins come in so many different forms and shapes that one gets the impression that living beings have a limitless repertoire of molecular jigs. The repertoire is the product of a long evolutionary search for complementary molecular forms, and that very complementarity also makes proteins good selectors; they can recognize and select one molecule out of a myriad. This is exactly, we recall, the capacity Maxwell had endowed his demon with. It is a capacity born of information. If we have information, we can select things; with three bits we can select one out of eight molecules (Fig. *1.2*) and with twenty bits, one in about a million. The information here resides in the three-dimensional configurations of the protein, their shapes. So, the selectivity is inherent in the molecular shape, we might say.

Crude selectivity devices are offered by proteins that form the pores in cell membranes. Some of these proteins only wield enough information to select molecules by their size and charge. They are electrically charged sieves. They can separate the small cellular inorganic ions from the larger or more charged ones and generate a degree of order if they are provided with an energy source that defrays the entropy cost of this cognitive act.

Some sieving of this general sort also goes on in proteins with more finesse. But there the sieving—a sieving of water molecules—is a sideplay, as it were, of the main cognitive act. The recognition sites of the proteins, which serve as jigs, often lie in pockets or narrow clefts that are lined with water-repellent atomic groups which keep the water out. Water, the most abundant molecule in living beings, is bad news for catalytic action. It is a dielectric, an insulator that keeps the charged molecules apart, hindering their electron-orbital interactions. Those clefts take the recognition sites out of the water, so to speak, allowing them to do their catalytic job.

But many proteins do much more than just sieving water. The catalytic ones, the *enzymes*, are capable of fine-tuned selection of substrate molecules. The enzyme pocket here selects the molecule with the complementary shape. There is quite a bit of information built into such a pocket, and the more complex the pocket's configuration, the lower the probability that the wrong molecule will fit into it. We are dealing here with a very different order of magnitude of information than in the case of the sieves. The selectivity is more like that of a lock and key. Indeed, the lock-and-key metaphor is often invoked by biologists in this connection. It is the brainchild of the chemist Emil Fischer—one of the great visions in science, dating back to before the turn of this century, long before one knew about the structure of proteins or about molecular information.

Enzymes can be of a loose fit, like a simple skeleton key, as well as of a high-precision fit, where every nook and cranny has its complementary counterpart on the substrate. Thus, as the molecules in the surrounds are jiggled by thermal energy against the enzyme, only the ones that fit the enzyme mold do attach, while at the same time they are forced into the optimal position for reaction.

All proteins which transfer information—and nearly all of them do—can select molecules. But what makes a protein an enzyme is the capacity of promoting a chemical reaction. Consider, for example, a reaction between two substrates. Such a reaction requires that the two molecules approach each other and collide in a particular orientation, and do so with sufficient energy. Very, very few among the possible substrate positions and states are suitable for that. It takes a lot of information to guide it into just the right position and to find the right states. And this is precisely what the enzyme offers. It has two regions built into its structure which are complementary in shape to the substrates and lie side by side, and so can bind the substrate molecules, fixing them in space and juxtaposing them for reaction.

One might score some success here also without the enzyme information by using the brute statistical approach of dissolving one substrate into the other at high concentrations. But the action of the enzyme goes way beyond what concentration could achieve. By constraining the possible substrate positions, the opportunity for reaction is immensely increased. And "immensely" is the right word, for to get comparable probabilities of reaction, one would have to raise the substrate concentrations to levels of the order of 10^{10} M, which are physically unattainable. What the enzyme does here is, in effect, to whittle down the activation energies for reaction. This is a tough row to hoe—ask anyone who

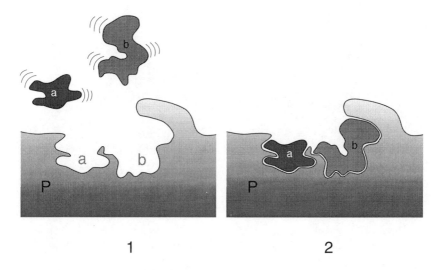

Figure 5.1   Two molecules, *a* and *b*, are selected and oriented in space for optimal interaction by their complementary nooks on the enzyme mold *P*.

earns his keep making chemicals! But enzymes have a whole bag of tricks for that: they can strain a substrate bond to near the breaking point by binding to the substrate's vulnerable molecular transition form; or they can destabilize a substrate bond by placing it next to a polar group on their own structure (a trick of the same stripe chemical engineers use in acid/base catalysis, though the enzymes do it better); or they can divide a reaction into several steps (each with a relatively small activation-energy demand) by holding all the participants in place and in the right sequence for a chain reaction; and so forth.

Information Is Transferred from Molecule to Molecule by Weak and Strong Electromagnetic Interactions

So, the pieces of our molecular jigsaw puzzle now are falling into place (Fig. *5.1*): the enzyme selects the pieces—the unit building blocks, the energy-delivering ATP, and other intermediates—out of a mix of thousands of molecular species and aligns them optimally for reaction. The heart and soul of all this is the information contained in the protein mold. There is a nook in that mold for every piece and enough information passes hand for three functions: (*1*) the recognition of each piece, (*2*) the optimal positioning of the pieces relative to one another, and (*3*) the weakening or splitting of certain bonds in the pieces or the making of new bonds for welding the pieces together.

Thus, in the making of a glutamine molecule—to return to our simple biosynthetic example of the preceding chapter, the pertinent enzyme (glutamine syn-

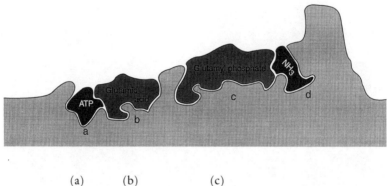

(b)   (c)
ATP + glutamic acid → glutamyl phosphate + ADP

(c)   (d)
glutamyl phosphate + NH₃ → glutamine + phosphate

Figure 5.2   A protein mold for sequential molecular interaction. A schematic representation of the nooks on the enzyme glutamine synthetase, which selects the ATP, glutamic acid, amonium (NH_3), and the intermediary glutamyl phosphate molecules and positions them optimally with respect to each other for electron-orbital interaction. As a final result, the amonium ion is welded to the glutamic acid lengthening that structure by one chain link, to glutamine. The ATP supplies the entropy flux that nurses the gain of information in this molecular growth.

thetase) presumably has four nooks engraved in its structure—each with the complementary shape of one participant—for carrying out function *1*. The nooks are so oriented and sequentially arrayed as to constrain the participants also appropriately for functions *2* and *3*, and this includes a nook for the ATP, which positions this energy donor appropriately on the welding spot (Fig. *5.2*). In short, the enzyme mold here contains the core program for a sequential chemical reaction.

We see here, in steps *1* and *2*, a basic mechanism of biological communication at work: transfer of information by intermolecular forces—a recurrent theme in this book. One molecule, an enzyme here, forces the other molecules, the substrates, to take up certain positions in space. These determinate positions represent a state with fewer degrees of freedom than their former random wandering in solution. So, clearly, by our definition (equation *2*), the substrate molecules have acquired information; or, looked at from the other side of the transaction, information from the enzyme structure has been passed on to the substrate molecules, orienting them. We might say that the enzyme molecule has told them the spatial coordinates they should occupy.

So much about information transfer. Now, to the force that implements it. From what was said in Chapter 1, we know what force to expect in general terms: the electromagnetic force, the gray eminence behind all molecular information affairs. But it's time now to spell out the particulars of the interacting

 1

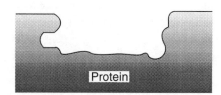

2

Figure 5.3 A flexing enzyme. The diagram shows the protein (*1*) before the encounter with its teammate, the enzyme substrate (*S*), and (*2*) after it. The protein snaps into its lock conformation upon interaction with the substrate, embracing it. (*P*) protein; (*S*) substrate.

electromagnetic fields between the molecules. In the case of step *1*, the information transfer is mediated by weak electrostatic interactions—induced dipoles (van der Waals), permanent dipoles, net charges, and sundry combinations. Besides, hydrophobic interactions may come in. This kind of interaction, more than from the protein itself, stems from the lack of attraction between the hydrophobic substrate and the water; the water expels the substrate and pushes it into the protein pocket.

Step *2* is by and large mediated by weak interactions, too, but in proteins of the precision-lock type, stronger forces enter in addition. Step *3*—a special function of enzymes—is, by definition, mediated by strong interactions. Here is where the narrow water-free clefts contribute their part. Dominated, as they are, by the relatively low dielectric constants of hydrophobic groups, these clefts provide an environment where strong electrostatic forces can be brought to bear on the reactants, rapidly breaking and making their bonds.

Molecular Relays: The Foolproof Genies

Their mold and matrix function notwithstanding, proteins are dynamic structures. Even many of those of the precision-lock type arrive at their lock form in a dynamic way. They snap into it on encounter with their teammate, embracing that molecule, as it were (Fig. 5.3). Such nimble behavior has its advantages. There are many kinds of molecules in cells that could enter an enzyme pocket and get in the way of the reaction—even the breakdown products of ATP or those of the substrate itself might do so. A rapidly flexing and unflexing enzyme sidesteps this problem.

The flexing enzymes belong to a class of proteins, the *allosteric* ones, that can adopt two or more stable conformations. All proteins have a stable form in their water or lipid habitats, the form their polypeptide chains spontaneously fold into. Like most polymers, the polypeptides can fold because many of the bonds that hold their atoms together can rotate; these covalent bonds are like hinges. But it is one thing for such a structure to fold and another to do so always the same way. An average protein of 500 amino acids has some 5,000 atoms, many of which can rotate. Nevertheless, the protein folds up in an invariant way and stably so.

This is a skill that is not easily come by. If you string up amino acids in random sequence in synthesis in the test tube, they may, with some luck, form coils. But they are likely to be loose and will sway and teeter and totter; generally, they won't even fold. Stable conformations are the products of a long evolutionary search among proteins and their polypeptide subsets. And as if that wasn't achievement enough, the allosteric proteins have two stable conformations. They can shift back and forth between these states with the help of a teammate. This teammate binds to the protein and stabilizes it in one of the states. The probabilities for the two conformations are generally not too different, and so these proteins can flip-flop with rather little input of information—they are molecular switches.

Such switches come in various styles. Some are touched off by weak interactions, others by strong ones, or by both. This depends on who the teammate is. That partner, for instance, can be a weakly binding polypeptide or a substrate, or a covalently binding phosphoryl, acetyl, or methyl group, or even just a little calcium ion (which binds to a set of oxygen atoms of the protein). But whatever the switch style, the teammate is always specific—only one particular molecule can pull the switch, the one that fits the protein's mold. *So, more than switches, the allosteric proteins are sophisticated relays that obey only ciphered commands.*

These protein relays are the keystones of biological information transfer. The principle of their operation is illustrated in Figure 5.4 . The flipping of the relay starts with the teammate binding to a site of the protein, disturbing the configuration there. This disturbance then propagates through the protein structure to another site, causing it to adopt another conformation. The protein behaves much like a solid here. Ordinarily, its structure is rather stably pent in by the interatomic forces; the bonds between the atoms along the polypeptide backbone are much like stiff springs. So, the individual atoms do not move far from their aver-

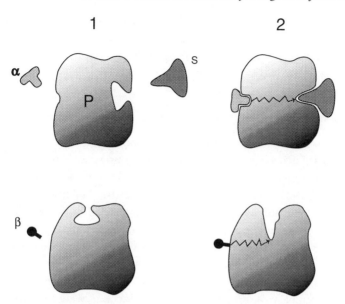

Figure 5.4   Molecular relays. The allosteric protein (*P*) flip-flops between conformations *1* and *2*. The *top* diagram represents an example where the protein has a small peptide (α) as a teammate. The interaction with α flips the structure to conformation *2* (in this conformation it can accommodate molecule *S*, say, a substrate). The interaction, a binding by weak electrostatic forces, causes a shift in the conformation at the interaction site, which propagates through the protein structure. The conformation flips back into *1* when α is absent. In the *bottom* diagram, the protein example has a covalently interactive teammate, a phosphoryl group (β).

age position of balance. But if a group of atoms here does get moved to another position, the whole lot must move, as in a domino. And this is precisely what happens when the group is thrown off balance by the electrostatic field of the teammate. The movement gets damped to some extent by the friction between the structural atoms and between these and the molecules of the surrounding liquid; still, it can spread far into the molecule and with acoustic speed, and the molecule may end up in a new configuration in thousandths or even billionths of a second. Thus, a cell-membrane protein pore can open or close in a flash, or an enzyme is turned on with one twitch and turned off with another.

These sprightly proteins are found everywhere in living beings, taking care of the transmission of information inside cells and between cells. They are the connecting links and switching stations in the vast cellular communication networks, and if we could peer with high enough magnification into an organism, we would see their diligent twitchings channeling the information everywhere, and their relentless hustle and bustle controlling and implementing everything in the three-dimensional sphere. They are the genies which, over the aeons, Lady Evolution bestowed on those who she favored. The powers thus granted were immense. But if the image of the Sorcerer's Apprentice should cross your mind, you can put a stop to it right here. These genies are foolproof; the lady saw to it that they can be turned off with ease.

The Molecular Machines for Coherent Transfer of Information

The preceding notions about protein relays took more than two decades to mature. The seminal idea of allosteric proteins was proposed by Jacques Monod, Jean-Pierre Changeux, and François Jabob in 1963 and was shaped in the following decade to the concept of flexing enzymes by Daniel Koshland. These days, the internal motions of proteins can be explored in computer simulations, and things even are advancing to the point where one can begin to see these molecules with nearly atomic resolution. So, a detailed picture of the workings of these prodigious structures may not be too far off.

But we don't have to wait for that to get a feel for their puissance. These nimble proteins can do what a rigid one never could: embrace another molecule all around. That partner, thus, can be fitted on many sides and this maximizes the power of recognition and information transmission of the protein. Indeed, from the information point of view, molecular evolution reaches its peak with these dynamic proteins and, not surprisingly, living beings use them in all transactions where large amounts of information pass hand.

Allosteric proteins can interact with a broad spectrum of molecules, organic and inorganic ones, but the most massive information transfers take place when they interact with one another. It is not uncommon that two or more such proteins band together to form a cooperative ensemble. Provided with a sufficient flow of energy, such ensembles can transfer information serially; one protein passes information to a neighbor, which in turn passes information to another neighbor, and so on. The ensemble itself is held together by weak electrostatic interactions. Such interactions by and large also implement the information transfers between the members when one of them switches to an alternate configuration; though occasionally the transfer is mediated by stronger electrostatic interaction. The flow of information here generates an ordered sequence of conformational changes, enabling the ensemble to perform complex coordinated actions. For instance, the ensemble may move a substrate from one enzyme site to the next and interact with it sequentially. And all this takes place segregated from the cell water—in the dry, so to speak—immensely raising the probability of molecular encounter and interaction.

The larger ensembles of this sort are coherent and largely self-sufficient machines, veritable assembly lines where the substrates are worked step by step into an end product. Many of the complex molecules in living beings, including the proteins themselves, come from such assembly lines. These allosteric ensembles may contain quite a few members. For example, the one that makes the proteins, the *ribosome*, is an ensemble of some sixty protein and RNA molecules (rRNA), which is large enough (about 250 angstroms) to be seen in the electron microscope. This ensemble has such awesome information power that it can sort out and link up hundreds or thousands of individual amino acids in a particular order.

The Proteus Stuff

Organisms make a myriad of different molecules. These productions are all variations on the same theme: energized modular molecular growth instructed by information-containing molecules. The making of glutamine from glutamic acid, we have seen (Figs. *4.1* and *5.2*), provides an example of such growth. Since molecular growth invariably entails three-dimensional organization, we find proteins instructing the manufacture of all biomatter: carbohydrates, lipids, proteins, and their blends—even DNA and RNA.

So, each type of synthesis has its protein. There are enough proteins around to afford such private instruction. Although named for *protos*, in Greek, "the first," they are more like Proteus, the sea god who could assume many shapes. Collectively, the sundry proteins in an organism can match this god's every shape, and his wonders to boot.

Individually, however, the proteins specialize in one or, at most, a few shapes; each protein only has one or a few sets of instructions to give. This specialization probably goes back to prebiotic times, to the first three-dimensional catalytic structures. Selection favored systems that could preserve the information contained in these structures in amalgamated one-dimensional form, systems that could record the information on a tape, so to speak. When eventually modular linkages came into being, that could pass on the information from the tape to the three-dimensional world—the direction of information flow we find today—the information was broken up again into structural units akin to the aboriginal three-dimensional ones. This strategy saved information and had potential for combining in the three-dimensional world the diverse instructions contained on the tape.[1] We have the benefit of 4 billion years of hindsight to know how enormous that potential really was. We see the structural materializations in every living being around us, in the total coherence of their information transfer and in their endless adaptability.

---

1.   An alternative, a strategy based on a manifold-instruction principle, conceivably might have worked too, from a certain stage on in prebiotic evolution—an interesting possibility to be pondered by those who follow Arrhenius's *panspermia* idea. However, if life on earth started from scratch, it is hard to imagine how such a vastly more information-costly strategy could ever have come off the ground in competition with systems in which information is parceled in modular unit form.

Much of this protean aptitude is owed to the modular makeup of these molecules. The sequence of their modules, the amino acids, is completely determined by the sequence of their genetic modular counterparts, the DNA bases. All proteins probably derive from a few ancestral genes which were duplicated and modified, and whose pieces got reshuffled, exchanged, and reamalgamated in the course of evolution. The variety of proteins that eventually emerged in this combinatorial way is so huge that there is a fitting protein shape for virtually every life-important inorganic and organic molecule. One could hardly dream up a more suitable stuff for transmission of information between molecules. Some RNA molecules of the looped sort, like the constitutive rRNA of ribosomes, have some talent for such molecular fitting, too, but the proteins have so much more.

Apart from this malleability—an adaptability in the developmental sense—proteins are adaptable in a dynamic sense; they can change shape in three dimensions. We have seen before how advantageous this can be in enzyme-substrate interactions or, in general, in molecular recognition and information transfer. But there is more. The changes of shape entail large displacements of the structure, allowing the proteins to do useful work, that is, work with a direction (it would do little good if the molecules went just randomly to and fro). So, these protean molecules can receive information and move on it; they can do work as they receive information. Since no other kind of molecule can fit that bill, everything that gets done in an organism or by an organism is done by proteins.

Transmission of Information by Weak Forces Hinges on Goodness of Molecular Fit

The weak electrostatic forces mediating the basic two steps of information transfer between a protein and its partner drop off steeply with the distance between the two molecules. Therefore, the fit between these molecules must be geometrically tight. This is the general rule for effective information transmission between molecules, a rule rooted in the tectonics of the attractive intermolecular forces.

Consider the most common forces between atoms, the van der Waals attractive forces. All atoms attract one another with these electrostatic forces. The forces arise from the attractions between the positive nucleus of one atom and the negative electron clouds of another. Individually, per interacting unit, the attraction is very weak. The van der Waals forces between a few atoms in one molecule and a few atoms in another would not be enough to produce a bond between the molecules that would resist the usual thermal jostling of the molecules. To hold two organic molecules in place even for a short time (say, one millionth of a second), one needs attraction energies well in excess of the thermal energy (on the order of 30 Kcal/mole). So, only if there are many van der Waals interacting units in parallel do they add up to that amount. This means that the number of interacting atoms in the two molecules must be relatively large or, put differently, the interacting molecular surfaces must be large. The attraction energy falls off with the sixth power of the distance; for organic molecules the distance for effective attraction is limited to 3 to 4 Å—the molecules must essentially be in contact. *Thus, in sum, for effective van der Waals attraction, the molecular surfaces must be large enough and their shapes must match within a few angstroms.*

This is the basis for the selection of molecules by molecules, a selectivity which in biology goes under the name of *molecular specificity*. Moreover, since the orientation of the interacting molecular structures with respect to each other is impelled by the same kind of forces, this part of the information transfer, in turn, hinges on a good surface match between protein and substrates, too. One local bump on the surface of the protein pocket exceeding the 4-Å limit, and the orientation gets out of whack.

A lock and key, adjacent pieces in a jigsaw puzzle, an enzyme and substrate are all examples of matching complementary shapes. But matches can obviously differ in accuracy. Armor, a suit of clothes, and a plaster cast are all complementary to a body, but they don't fit equally well. On the other hand, the available X-ray diffraction data show that enzyme and substrate tend to fit closely. Millions of years of evolutionary tryouts have seen to that. The fitting surface areas range widely in different types of proteins, but the fit is always good at the site where the making and breaking of bonds takes place. For example, in the case of trypsin, a smallish protein-splitting enzyme secreted by the pancreas into our intestines, the fit is so close that it can even form a covalent bond with its protein substrate. And this is why it is as adept at splitting peptide bonds as some of us are at splitting infinitives. The structures of the bulkier biosynthetic enzymes have not yet been worked out, but at least in those cases with high catalytic power, one suspects the fit to be at least as tight. A misalignment of a fraction of the length of a carbon-oxygen bond (about 1.3 Å)—the difference between positioning a catalytic group next to a carbon atom instead of an oxygen atom from which it can pull out electrons—could foul up the catalysis.

Weak Forces Are Good for Da Capos in Information Transfer: The Useful Nook in Interacting Atomic Fields

Why are weak forces used in the two basic steps of information transmission? Why not covalent bonds? Indeed, the covalent forces, one might think, could permit a better information transfer; they could keep the information recipient molecule, say, an enzyme substrate, on a shorter and stiffer leash, making its positions in space more determinate. The reasons lie in the kinetics of the interactions. Van der Waals energies allow the substrate to come off the enzyme more easily after reaction, permitting higher reaction rates. For fast reaction, it is as

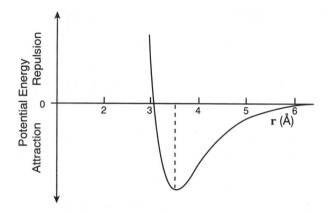

Figure 5.5   The serviceable nook in the interatomic energy field. The van der Waals energy of interaction between a carbon and an oxygen atom as a function of interatomic distance (*r*). A peak of attractive energy here is at a distance of 3.4 Å.

important for a molecule to dissociate from its binding site as for it to bind readily; large equilibrium constants for association lead to excessively slow rate constants for dissociations, even if the binding is diffusion controlled. Evolution, therefore, preferred this sloppier way of information transfer, whenever speed was of the essence.

Evolution here cleverly took advantage of a sharp transition in the energy fields of interacting atoms. The total potential energy of a van der Waals interaction is the sum of an attractive and a repulsive energy term. The first term represents the attractions between the positive nuclei on the one side and the electron clouds (their fluctuating local charges) on the other; and the repulsive energy is due to repulsion between electrons in the filled orbitals of the two sides. The repulsive component falls off even more steeply with distance than the attractive component does. Thus, for C, H, O, N, S, and P atoms, it works out that, at more than 4 Å between atoms, both components are negligibly small, at about 3 to 4 Å only the attraction is important, and at closer distances the repulsion is dominant (Fig. 5.5). *There is then only a 3–4 Å distance in the energy field around atoms which is useful for attractive interaction. And it is this little nook that Evolution shrewdly seized upon for fast repeats of information transfer.*

Why Proteins Do Not Run Out of Wit

There is some similarity between the foregoing enzyme-substrate scheme and certain toys which one shakes to guide two or three thingumajigs into place (Fig. 5.6). The complementary shapes of the thingumajigs are engraved in bas-relief on the toy's surface and if the toy is jiggled long enough, the thingumajigs

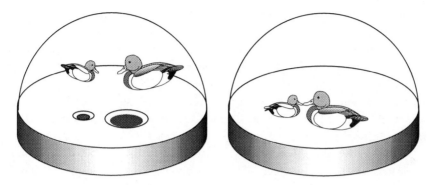

Figure 5.6 Jiggle toys are instructed to adopt determinate positions in space by the oriented complementary shapes in a bas-relief. The information for the kiss-kiss position is transferred from the bas-relief to the thingumajigs. The bas-relief is the only source of this information, just as the enzyme in the preceding figure is the only source instructing the two substrate molecules to the position for electronic interaction.

may eventually fall into place and be held there (the jiggling should be done blindly here to simulate the random thermal movement of molecules). Part of the information contained in the bas-relief structure has been transferred to the bobbers which now "know" their place. For replay, one would not want the interacting surfaces to be too sticky and certainly not to be covalently cemented to each other.

For enzymes of the flexible kind, where the fit is induced by interaction with the substrate, this model would need some elaboration. Not to wear the analogy too thin, the bas-relief surface would have to be made of little, elastically hinged pieces which, on contact with the thingumajigs, move a little, adapting the surface to the shape of the corresponding thingumajig. With the right snap and spring, such a contrivance could be extremely fast in play and replay.

Where does the negative entropy come from in these protein games? In many enzymes the selection step is readily reversed; the substrate is trapped and lost again. Thus, what the substrate had gained in information during the transfer must be precisely counterbalanced by an increase in entropy elsewhere. Consider the logic of the selection process. The recognition of the substrate by the enzyme—a grasping in the two senses of the word—is tantamount to a physical measurement in which the molecular contours are surveyed. The survey is done in successive measuring steps in the course of the molecules' buffeting, using the contour lines of the enzyme pocket as the standard. In every step, the enzyme compares a piece of contour with itself as well as it can, given the thermal molecular jitter. When a certain proportion of contours match on integration, the enzyme says, "Aha, you are my type!" As with everything, this measurement is subject to the conservation law: the gain in information demands an increase in thermodynamic entropy elsewhere. So, in the course of the buffeting, the enzyme system must be compensated with k ln 2 entropy units (1 bit) for each recognition. For a capable enzyme which manages to raise a substrate concentration to an effective 10^{10}M (p. 64), this amounts to a tidy number of bits.

The compensation can happen in a number of ways. For instance, it can be brought about through an increase in entropy in the water around a somewhat hydrophobic substrate: when the substrate leaves the water for the enzyme pocket, the ordered arrangement of molecules in the substrate's water shell decays. Or, to cite another common instance, the compensation can come from a well of potential energy created by the attractive forces between enzyme and substrate: when the substrate binds, this energy dissipates into molecular motion—an entropy increase that spreads with the motion.

While this foreplay in the enzyme action can go back and forth, the actual reaction catalyzed by the enzyme normally will proceed in one direction. Trypsin, a digestive enzyme, will break the polypeptide chains of the substrate, but normally it will not paste them together. Most enzyme reactions go, for all practical purposes, one way, because they are far away from equilibrium. Yet the enzyme comes out of the reaction unscathed. This may not be surprising in those cases where the protein only serves as a mold; it just serves to hold the substrate molecules by their cold ends, so to speak. However, it is truly prodigious in those cases where the enzyme engages their hot ends and goes even as far as to form covalent bonds. An enzyme like that could get quite a dent or bump.

How, then, does the enzyme snap back to its original shape, and how does it rid itself of the reaction products (the transformed substrates)? This it certainly must do if it is to repeat its action. And that admits no sloppiness; the enzyme must be returned exactly to its original configuration of atoms and energy fields (and its pocket must be clean). Thus, the question is, where does the information for this restoration come from? The pertinent mechanisms took a billion years to develop, but the answer can be given in a few words: *the enzyme taps the information from the very chemical reactions it catalyzes;* or, put in the chemist's language, *the enzyme uses the Gibbsian free energy of that reaction for its restoration.*[2] Together, the stages of substrate selection and orientation, catalytic reaction and enzyme restoration form a cycle. The cycle can be repeated with perfect reenactment of all stages, and this can go on and on, so long as the system as a whole—the system consisting of enzyme, substrate, and environment—can pick up the energy tab and balance the entropy books, as prescribed by equation 3.

2.   With information expressed in entropy units, the Gibbsian free "energy" (ΔG) is obtained by scaling the information by the absolute temperature. Thus, ΔG expresses information in terms of an equivalent energy, namely, the energy of one quantum whose degradation into thermal energy would wipe out the original information. The original information can come from order in matter or in actual energy.

The Recharging of the Protein Demons

The enzyme, thus, is as good as Maxwell's demon at ordering its little universe and, like the demon, it pays for each cycle of intelligence. It pays as it goes—there are no information credits in our universe—and it does so at the end of a cycle. It is then when the enzyme defrays the entropic cost of resetting its structure—the basic fee for forgetting the previous state, we might say, the minimum thermodynamic payment all lawful demons must come up with (Chapter 1). The enzyme's restoration here is equivalent to the resetting of the memory of Maxwell's demon, and in both cases the overall entropy increases as the previous information is discarded.

So, each cycle of action requires that the protein be supplied with information anew. The protein emerges unscathed from the innumerable cycles only because it can draw negative entropy onto itself. For this ethereal act proteins have earthy recipes. For example, they transform their substrates into new shapes which no longer match their mold; thus their remaining in the pocket is made unlikely by the reaction itself (a negative "aha!" event). Or they draw the high-energy bond from the ATP or from another substrate onto themselves and, with it, information that serves for their restoration. This is what recharges the demon.

By looking at things this way, we cut short the chemist's usual approach of considering Gibbsian free energy. The chemist's approach has its uses in energetics, but it can distort the perspective, giving the impression that it is energy that rules the information game. Quite the opposite is true: the distribution of energy is ruled by information. Had we canvased the problem of information transmission in terms of energetics alone, the loss of energy through friction in the molecular buffeting would have been evident, and the numbers would have come out by appropriately manipulating the arithmetic digits. But missed in the shuffle would have been the forest for the trees. The wonder of it all is not that all of the information in the bonds is lost, but that the enzyme retrieves some of it for its information transactions.

To end this topic, we have one more go at our jiggle toy. The measuring operations in this toy are essentially like those of the enzyme, but the reader may wonder where the negative entropy comes from. One should not be flim-flammed by the jiggling image into thinking that the analogy with thermal molecular motion is complete. Fair jiggling, like thermal movement, is random, but the system is not at equilibrium. Our jostling raises the energy in the jiggle box high above the level for that temperature, more than enough to power a few flashlights for information demons. This muscular activity is driven by ATP, and so, in an underhand way, bond energy here is the source of the negative entropy, too.

The Limits of Biological Demons and Demon Watchers

Although proteins are highly capable demons, they are clearly not designed to do the impossible: to decrease the *overall* entropy. At best, an enzyme may break even, and usually it will increase the overall entropy and thrive on it.

By the same law, it also cannot control the direction of the reaction; it cannot drive the reaction away from equilibrium. Hence, it must be equally good at speeding the reaction forward or backward. In other words, the equilibrium of the reaction is exactly the same with or without the enzyme.

While we can be certain that any piece of information that an enzyme invests in a catalyzed reaction is re-collected by the end of a cycle, we find it difficult to trace the whereabouts of this information. By looking at any enzyme as an isolated piece of a biological system, one can easily be misled into thinking that it carries *the* information (meaning all information) for a function; it is not uncommon to read in biological textbooks that such and such molecule contains the information for such and such function (or structure). What is really meant there is that the molecule contains a part of that information—what we called the *core part* (and this also holds true for the molecules of the next hierarchical levels, the RNA and DNA)—though this need not always be the larger part. Biological molecules form integrated systems with nonbiological ones, holistic systems in the operational and evolutionary sense. A protein operates in a system of which the substrate, the water, the inorganic ions, and other factors are as much integral parts as it is itself; and in this holistic way the protein originally came into being and matured. The parts in biological systems are so many and so intertwined that it is fiendishly difficult to see precisely where the information comes from and where it goes. But, c'est la vie!

The Growth of Biological Information

How Sloppy Demons Save Information

We have seen that proteins use mainly weak electromagnetic energies to implement their information transactions and that, as Maxwell demons, they rely on a sharp, 3–4 Å transition in their energy fields. This gains them speed in molecular interactions. But there is more in that penchant for weak energies than meets the eye: it actually works to their economic advantage not to use stronger interactions. A little digging into the economic information aspects of the demons' sleights of hand will bring this into the open.

All Maxwell demons require an energy source that can supply the information for their cognitive acts (Chapter 1). Protein demons don't carry that energy in themselves, but they have a donor, usually a nucleotide with high-energy phosphate bonds, who supplies it—he dishes it up, we might say, for he presents it right to the spot of cognition (Chapter 5). Thus, demons of the biosynthetic sort manage to steer the donor to the molecular welding sites, and there, right on the spot, draw their drop of blood. In fact, they manage to draw off so much information that the probability for synthetic reaction skyrockets, virtually making reaction a certainty. Some demons, the allosteric ones, contrive with that information even to metamorphose into another gestalt which matches those of the substrates. These demons squeeze quite a bit out of a high-energy bond. But even they, efficient as they are, cannot do so to the last entropy drop. The laws of thermodynamics forbid that; inescapably, the cognitive act itself produces (positive) entropies which, together with the rest, must be balanced against the information gain in the enzyme-substrate interaction (equation 3). And here is where the information economy comes in, for the burden of the expenditure of the high-energy bond depends on the fit between enzyme and substrate.

Consider the probabilities for the binding between the two molecules at the sites of interaction, the fitting domain on the enzyme and its counterdomain on the substrate. When the fit is too tight, there may only be one mutual position (or very few) that is compatible with the binding of the two molecules; the probability for binding is low and, hence, the information required for finding that position is high (equation *1*). On the other hand, when the fit is too loose, we again get a low probability for binding. Somewhere in between lies the optimal, sloppy fit where the binding probability is highest and the information required, lowest.[1]

Suppose, then, we have a sloppy fit where the same bonds are made between an enzyme and a substrate, as in the perfect fit, but where, instead of only one position, the two molecules can adopt a thousand ever so slightly different positions to form these bonds. The probabilities for binding are a thousand times greater in the wobbly situation than in the tighter fit and, what makes this the perfect bargain, the information cost is lower. Now, it is precisely this wobbly situation which the protein demon slyly exploits to pull his card from the sleeve. He acts precisely where the fit is neither too tight nor too loose, and so brings off his cognitive act with a minimum of information.

Thus, a little sloppiness is not such a bad thing in demon circles. Well, each culture values what is in short supply. If with us humans this is fickle, with Maxwell demons it's always the same thing: information. So, with them a little wobble is worth its weight in gold—without it, there would be no sleight of hand ... and the twig of life would be parched and withered.

Chemical Performance Guides the Selection of Molecular Complementarity in the Laboratory of Evolution

Now how does that optimal wobbly molecular fit come about? How is a fit which is neither too tight nor too loose selected in evolution? Natural selection, in Darwin's sense, results from the comparison of a trait among competitors—a comparison without a fixed standard—by an endless testing of performance in the laboratory of evolution. Looked at from today's vantage point, the critical performance or the criteria for its measurement are not always readily discerned in the case of whole organisms, the classical evolution of species. But no such problem generally attaches to evolution at the molecular level, the performance of macromolecular species—the speed of the reaction which is catalyzed by the macromolecular fit is a "natural" here. This performance, say, the rate of synthesis in a chemical pathway, could serve to cull out a wide range of evolutionarily

---

1.   The reader versed in thermodynamics may wish to work this out the traditional way of the chemist, in terms of Gibbsian energy instead of straightforward information, bearing in mind that the positive Gibbsian term represents information.

unfavorable things: excessive accumulation of energy intermediates in the reaction paths, excessive desolvation, excessive electrostatic repulsion, excessive mechanical strain—all of which at once would be selected against by the highest rate of synthesis.

Just as natural selection prevailed in the evolution of whole organisms, so it did at the level of their macromolecules—that's where the whole thing started. The competitors here, we may imagine, were the first self-reproducing molecular systems—cooperative ensembles of two or more molecules forming closed information loops (Fig. *3.2*). Eventually such loops constituted linked systems, as we discussed before. Operating in positive feedback mode, such systems can generate molecules round the clock, and the loops which generated the most— the systems with the highest synthesis rate—won the contest in the playing field of evolution. The optimal wobbly fit between their member molecules thus would mark the winners of the primordial competition.

An Evolutionary Hypothesis of Conservative Selection

Closed information loops of the positive feedback sort are found in all living beings. In today's complex creatures, such loops are integrated with loops of negative feedback which stabilize them. The two kinds of loops form integrated circuits where those with negative bias often appear to have the upper hand— they keep the positive-biased ones in rein—but this should not distract us from the fundamental role and importance of the latter. These loops drive all macromolecular genesis—they are the heart, hub, and pivot of procreation.

One can get a feel for their developmental power from experiments in which the loop components are pried loose from the cellular circuits, and the basic information circle is reconstituted in the test tube. Of special interest to us in the present context are the results of experiments in which RNA was looped with protein. Such uncomplicated bimolecular concoctions gave rise to self-reproducing systems when they were furnished with a suitable energy donor; molecule here begat molecule, until the supply of building blocks ran out. Thus, for example, a mix of viral RNA and replicating enzyme—as Manfred Eigen and his colleagues have shown and penetratingly analyzed—will give birth to a procreating molecular system, if it is supplied with nucleotide monomers containing high-energy phosphate. Even natural selection then will take place here as the building-block supply gets constrained: in due time, molecular mutants arise which will compete for that supply, and the mutants which pull off the highest synthesis rates carry the day—an evolution in the test tube.

Central to all of this is the circling information (Fig. *3.2*) which makes a system out of what otherwise would be just a batch of individual molecules. And what is more, the system is a self-reproducing one—a *species* in the Darwinian sense. Individuals cannot evolve; only species can, and only molecular systems with positive feedback loops can foster their own progress. Thus, we now will move this information loop to the center stage and build an evolutionary model around it. The thesis that I wish to develop here is by and large a Darwinian scheme which operates by natural selection, that is, standard-free selection. But to deal with molecules, I take leave of the precept of the "survival of the fittest" designed for whole organisms, in favor of a more conservative selection mode where not only the fittest, but also something of the not-so-fit survives: its information.

The hypothesis will shape up in the following sections of this chapter—I shall build my case on a principle governing the information economy of self-organizing systems. But to show the tenor of the way, I will set here the conservative scheme briefly against the classical Darwinian one. In either scheme, selection results from the termination of the evolutionary progress of the losers—this is sine qua non for natural-free selection—but the evolutionary games are played for very different stakes: in the arena of the survival of the fittest, the players wager everything, their genetic and bodily possessions; in the conservative arena, they only wager the latter. Thus, while in "the survival of the fittest" the losers are forever beyond recall, in the conservative game their information gets a second chance.

A word also about the winning hand. This, in the classical game, is a crucial ability of whole organisms, which may vary from case to case—a talent for killing, outeating, or outlasting the rivals, or something along those lines. In the conservative game, on the other hand, it is one thing pure and simple: the talent for generating information loops. And this takes us smack back to the protein fit, for this is what closes the loops and determines the winning hand. Thus, in building our models for a primordial evolutionary playground, we will start from the optimal wobbly fit and track the flow of the circling information, working ourselves forward in time, as it were.

An Evolutionary Thought Experiment

Let us reconnoiter, then, with our mind's eye, an evolutionary playground—a world of aboriginal RNA molecules that can adopt one-dimensional and three-dimensional configurations—where a protein with some talent for closing information loops has arisen on a three-dimensional RNA template (Fig. 6.1). This will be our point of departure for the growth of macromolecular information—a model system that grows by itself. For the choice of the aboriginal molecular species in the model, we take our cues from actual experimental work simulating the prebiotic conditions on Earth, and from work with RNAs that can catalyze their own synthesis, especially the experiments of Leslie Orgel and Thomas Cech, whose core of facts is presented at the end of this chapter. Working our way through the model may help us see how circling biological information could have grown autonomously through information-conserving selection. For the validity of the hypothesis, however, it is not important that the model be correct in detail. (Those who have no fancy for modeling may wish to skip to page 88).

Follow, then, in Figure *6.1 c* the information flux past that protein to where the information stream turns onto itself. The instant the loop is completed, the system has attained cybernetic capability: it furthers its own growth. We freeze the system at this point in time and make it our system unit.

It helps to use an early stage of evolution for our thought experiment, because the system then may be expected still to be simply of the straightforward positive feedback sort and not yet tangled with connections with other sorts of loops. In Figure *6.1 a*, the system is represented by an RNA molecule (R), a primeval three-dimensional one, of the kind of today's tRNA and rRNA, that assembled from its subunits, assisted by a low-power catalyst (C_R). In the presence of another low-power catalyst (C_P), R manages to pull off the difficult trick of polymerizing a modest protein molecule (P) from the amino acids available in the environment. Mutants arise and one among them (P_α) happens to have the right fit for catalyzing the assemblage of its informational progenitor (R_α). The reasons for choosing RNA as a starter are given further on. However, for our argument it does not really matter what these molecules precisely were. What matters is that P_α be a better catalyst than C_R —not something difficult to imagine if P_α had only a modicum of the steric talents of today's proteins. Indeed, a protein with two nooks, one fitting R and another fitting an energy-donating polyphosphate, would be hard to beat in catalytic power by anything in the less-specific catalyst world, be this clay, platinum black, random polypeptide, nucleic acid, or what have you. As soon as the loop is closed, we have a self-reinforcing system where P leads to more P—and exit C_R.

This, then, would be the crucial point in evolution from the point of view of our hypothesis. As soon as this information loop is established, the system, interacting with its environment, could advance to higher and higher complexities.

A Cybernetic Beginning

Consider the start (Fig. *6.1 c*). The loop between P and R can close only if there are energy donors capable of supplying the entropy flux necessary for the transactions at the two stages of information transfer $(P_\alpha/R_\alpha$ and $R_\alpha/P_\alpha)$. Assuming an abundance of energy-rich polyphosphates and amino-acid building blocks in the medium, the probability for a loop to be closed should increase with the P and R concentrations. And once the loop ticks (c), the number of P and R molecules will increase rapidly under the positive feedback. The stage is now set for the appearance of new catalytic loops—the first competitors on the playing field.

Nucleic acids undergo intramolecular rearrangements (mutations) by radiation and by errors in replication due to loose information transfer at their tem-

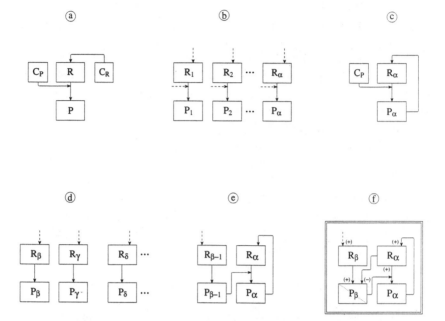

Figure 6.1 An evolutionary model. R, the central actor, is an RNA of the sort that can adopt two stable conformations: a folded three-dimensional form (stabilized by hydrogen bondings between some of its own base pairs) and a linear form (where the three-dimensional conformation unfolds as those bonds are broken). In the linear form it is easily copied and in the three-dimensional form it serves as a catalyst. R was assembled from nucleotide units abounding in the primordial soup, assisted by a low-power catalyst, C_R. On this background, largely a world of RNA, cybernating systems arise, whose main developmental stages are pictured in *a–i*:

(*a*) The first protein molecules, P, form on R as a template, assisted by the low-power catalyst C_P. (*b*) Mutants R_1 . . . R_α arise which are capable of engendering other proteins, P_1 . . . P_α. (*c*) P_α, with a fit suitable for catalyzing the formation of its progenitor R_α, displaces C_R and establishes the first self-reinforcing R-P-R information loop.

(*d*) R-mutants now arise faster, and more and more P lineages pile up at the first self-amplifying system. (The various P lineages are informationally interconnected, as discussed in Chapter 3; for simplicity, these connections are not shown here.) (*e*) Among those lineages, $P_{\beta-1}$ is suitable for catalyzing the formation of P_a. This mix is explosive and consumes the resources of the compartment; the system reaches a dead end. (*f*) In another compartment, an allosteric lineage has emerged, P_{β} which has a cognitive site for R_a which negatively regulates the catalytic action; as R_a binds to that site, the catalytic information flow to P_a is switched off—a negative feedback that keeps the original self-reinforcing loop R_a-P_a-R_a in rein. The system now is self-regulating and is ready to grow on both ends.

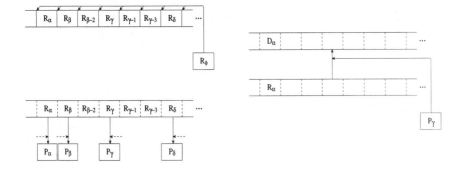

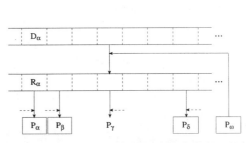

(i)

(g) Conservation. The family of Rs, the successful ones at engendering Ps along with some of the unsuccessful ones, are linked, forming a continuous RNA string. The phosphodiester linkage is catalyzed by an RNA, R_ϕ. (g, bottom) The former successful Rs now have become integrated informational units which continue to engender the respective Ps. The information loops from stage f by now have multiplied; they are indicated by the dotted short arrows. (Ever new Ps eventually arise through mutation and recombination of the informational units; that development is not shown.)

(h) Assisted by P from the γ lineage, RNA serves as template for the assemblage of a molecule from deoxyribonucleotide units, the first DNA. This molecule is a simple transform of the RNA that contains the conserved ancestral R information in deoxyribonucleotide-unit form (D). (i) Another lineage, P_ω, furnishes the catalyst reversing the direction of the information flow. This marks the end of the primeval RNA world. DNA now has become the information depository, and the flux has taken on the direction we find in living beings today. The information loops from stage f, which by now have grown to a circumvoluted network, are indicated, as before, by the dotted short arrows.

plates. Errors of that sort are kept low in today's cells by sophisticated enzymes, but the errors must have been more frequent when nucleic-acid formation was still run by more primitive catalysts. Thus, the variety of R and P molecules would steadily increase (Fig. *6.1 d*). And in the enlarging pile the competition is now on for optimally wobbly P-fits for catalysis of P_α and for catalysis of reactions extending the information transfer (unlooped) in other advantageous directions for the system. The next diagrams (*e* to *i*) outline five stages of such system growth in the order one might imagine them to have occurred in evolution; each stage is the outcome of the competition of a P lineage.

One lineage (β-*1*), the first in a series, is taken to provide a good fit for the catalysis of P_α. It replaces the primitive catalyst C_P and sets the R-P-R loop agoing. R-P-R is now a self-amplifying system which can reproduce itself over and over (Fig. *6.1 e*).

At this stage, the system still has not much evolutionary life expectancy. It sooner or later runs out of steam. How soon will depend on the amount of building blocks available in the environment and on the power of catalysis at the intervening steps of information transfer. It might work for a while if the catalytic power of $P_{\beta-1}$ is not too high and the compartment of primordial soup is large. But amplifier systems whose evolution is geared to selection of catalytic power will unavoidably get better and better at self-catalysis, and must eventually consume their own resources.

We need thus a mechanism that can hold the system in check: a brake. This wasn't much of a problem before the P_βs came along; their predecessor (C_P) was a poor catalyst and, in effect, served as a brake (and this is why—in case you wondered—the system did not go up in smoke). But now, whipped on by the powerful P_β enzyme, the ensemble must be reined in. Perforce, the controlling information here must come from within—on the evolutionary playing field all players are self-reliant. So, information about the R_α or P_α quantities somehow has to be negatively looped in. Most simply, the R_α concentration could serve this end by switching off the catalytic activity of P_β. All that would be needed here is an R_α-sensitive relay. This is not too tall an order to fill when proteins are around. Allosteric proteins are natural relays that can switch between two stable configurations. An allosteric P_β with two cognitive sites—a catalytic one for P_α formation and a regulatory one for R_α—could turn the trick all by itself. So, with such a catalyst-*cum*-relay in our model, all elements are now in place for a negative feedback loop which will shut the primary R_α-P_α-R_α loop down when R_α reaches a concentration threshold (Fig. *6.1 f*). The system now is not only amplifying, but also self-regulating, and can move on to bigger things.

The Consolidation and Conservation of Information

In due time a number of such systems, all of the same pedigree (R_α, R_γ, . . .) and informationally interconnected, ripen to the self-regulating stage. The scene is now set for the conserving of that *R* information in linear form, for splicing together the host of *R*s—those *R*s that engender useful proteins along with some of those that don't. Three-dimensional *R* can unfold to linear form (a transformation which can be catalyzed by certain *R*s themselves that are found even in some organisms today where catalysis ordinarily is dominated by proteins); in the linear form they can easily be replicated, and in the three-dimensional form they can catalyze the splicing reaction. So, when *R* matures to the self-splicing stage, the host of *R*s are bonded together by their ends, consolidating the aboriginal RNAs into a single molecular string, the first repository of information (Fig. *6.1 g*).

This is the stage of our postulated conservation of information, the garnering of information from the winners as well as from some of the losers. The consolidated winners continue to be informational units, albeit on a continuous RNA band, engendering their respective proteins. The integrated losers transfer no information to the three-dimensional world, for the time being; they stand by, as it were, for future recruitment.

We could leave our thought experiment at this point, but to better perceive the consanguinity of the model with the information system in today's cells, we follow it through to where it unfurls its full growth potential.

The Growth Phase

Once the system reaches the stage of amalgamation of the *R*s and sets multiple, parallel *R-P-R* fluxes going, it is over the hump. It is now immensely more powerful than its underdeveloped neighbors and makes headway ever so much faster. The system now can advance not only through mutations, but also through recombination of useful *R* information. It only needs to continue doing its thing, testing novel protein fits, and information arms will grow everywhere. It is now Hydra and Medusa combined, with arms reaching out everywhere, growing and regrowing, fed by its amalgamated and reamalgamated heads.

One of the novel *P* lineages (P_γ) completes the deployment by catalyzing the transcription of a descant (*D*), a simple information transform, from the consolidated RNA (Fig. *6.1 h*). That transform, like the RNA, is a nucleic acid, but it contains thymine instead of uracil, and its sugar backbone is made of deoxyribose instead of ribose; it is a deoxyribonucleic acid—*DNA*. This molecule now becomes the final depository of the accumulated ancestral *R* information, a stable and specialized repository unburdened by three-dimensional tasks.

One more *P* lineage (P_ω) eventually furnishes the catalyst for reversing the information flow (Fig. *6.1 i*). The stream now takes on the familiar direction DNA → RNA → protein we find today, and the RNA steps down to the role of intermediary in the information transfer. This marks the end of the primeval RNA world. We will pick up again the threads of the successor stream in Chapter 7.

The Game of Chance and Savvy

The conceptual skeleton won through our thought experiment may help to steer us through the perplexities of evolution. The main perception is this: a suitably looped molecular system with just a dash of information can grow by itself in the earthly environment; it can wrest information out of its surroundings and, by dint of its circling information and molecular mutations, propel itself forward to more and more information gains. This was worth, as the Michelin guide says, the detour.

The circling information in positive feedback mode here is the prime mover. It provides the crucial element of amplification that pulls a macroscopic system out of the quantum soup. Without it, there wouldn't be enough hard molecular copies around and not a snowball's chance in hell for molecular interaction. The spontaneous fluctuations in the constitutive information, the molecular mutations, furnish the essential element of chance for the system's evolutionary advance. The system rolls dice, as it were, and its line of advance may be represented by a series of forkings where binary choices are made by the fall of the dice (the mutations). But this is no ordinary game of chance, and certainly not a stupid one. The system has a memory bank—an information repository—where it can store the results of the good throws, and so it doesn't need to repeat them over and over again.

Thus, chance becomes less and less important as the system mounts the ladder of success and grows richer in information. It then makes ever more efficient use of the throws of the dice and plays an ever-improving hand. The chance events never leave the wellhead of growth, not even at the highest rungs, but they are eventually overshadowed by the élan of the accumulated information. By the time we get to the first modern cellular beings, the game has become one of subtlety and cunning where the information literally runs rings round fickle chance, and the system well-nigh steers its own course. It is the course of the consummate gambler who plays it safe, the one who sticks to the "saddle points" of mathematical game theory, as we shall see (Chapter 8). Among the many fascinating sides of this game, the gradual shift from chance to savvy, perhaps, is the most tantalizing one.

The Why and Wherefore of the Conservative Selection

We turn now again to the evolutionary hypothesis of conservative selection to see the basic reason behind the postulate of conserving the information of the losers along with that of the winners in the primeval biomolecular contest. The immediate question before us—and I'll put it in the plain terms of Figure *6.1 g*—is this: What is more likely, the amalgamation of all *R*s or that of only the successful ones?

It pays to take a look here at the amalgamating molecular regions, the *R*-ends. These ends all have the same chemistry—on one side there is a hydroxyl group and on the other, a phosphate group—the ideal situation for wholesale catalysis. The linking of the various *R*s, thus, is highly probable once one good catalyst is on hand; one catalyst, a three-dimensional RNA fit (R_ϕ), or a protein fit suiting the (diester) linkage of one R-pair may suit them all. Or put in terms of our game metaphor: *the wholesale conservation of Rs is a more probable event than a conservation of only winners.*

Under the looking glass of information theory, this conclusion casts its most alluring spell. By virtue of equations *1* and *2*, the more probable event (the one closer to thermodynamic equilibrium) requires less information. *Therefore, the wholesale conservation of Rs is less information-costly than a conservation of only winners. The operation here that conserves the most information is also the one that costs the least information.* And this is the rhyme and reason for the proposal of a conservative selection.

The Principle of Evolutionary Information Economy

Perhaps, despite its cogency, the notion that it should cost *less* to conserve *more* may look offbeat. It seemingly clashes with our daily economic experience. But this is mainly an experience with matter and energy costs, which teaches us nothing about information. Not even our daily dealings with information apparatus, computers, are of much help here. These machines do not evolve on their own as biological systems do. Developmentally autonomous systems have their own peculiar information economy, and there is in truth no common experience that could nurse our intuition in that respect. I do not wish to imply that developmental autonomy is inherently unique to living beings. Though it presently is found only in the living world, I hesitate calling that autonomy life-specific because I am afraid that one day an artificial cybernetic machine might thumb its nose at me. However, we should be in the clear if we broaden our circle of things and include any system that can grow by itself. And this gets us right to the heart of the evolutionary game, the hard and fast rule of its information economy.

First, let me spell out what a "system that grows by itself" means in information terms. It means a system that adds new information to itself without outside help. This tells us right off half the story. We only need to look at equation *3* and consider how fresh information is won: since there are no information loans to be had anywhere in our (macroscopic) universe, such an autonomous system can only acquire fresh information through the painstaking entropy transactions prescribed by the equation.[2] It goes without saying that no one ever makes a fast buck under that austere rule. To come out ahead in the entropy balance, to gain information, the system already must have some to call its own. This information

---

2.   However, raiding, the takeover of information from another system, is permissible under equation *3*. Living systems resort to this practice in evolution (as they do in the corporate world, though not as often). For example, the mitochondria in our cells may have originated that way when an ancestor of ours swallowed a suitable bacterium, body and soul. The information in either system had been correctly paid for before the merger.

capital, we know, is the intrinsic information of the system's molecular Maxwell demons, and the amounts of new information it can add to itself will depend on how much information it has on hand. This is the growth formula of the most conservative banker: the gain is commensurate with the current working capital. The developmentally autonomous system thus is of necessity information-thrifty; it must virtually pull itself up by its own bootstraps.

As if that were not parsimony enough, there is more to come in due time when the system enters into competition with others. Then, as the various systems vie with each other for the informational resources in the environment, their growth gets naturally streamlined, and the contest eventually ends up with the selection of the most information-economical one. This outcome is inescapable in biological systems, as we shall see, and leads us to an evolutionary principle.

To get there, let us consider an early generation of biomolecular systems that have just come of cybernetic age (the stage in Fig. *6.1 f*) and are beginning their contest. Each system then advances step by hard-earned step, as it rolls dice at the forks of the road. One could not predict its individual progress—the stochastic understructure precludes that—but one thing is as good as certain: the lines of advance of the various systems won't be the same over the long haul; it is extremely unlikely that the information increments at the various forkings or the direction of the steps there in the functional (three-dimensional) world would be the same for all competitors. The different lines of advance provide a basis for natural selection, though a somewhat scrappy one, on their own. Were the selection to rest on this difference alone, it would hardly be sharp-edged; despite their different developmental paths, some systems might well end up having comparable functional capabilities. However, another element here enters the game, the *developmental time*, and this narrows the constraints and offers a second basis for natural selection. This element ensures a clear-cut end to the evolutionary contest, for the system with the shortest developmental time, the one which attains the critical functional capability first, has the unmistakable edge in the competition for the available informational resources: it corners the information market, stopping the laggards dead in their tracks—literally choking the competition.

One more inferential step concerning the developmental time, and we arrive at our goal. Apart from what was already said, that time enters elsewhere the chance part of the game: the mutations. These molecular events occur randomly but, over the long run, their average frequency in kindred systems in the same environment is about the same. So, the system with the shortest developmental time undergoes the fewest mutations—or, falling back on our metaphor, it makes the fewest throws of dice. Now, in the information realm, this gambling efficiency has a fateful spin-off: economy of information. Each successful throw of dice, each step of information gain, entails an investment of information. So, the winning system, the one that reaches the critical functional capacity with the fewest throws, makes the smallest total investment; though the investment for

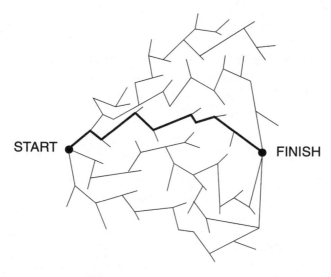

Figure 6.2   The labyrinth sport, a mass game of chance and endurance, played by the following rules: participants trot at fixed speed and rely on chance in their choice of forks. The track of the lucky winner, the shortest route between start and finish, is shown as a thick line.

each individual step need not be very different from that of the laggards, the sum of the investments will be smaller.

And so it comes out that the winners of the evolutionary contests are the most information-economical systems. We may formulate our finding in two statements: *(1) biomolecular systems tend to minimize information expenditure; (2) they evolve along paths maximizing information efficiency.* And this, I propose, is the guiding principle of biological evolution.

The first statement embodies the rule for autonomous growth, a basic time-independent rule; and the second statement incorporates the rule for competitive autonomous growth which depends on time, the developmental time. In the old veterans of competition now populating the earth, the two rules are inseparable; they have merged. But at the dawn of evolution, when biomolecules were still too diluted for deadly competition, the first rule probably was all there was.

It would be helpful to lean here on an analogy from daily experience—it often is when things reach some level of abstraction. But it is not easy to find a simile for the evolutionary game—that game is unique. However, if we are willing to ignore for a moment the subtle interplay between chance and cunning (alas, the best part of the game!), we can make up an analogy for the purely-chance part of the evolutionary contest, without laying it on too thick. Consider a game in a maze of forks with a single exit (finish line) where scores of runners try their luck and test their endurance under these ground rules: jog at a fixed speed and choose your way at the forks by blind chance (Fig. *6.2*). We could not predict *who* will be the first to arrive at the finish line, but we can predict *what* will be the winner's relative performance: he is the one to take, as luck would have it, the shortest route.

In the statistically equally sporting but not-so-friendly evolutionary game, the less efficient laggards get choked along the way. So, the few who made it to the Three-and-a-Half-Billion-Year marker took the shortcuts—information penny-pinching runs in their blood, we might say.

Keeping Things in the Family of Adam and Eve Has Its Advantages

We shall run into this evolutionary principle quite often as the story of this book unfolds. In fact, we tacitly have stood on it already just a moment ago when we weighed the probabilities of R conservation during the consolidation stage of primordial information (Fig. *6.1*, stage *g*). That stage also gives us a chance to get acquainted with the principle's implications in evolutionary design.

So, let us take another look at the postulated amalgamation of primordial information and take stock of the benefits of R conservation. We recall that the Rs have a common genealogy; they stem from the first cybernetically successful RNA, R_α, the Adam in our thought experiment. This Adam had two Eves, P_α and P_β, who got more than a rib from him; the whole basic information skeleton of P_α derived from his template, and P_β is a close cousin of P_α (Fig. *6.1 e, f*). *So, all Rs bear a family resemblance among themselves, a certain congruousness of information that stretches all the way from the one-dimensional to the three-dimensional realm.* This congruousness is as momentous as it is unique: *it permits the kaleidoscopic blending of R information; it makes the Rs suitable for combinations with the yoretime winner and each other, and it makes their three-dimensional transforms suitable for combinations in the physiological world.*

The cat's now out of the bag. A system that conserves kindred R-information can "recall" pieces of that information in the course of evolution and recruit them for new functions, *with minimal expenditure of information*; the same old tested pieces (and entropy donors) can be used over and over again in different combination. But not only that—a conservative system goes one better. It can enhance the combinatorial potential further by active variegation, namely, by duplication of the old information and subsequent mutations in the duplicates. Duplication is an amplification of existing information. This costs relatively little information if loops of the positive feedback sort are on hand, and even less when such loops are interconnected. Living systems have taken naturally to such loops ever since the evolutionary beginnings. So, under the selection pressures operating in the three-dimensional realm, duplicates of the Rs with the most winning features may arise and their mutants be added to the R-arsenal.

Thus, to sum up, the conservation of Rs is information-saving not only during the primordial act of consolidation, but also over the long haul. Compared to a nonconserving system, which for each new function must generate all information from scratch, the conserving system advances at an immensely faster pace, virtually ensuring its evolutionary ascendancy.

The Vestiges of Ancient Information

So far so good, but is there anything to show for the large-scale conservation in today's DNAs and RNAs? One would not expect that all ancient information be retained. The selection pressures on the three-dimensional realm would see to it that the detrimental pieces be eliminated outright and that further stretches be eliminated if there is a bonus on replication speed. The latter may well be the case of bacteria and other fast-growing creatures without cell nuclei, the *prokaryotes*, but it may weigh much less in the slow-growing nucleated organisms, the *eukaryotes*. Thus, it is in eukaryotic cells where one stands the best chance of finding vestiges of old conservations.

Indeed, the marks are there for all to see in the sequence of the eukaryotic DNA and in the processing of its information. This DNA, we shall see in the next chapter, contains undecoded stretches called *introns*, which are dispersed between the coding sequences, called *exons*. The number of introns varies from gene to gene; some genes have just a few, others, as many as fifty. We don't know the age of all introns, but those in the genes of old enzymes, like the dehydrogenases and kinases of carbohydrate metabolism are old enough to qualify as R relics; they have been there since before the divergence of plants and animals, which occurred more than a billion years ago. While most introns seem functionally silent (their RNA transcripts are cut out of the exon transcript and destroyed in the cells), some introns can enter into combination with exons and the amalgamated information is expressed in the three-dimensional protein world. For example, the information of an intron of our immunoglobulin gene can be processed in two alternate ways: it can be excised from the RNA transcript; or it can be partly retained and translated together with the contiguous exon into a different protein. And the latter is the option that interests us here, for, in terms of the conservation hypothesis, its exercise amounts to a recruitment of ancient information (a recruitment of $R_{\beta\text{-}2}$, $R_{\gamma\text{-}1}$, and their likes; Fig. *6.1 g*). Intron recruitment and intron-exon sliding of this sort may have given rise to today's spectrum of complex exons that originally may have started as relatively short stretches which encoded perhaps no more than some twenty amino acids.

As for amplification of ancient information, there are telltale signs of duplications of old DNA stretches in the structure of many basic proteins — hemoglobins, immunoglobulins, albumins, collagens, oncogenes, to name a few. There is a molecular archeology now in the making. It is still in its infancy, but interesting relics are turning up. By attending to homology of DNA sequences, the history of the globin chain in hemoglobin, for instance, can be traced in marine worms, higher fish, mammals, and primates, over about 700 million years. Many genes share structural features suggesting that they originated by modular assembly. And the genes of the epidermal growth-factor precursor, the low-density lipoprotein receptor, and certain blood clotting factors, which share a number of exons, tip us off that they may have been built up by reuse of modules. The molecular biologist Walter Gilbert has advanced the hypothesis that such a shuffling of exons gave rise to the current diversity of genes; exons from an ancestral gene were duplicated and subsequently recombined wholesale in the DNA into different transcriptional units. This way successful exon motifs may have been used again and again, cutting short the testing of three-dimensional transforms and quickening the pace of evolution.

The information-conservative genome, thus, offers fertile ground for the sleuth of biological evolutionary history. The new tools of molecular genetics allow him to penetrate far into the distant past. He is about to enter territories that seemed inscrutable only a few years ago, finally catching up with his colleagues of cosmic evolutionary history—and we have seen in Chapter 2 how these investigators managed to bring the origin of the universe within their grasp. Cosmologists and biologists here are both concerned with the same thing, a fifteen-billion-year-old continuum which comprises our own lowly origins. And those origins have beckoned ever since a master sleuth presciently wrote a century ago:

> Man with all his noble qualities, . . . with his god-like intellect, with sympathy which feels for the most debased, which has penetrated into the movements and constitution of the solar system—with all these exalted powers—Man still bears in his bodily frame the indelible stamp of his lowly origin. (Charles Darwin, *The Descent of Man*, 1871)

The First Commandment of Evolution

To fix ideas, let us pull together what has transpired so far in our inquiry about biological evolution. First of all, there is the trend of ever-increasing information as life climbs the phylogenetic rungs—what actually evolves is information in all its forms or transforms. Thus, if there were something like a guidebook for living creatures, I think, the first line would read like a biblical commandment, *Make thy information larger.* And next would come the guidelines for colonizing, in good imperialist fashion, the biggest chunk of negative entropy around.

To abide by the commandment, the covalent bond was the best thing around. Energetically it was a good match for the quantum in sunlight, and so could capture this most abundant energy form, and it was the best instrument for holding onto molecular information once some was gained, or for adding new molecular information to old. So, the covalent bond became a tool for creation and procreation.

Carbon was the most talented atom for bondings of this sort. This atom has four valency electrons and four valency orbitals; it can make as many as four covalent bonds, including a bond to itself, which is as strong as a bond to oxygen. It is thus beautifully suited for the accretion of large and stable molecular structures, not only those of monotonous design but also those with ever-fresh diversity, structures with branches and cross-links. And as if this were not enough, this atom has even more cachet: its mass is small and its entropy, low. (The lower the mass of a molecule, the fewer and more wide apart are the quantum energy levels; hence, the low entropy.) This low inherent disorder, no doubt, is important for storing information, but other low-mass elements have this feature, too; among them is silicon which is no less abundant. What must have decided the choice of carbon for information-carrying structures—one-dimensional as well as three-dimensional ones—are the extraordinary bonding abilities of this atom, which put all its neighbors on the periodic table to shame: it can make straight chains, branched chains, cross-links, double bonds, and ring structures with delocalized electrons. Silicon can do some of these things, but it cannot do them all—no other element can.

The very first carbon chains, we may imagine, obeyed the commandment, without help from Maxwell demons, under the direct energy input of sunlight. The necessary negative entropy may have been produced in the process of the capture of light quanta by a molecular intermediate, and some swirling and seething may have helped things at one time to throw them off thermodynamic equilibrium. However, as the molecules grew beyond their initial modest size, such a method of direct energy input was no longer feasible because of limitations in the stability of the covalent bonds.

The kowtow tactics then changed. Informational molecules, like RNA, now began to generate three-dimensional information transforms of themselves. These new structures were Maxwell demons which served the old ones in the processing of information. To keep these demons thermodynamically happy at their processing-*cum*-organizing job, the sunlight energy was now conserved in the covalent bonds and doled out to precise molecular spots where needed—a splendid variation of the old capture-of-sunlight theme.

Thus, step by parsimonious step, the information garnered by the evolving systems grew enough to form cybernating loops, first amplifying the information, then self-regulating the systems.

So much for the rewards for obedience of the first commandment. As for disobedience . . . well, that was punished with oblivion—something that ever since has worked wonders in spurring on all sorts of creative activities.

Rolling Uphill

If growth of information is the essence of evolution, so is cybernation. Any information loop is inherently information-saving, and the self-reproducing loop achieves the ultimate saving: total conservation. Thus, more than a cornerstone as we have considered it all along, the information loop may be in a deep sense the immanent cause of evolution. The number of cybernating loops increases at every evolutionary stage, and the loops form loops within loops within loops. In organisms there are millions of such parallel and coupled loops. Awesome? Yes, but perhaps a little less than only a few years ago. In the age of computers it is no longer so unfamiliar for us to think of large numbers of parallel and consecutive operations programmed into a system. We can thus fathom a system that, fed with enough information, moves itself more and more away from thermodynamic equilibrium; it virtually *rolls uphill.*

If rolling uphill sounds weird, it is nonetheless thermodynamically correct (equation 3). There is even something inexorable about it, although not in the same deterministic sense we are used to thinking of the events in macroscopic physics which, in the downhill direction, follow time's arrow purely and simply; the rolling in biological systems covers a very different matrix of constraints and a very different order of probability. Once the first successful self-reproducing amplifier loop is formed and stabilized, in the right environment, the system perforce continues to grow in complexity, until it is annihilated or incorporated into another system. Such development of cybernating loops—and I am using "development" exchangeably in the phylogenetic and ontogenetic sense—is only a matter of time. But there were aeons of time and oceans of material.

The Basic Loop

The systems that made good in this fateful game, the organisms around today, exhibit self-reproducing information loops that in the main have only a few molecular members. The number of loops, their interconnections, and attendant adaptor molecules have proliferated immensely, and RNA now has become an intermediary between DNA and protein, but the basic feedback loops impinging on the linear information depository have remained simple. Evidently, evolutionary advance here was made by multiplication of loops, rather than by loop elongation. Why?

The reason is rooted in the economics of mutational information transmission. A little excursion into the far-fetched will help to make this clear. Suppose it were possible to generate protein with protein, that is, to instruct protein synthesis, using a protein catalyst instead of a nucleic-acid template. Suppose, further, it were possible to build a loop with such proteins, a self-reproducing cascade—P_1, $P_2 \ldots P_n$—where one P catalyzes the generation of the next by way of informational side branches (say, a cascade where the Ps are of increasing size, have multiple recognition sites, and split off catalytic subunits as side branches, Fig. *6.3*, *bottom*). Now, compare such an imaginary loop with a loop that is made only of one RNA *(R)* and one *P* species, as in the model we considered before. A single mutation (say, a substitution of one nucleotide for another) that can start the simple *R*-*P* loop rolling on the way of evolutionary advance would

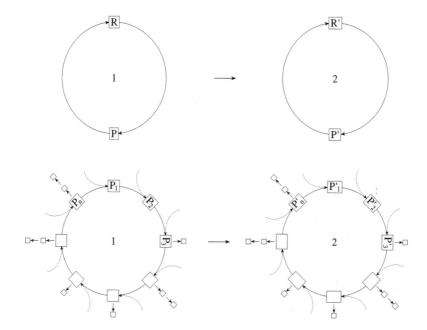

Figure 6.3 The winning features of cybernetic loops of nucleic acid and protein. The simple self-reproducing information loop made of one RNA *(R)* and one protein *(P)* species advances when a mutant *P'* arises that is a better catalyst than *P*. Loop *2* out-performs *1*, and replaces it; progress *1* → *2* results from a single mutation. This basic loop, where information flows from a nucleic-acid template to protein and back to nucleic acid, is the mark of all living beings. The ability to transfer the entire structural information, including the mutated one, from the one-dimensional template gives the loop evolutionary agility. In today's living beings, DNA (not shown) comes between *P* and *R*, but the loop does not lose its agility for that, because, mutatis mutandis, all information is faithfully transferred between the loop's members.

Below, a hypothetic self-reproducing protein loop. The loop consists of a cascade of *P*s of increasing size, where catalytic side arms (small boxes) from each *P* instruct the formation of the next *P*. The dashed arrows looping into the *P* circle indicate the information flows from the various side arms. Here only a fraction of the information in the three-dimensional structures can be transferred, and so only very few mutations propagate in the cascade; a mutation in, say, P_1 alone *(P)* merely reproduces the original P_n in the loop. For the system to progress from stage *1* → *2*, *all* members in the cycle must undergo compatible mutations (P_1', P_2' ... Pn'), a highly improbable event. Although high in information, such loops are evolutionarily inert; they are not found in living beings.

do no good in the complex *P* loop. That loop merely would continue making more of the old *P*, if a mutant (P'_1) occurred that is a better catalyst than one of its members. For such a loop to progress, compatible mutations would have to arise simultaneously for all its members. The probability for such coupled events is dismally small; it gets smaller and smaller with each added member. Therefore, although tacking on *P* members in a cascade of increasing molecular information might seem a possible way for the system to grow (and even a rather plausible way from the chemical point of view), it would require an immense amount of information to send mutational information through the loop. The evolutionary information cost would be unaffordable.

Thus, from the beginning it was in the cards: the simplest self-reproducing loop, the one advancing with the least information cost, is the winner. Although it may contain less information, the basic loop with few members is evolutionarily more agile than any loop with more.

Why Mutations Don't Rub Off Folded Molecular Structures

Evidently, it takes more than just the blind obeying of the "first commandment" for a system to make headway in evolution. You may cram as much information into a system as you might, but if that information gets into the wrong sort of molecule, as in the case of the top-heavy protein loop we dreamed up, the system won't come off the ground. Such a system might amass information for a while, but it has little ability to improve itself through mutations. Not that mutations don't occur there—proteins by and large are as prone to be altered by statistical-mechanical events as the nucleic acids are—but the problem is that these mutations, unlike those in nucleic acid, don't necessarily rub off.

This immediately raises another why. The key to such things never lies far from the electromagnetic force, the mother of molecular information transfer (Chapter 3) and, with a little more digging, we find the answer in the geometric constraints on that force. The effective electromagnetic field in organic molecules, as we have seen, extends only a very short distance away from their atoms. Thus, transfer of information between such molecules demands that the interacting sets of atoms have closely matching shapes (Chapter 5). But we might as well ask for the moon if we expect coiled-up molecules, like proteins, ever to be matching over their whole body; their own convolutions stand in the way of such total complementarity. Protein structures, to be sure, can be pushed a little; they are not entirely without give and take, but you can only squeeze them so much without ruining them.

If we look through a higher lens, we see where the shoe pinches the toe. For a set of extrinsic atoms to receive information from a protein, it must approach it within 3–4 Å (Fig. 5.5). Thus, if the potential donor atoms are in a tightly involuted protein region, the recipient atoms would have to occupy a space that is already filled by intrinsic atoms. Intuitively we feel that this is a no go; from daily experience we know that two things can't be at once in the same place. However, we also have learned to be wary of intuitions in our dealings with small-scale things bordering the elementary-particle realm. We have seen in Chapter 2 how misleading our macroscopic experiences can be when they are extrapolated to that realm. And as for two being in the same place, *bosons*, particles associated with force, like to do just that; any number of them can be squashed into the same space. Nevertheless, we need not worry on that account; in the issue at hand our intuitive feeling is OK, because the relevant particles, the electrons, protons, and neutrons, are not in the least as gregarious as the bosons are. So, short of altering the structure of the protein, there is no way how information can be directly transferred from inside a protein to another molecule when the quarters are too tight. Electric repulsion and, in an absolute sense, Pauli's exclusion principle, see to it that potential recipient molecules are shut out from such quarters. The exclusion principle, a basic tenet of quantum physics, states that two electrons (or two protons or neutrons) cannot be in the same state, which means, they cannot have both the same position and the same velocity, within the limits given by the uncertainty principle.

So, we are on safe ground to say that tightly folded three-dimensional proteins cannot transfer information from their entire structure—and this includes mutated structure—to other molecules. Therefore, in protein circles, mutations do not rub off, unless they occur at "active sites," the sites where the information transactions normally take place. Mutations occurring elsewhere are informational dead ends which, even in the most amplifying feedback loop, lead to nowhere in evolution.

And Why They Rub Off Linear Structures

All this becomes a different kettle of fish when nucleic acids are blended in. The structure of nucleic acids is strung out in one dimension, and not too many constitutive atoms get into each other's way during information transfer. Thus, to achieve a transfer that embraces the whole molecular structure is not the perverse catch-22 it is with protein. Quite the contrary, some RNAs and DNA are able to do just that.

This aptitude comes with their long and slender figure. Such shapeliness is not common among molecules. All molecules are necessarily three-dimensional, and most natural ones are rather stocky or roly-poly. But DNA and many RNA molecules are so much longer than wider that they pass for linear. Take, for example, the human DNA; the average molecule is some ten million times longer than it is wide or deep. Such a lanky figure is the result of a long search in evolution for molecules whose atoms are so distributed as to make efficient use of their electromagnetic fields for information transfer; in other words, structures

The Exclusion Principle

The exclusion principle, formulated in 1925 by Wolfgang Pauli, was an extension of Bohr's old quantum theory, which stated that an electron in an atom could only move in one of a discrete set of orbits. Pauli added that there could be only one electron moving in each orbit. When Bohr's theory gave way to modern quantum mechanics, where the orbit of the electron is replaced by the more abstract notion of "state," the exclusion principle transformed to the statement that two electrons cannot be in the same state (see text). In this form the principle also applies to other particles of matter, like the protons and neutrons.

The exclusion principle explains why the electrons in heavy atoms stack up in an orderly fashion, filling each energy level, rather than all trying to dwell in the lowest energy state. This accounts at once for the organization of the Periodic Table of chemical elements, the basis of all chemistry. The exclusion principle is at the root of all organization in the universe. Without it the universe would be a formless soup: the quarks would not form separate differentiated neutrons, and neutrons would not form separate organized atoms.

Bosons are particles of zero spin or of whole-number units of spin, which represent quanta of field energy. They include the photon and the graviton. Particles of half-number units of spin are called *fermions*. They are usually regarded as particles of matter, and include the electron and quark. Fermions, but not bosons, are subject to the exclusion principle. Thus, two or more bosons can be in the same state; for example, vast numbers of photons can merge to build up an electromagnetic wave, and so pull a macroscopic effect out of the quantum world, which we can perceive directly with our eyes as light.

The universality of the principle has been put to the test. Two sets of critical experiments done in 1900 showed that, if the principle ever is violated, it happens extremely rarely: less than once in 10^{26} times, as tested for an electron-electron interaction, and less than once in 10^{34} times, as tested for a proton-electron interaction.

that can easily donate their information to others and serve as organizational templates. And in this regard (as well as regarding the ability to hold information) these "linear" molecules are peerless; they can transfer the information contained in the entire sequence of their unit building blocks.

Their proficiency to go the whole hog enables these nucleic acids to transmit information from any mutated place of their structure. Therefore, loops made of nucleic acid and protein succeed precisely where the protein-protein loop fails: they efficiently pass mutations through the whole information circle (Fig. *6.3,* *top*), and so can change for the better and make progress in the world.

Mutatis Mutandis

Those self-reproducing loops that made the grade, the ones we find in living beings today, send information through with whacking efficiency. Here are some error-rate figures for the various stages of information transmission: DNA to DNA (DNA replication), 1 : 1,000,000,000; DNA to RNA, 1 : 10,000 – 100,000; and RNA to protein, 1 : 10,000. These figures are the number of mistakes (E) made per number of nucleotides or amino acids (n) assembled at each stage. They readily translate into probabilities $[p = (1 - E)^n]$; so, for a protein of 1,000 amino acids, for example, the probability of error-free transmission is 0.91. Since the information transfers in the loops are repeated many times, virtually all mutations make it through (1,000,000,000 repeats in the DNA-to-protein segments are not uncommon in the lifetime of a DNA molecule). Thus, when one finds a nucleotide mutation in the DNA, one almost can take it for granted without further checking that there will be an exact informational counterpart down the line in RNA and protein.

Such a fidelity of information transmission is, at first blush, hard to believe. After all, these are only molecules. For all their good looks, informational wealth, and concerted interplay, they are just atoms strung together and, like all chemical things, they are subject to the laws of statistical mechanics. Yet somehow they manage to beat the statistical odds. I can well imagine the reader brought up in physics or chemistry shaking his head when he sees the above error figures, but they are no misprints and a perfectly earthy explanation is forthcoming in Chapter 7. The key, we shall see, lies in the skill of the biomolecular loops to attract information from the outside and to loop it in with themselves—to make loops within loops.

Meanwhile, we keep this bald fact on tap: a molecular alteration arises in one member of the loop and, *mutatis mutandis*, turns up at the others downstream.

This is so extraordinary from any physics point of view and so arcane that the above statement rated a paragraph by itself. What we have here is a series of chemical transmutations which bear an exact relationship to one another—they are mathematical transforms (information transforms) of each other—and their error rates are far below those in the familiar (linear) physical or chemical systems that obey the "√n" error law where the mistakes are proportional to the square root of the number of cooperating molecules. Any major deviation from that law constitutes a wonder, but all such wonders so far, when genuine, have proven to be the products of nonlinearity in the systems, and this one, as we shall see, is the result of the very acme of nonlinearity.

We are ready to return now to our evolutionary playground. We can trace the history of the basic loops thoughout phylogeny—we survey them in organisms, going up the phylogenetic ladder. Thus, we see their fidelity—the fidelity of their information throughput—improving over the ages. But even far down, at the lowest rungs, the fidelity is nothing but prodigious—and by taking in also the fingerprints left in the DNAs of the various organisms, we can go back as far as 2 to 3 billion years. Now, from there we venture into the fuzzier past, to when DNA originally made the scene. We take that historical record to extrapolate—and 2–3 billion years would seem long enough for an extrapolation—that, by the end of the RNA Age, the fidelity of the ancestral loop was then already good enough for mutational information transfer; that is, the information in mutations accumulated in the primeval RNA could have been transferred (along with the rest of the RNA information) to the emerging DNA.

Thus, what we see in our loops today would be a recapitulation of very ancient history—a perpetual *mutatis mutandi* going back to well before the time when the mainstream of biological information adopted the direction it has today. The old Greeks and Romans would have been dazed to learn that someone knew the formula for "mutatis mutandis semper et ubique" long before their gods did. Well, they are not the only ones put to shame by Lady Evolution.

Cybernetic Loops and the Mass Capture of Photon Information

Let us now go back somewhat farther in time for a more panoramic view that includes also the pre-biological molecular evolution. We have seen (Chapter 3) that after the earth got cool enough, small organic molecules were organized from the available carbon, nitrogen, oxygen, and metal atoms. These modest asymmetric molecular geneses were nursed with information from photons coming in from the cosmos. The nascent molecules suckled photons, as it were. That suckling was stepped up when the larger organic molecules emerged and, while eventually there would be more mature ways to bring through the organizational entropy quid pro quos, it became a habit from which living beings would never be weaned. But it truly started in earnest with the advent of the cybernetic loop. Then, the photon quanta were sucked nonstop into a circulating flow of information that instructed the mass-production of organic molecules, small and large. It is this cybernetically driven uptake of photon information that is ultimately behind all organization of living matter, including its mutations.

Thus, after all is said and done, what matters is how well a system uses the photon information that is showered onto earth—the efficiency with which the system extracts information from that seemingly inexhaustible cosmic store. Living systems are not the only ones to avail themselves of that cosmic cornucopia. Nonliving molecular systems do so, too—and that's how the first organic molecules on Earth emerged. But living things do it so much more efficiently. Their molecular systems are hooked up in information circles that were naturally selected over aeons for their high efficiency of information uptake. Those circles are relentlessly pushy in their evolutionary advance . . . and the pushiest are right inside us.

The Indispensable Insulator for Natural Selection

Implicit in our evolutionary musings is the notion that from early on there was a barrier that prevented the components of the developing molecular system from drifting apart and undesirable outside elements of the environment from drifting in—a peripheral insulator, a sort of skin that kept the system's component apart from other systems and from the environment. We are dealing here with things that largely are in the liquid state—the medium for the molecular development was water. Thus, only by surrounding the pertinent ensembles of molecules with a skin that is impervious to water can those ensembles acquire an individuality—a unit quality that is essential for natural selection.

What precisely that skin was in primeval times, we can only guess. Today it universally is a thin film made of lipid and protein, a membrane that bounds every cell in an organism. Such a structure costs a lot of information—its organization is achieved with the input of macromolecular information. But the early stages of biomolecular evolution, like the stages in Figure 6.1, could possibly have occurred inside coacervates or under a cover of proteinoids—structures that are less information-costly.

Coacervates are droplets of polymer concentrate that segregate from the water in colloidal solutions; and proteinoids are amino-acid polymers that form when mixtures of amino acids and polyphosphates are heated. Neither one of these formations requires input of macromolecular information. But what exactly the early peripheral insulator was is not important for our argument.

Natural Selection Is Not a Gentle Game

To end this account of the evolutionary hypothesis, it may be useful once more to set the conservative concept we applied to primeval biomolecular development against the classical one. The classical concept is best summarized in Darwin's own words:

> This preservation of favorable individual differences and variations, and the destruction of those which are injurious, I have called Natural Selection, or the Survival of the Fittest.

Here selection ensues by the destruction of the less fit together with the information that brought the less fit into being. In the conservative view the information of some of the not-so-fit is stored and may be used later on in evolution. Or to put this in terms of our game metaphor: in the first game, the loser loses body and "soul"; in the latter, he loses only his body, and there is hope for resurrection.

We have seen that this hope is not unfounded; ancient information may eventually become expressed in the functional world and such close-fisted management of molecular information can pay big dividends in cellular differentiations. We shall see more of this when we get to the cellular communications.

However, the reader should not be left with the impression that the conservatism proposed is entirely indiscriminate or that it need prevail in all selections. It certainly entails destruction of some information, namely, of R mutants and their descendants which are chemically unstable, replicate too slowly or too inaccurately, are too harmful, or are otherwise unfit to make it to the amalgamation stage. As we get to the higher organismic level—the level Darwin had intended his theory for—selection by total destruction of variant organisms, indeed of whole species, does occur, of course.

On a humanitarian scale, then, the conservative view is only a little less harsh than the classical one, but there is no reason to expect otherwise. The going has always been rough out there. Only in very recent times has the life for humans and their pets become sheltered enough to abet humane behavior. As the old saying goes:

> When Adam delved and Eve span,
> Who was then a gentleman?

The Garden of Eden

With the conservation hypothesis in tow, we turn now to the actual experimental work behind our thought experiment. It was not without motive that we started the evolutionary diagrams with three-dimensional RNA, working ourselves backward, as it were, against the present-day streams of information. The choice of RNA was guided by considering what kind of chemical structures, among those possibly present in the prebiotic environment, were most likely to have generated closed information loops. To begin with, we needed a structure with a modicum of information for the building up of other structures with the blocks in the environment; and we needed a closed information loop for getting amplification. Only in this way could one expect an organization to arise that would not soon relapse into equilibrium. Without an informational structure to direct the assemblage, it is impossible to achieve a strongly bonded macromolecular organization. And without amplification, that is, without many copies of the assembled structure, the probability for closed-loop formation in liquid-state systems is virtually zero because molecular concentration falls off with the square of the distance from any molecular source.

All workers in the field of prebiotic chemistry agree that the basic building blocks for the macromolecular organizations, the amino acids and nucleoside bases, and the phosphate energy donors, could have formed in primeval conditions in the absence of macromolecules. These building blocks were shown to assemble "spontaneously" in experiments simulating the early environment on Earth (such molecular syntheses are ruled by the principle of directly energized short-chain elongation, as discussed in Chapter 4), and nucleosides and nucleotides might have formed from prebiotic precursors through mineral-catalyzed reactions. It also has been shown that from these building blocks protein-like molecules and short RNA chains can be polymerized in the absence of macromolecules: polypeptides of up to about 50 amino acids form on clays, proteinoids of 200 amino acids form in the presence of polyphosphates, and oligonucleotides of 40 nucleotide units with RNA-like linkages (3'–5') form in the presence of polyphosphoric acid and zinc. Our question is then, which of the three classes of molecules, DNA, RNA, or protein, started the cybernation? We ask what class or classes, not what particular molecule, because it is thermodynamically impossible to generate an information loop with a single molecular entity. To those who thought that life could have sprung from a DNA molecule, the physicist John Bernal once aptly said: "The picture of the solitary molecule of DNA on a primitive seashore generating the rest of life was put forward with slightly less plausibility than that of Adam and Eve in the Garden."

So, let us examine what mix is more auspicious. To begin with the proteins, this class could conceivably serve for molecular assemblage. Indeed, proteins can instruct the formation of other proteins without the intervention of nucleic acids. The cyclic peptide gramicidin S (an antibiotic), for example, is made of 10 amino acids with the information from a 280,000-dalton protein. It is one thing, however, to make a single peptide, and another to get a cybernetic loop. A loop would require a whole string of peptides. Only a cascade of proteins could embody enough information in members and side branches to generate an autocatalytic loop (and this would need some catalytic priming). A chain of eight, as pictured in Figure *6.3* , would barely suffice to build up to a protein of modest size.

Pure protein loops, thus, are of necessity multimembered, and such loops, we have seen, cannot compete in information economy with simpler loops, such as those that can be fashioned from nucleic acid and protein. There may be chemical reasons besides, but this alone is sufficient to ban protein loops from the evolutionary scene.

In choosing between DNA and RNA, we must consider that information is ultimately to be transferred to the three-dimensional world where all life's actions take place. The DNA class is evidently the weaker candidate for such transfer. Not that DNA is incapable of making amplifier loops—in fact, it is superlatively suited for that through the complementary pairing of its bases in the double helix—but in single-strand form it simply has not much talent for passing on information to a three-dimensional structure. It has much potential energy for short-range molecular interaction by hydrogen bondings, but very little of this can be actualized in pairings with three-dimensional molecules. Whichever way we turn the DNA with respect to organic molecules (that are not nucleic acids), the intermolecular distance most everywhere exceeds the critical 4-Å nook for information transmission.

We are thus left with the RNA class as the safest bet for starting amplifier loops. This class represents the best of two worlds, in a manner of speaking. Single-strand RNA can fold up to various shapes, depending on the sequence of its bases. The three-dimensional structure results from hydrogen bondings between the usual complementary bases (guanine-cytosine, adenine-uracil) and between other bases (guanine-guanine, adenine-adenine, and adenine-cytosine), forces that twist the strand into a partial double helix with a tertiary structure. When certain strategic bonds are broken, this usually stable three-dimensional structure untwists to a one-dimensional one. In the one-dimensional form it is well suited for transferring information by complementarity, and so can serve as an information store that is readily copied; and in the three-dimensional form it can transfer information to a three-dimensional structure. RNA, thus, makes an ideal bridge to the three-dimensional world of action.

There is experimental evidence that three-dimensional RNA can reproduce itself in the absence of proteins. Orgel showed that chains of up to about 40 guanine bases can form from guanine monomers containing triphosphate (GTP) on cytosine chains as templates, in conditions where zinc ions are the only catalysts. Such RNA replication is more error-prone than that catalyzed by today's sophisticated RNA polymerizing enzymes (which interestingly all contain zinc), but the fidelity is good enough to sustain an amplifier loop. The good fidelity is due to a particularly high interaction energy in the hydrogen bonding between guanine and cytosine (high enough even for transferring information directly from pieces of RNA to protein, that is, for translating the RNA in the absence of a ribosome machinery). There is evidence, furthermore, from Eigen's laboratory that RNAs can make self-reproducing loops with proteins, which undergo natural selection in the test tube.

Particularly revealing here is Thomas Cech's discovery of an RNA in the protozoan *Tetrahymena*, which replicates itself in the absence of protein. That RNA molecule has all the necessary information built into itself to assemble RNA. It accomplishes this by means of an amazingly versatile part of its own structure: an untranslated sequence of 400 nucleotides—an intron. This intron can unlatch itself from the molecule's main body and adopt a three-dimensional configuration capable of catalyzing the synthesis of short nucleotide chains (polycystines). This little RNA holds enough information to make RNA, using itself as template.

There are a number of such relics from the ancient RNA world still around in cellular genomes. A whole group of introns (*group I*), which can act both as catalysts for RNA replication and transsplicing, has been discovered.

These were the reasons for choosing RNA (R) for the start of the evolutionary models of Figure *6.1*. We still need to specify the mechanism by which the Rs form and reproduce themselves. For the earliest stage, we may assume a slow noninstructed assemblage of the Rs from the available nucleotides. Such assemblage of small RNA molecules from scratch does, in fact, occur in nucleotide solutions with high salt content in vitro. For a little more advanced, though still primitive, stage (Fig. *6.1 a*), one may assume a predominantly instructed mode of R assemblage, where the Rs formed in the earlier phase serve as templates, after the group-I intron model. (Interestingly, even in experiments with solutions of nucleotides and polymerizing enzymes, the RNA assemblage becomes more and more template-instructed, as RNA molecules become available in the solution; RNA-instructed enzyme formation here is looped into RNA-instructed RNA formation.)

As to the first catalysts for R formation (C_P, C_R), there is not much to guide our choice. These may have been organic polymers, clays, or other mineral surfaces. Polynucleotides adsorb strongly to positively charged surfaces of minerals, such as hydroxylapatite. Such surfaces might have oriented nucleotides in stereospecific reactions. However, from a certain stage on, and well before complex proteins were on the scene, the catalysis may have been carried out by RNAs. It turns out that proteins do not hold the monopoly on enzyme activity as was generally believed. Several three-dimensional RNAs have been found that possess considerable catalytic power; they can catalyze a number of chemical reactions, including protein synthesis. The ribosomal RNA (rRNA) is a case in point. This rRNA seems to be another relic of the old RNA world; it can catalyze the peptide-forming step of protein synthesis, as Harry Noller and his colleagues have recently demonstrated. Indeed, the all-important peptidyl transfer reaction in our ribosomes, it now turns out, may well be mediated by rRNA, rather than by the ribosomal protein as was formerly thought.

Thus, early on RNA molecules conceivably performed the catalytic activities necessary to unfold from three-dimensional to linear form, and to assemble themselves in a nucleotide soup. Such RNA molecules also may have been instrumental in the conservation phase (Fig. *6.1*, stage *g*), as catalysts of the first Rs' amalgamations. Even today a certain amount of RNA splicing occurs in protozoan and neurospora cells, where the RNA itself supplies the entire information without the help of protein. The amalgamations became more efficient with the advent of the modern splicing protein enzymes, but in the primeval RNA world catalysis by RNA molecules may have been sufficient; the ribo- or the deoxyribonucleotides are structurally all alike at their ends. So, one molecular species with the right fit may have brought about the diester linkages of all 16 possible combinations.

Polypeptides of modest size may have arisen, as already said, before other macromolecules were on the scene to serve as templates. But the making of protein probably only started in earnest in the RNA world, when molecules developed that served as adaptors to mediate the transfer of information from linear RNA to amino acids, which do not fit well enough together to transfer information directly to one another. (Molecules so conformed as to match two unmatching structures for information transfer are called *adaptors*.) The first adaptors for protein synthesis may have been primitive three-dimensional RNA molecules, precursors of the present tRNA, that could link (acylated) amino acids to an RNA template.

That the direction of information flow in the model at one time went from RNA to DNA (Fig. *6.1* , stage *h*) will not seem far-fetched. Even present-day RNA viruses are capable of generating flow in that direction by means of the "reverse transcriptase," an ancient enzyme, discovered by David Baltimore and Howard Temin.

The Oldest Circus on Earth

We end this chapter with a thumbnail sketch of the evolutionary scenario:

Some 10 to 20 billion years ago the four elementary forces burst forth from the primordial *Unum*, starting the cosmic march of organization from energy particles to atoms to molecules. Here on the cooling Earth, that march went on from symmetric molecules to asymmetric ones, ushering in, 4 billion years ago, the first exchanges of information between molecules of complementary shapes, exchanges where information ran in circles, instead of straight lines — the first molecular cybernetic loops.

Such cybernation began in earnest with RNA molecules, asymmetric structures of enough versatility to serve both as one-dimensional information repositories and three-dimensional actors. Successfully looped, such RNAs can seize as much information as the laws of thermodynamics will give them leave to — a "pulling themselves up by their own bootstraps," where the systems as a whole manage to stay always a little ahead of the game.

Thus, RNA begat RNA. In due season, some structures appeared in this world of RNA, which had features complementary to amino-acid groups. These RNAs could serve as adaptors to get information to flow to amino-acid chains; and so the first proteins came into being on RNA as a template. Eventually, some of the proteins, better catalysts than the RNA, took over the three-dimensional actions. The families of RNA molecules — the successful ones at engendering useful proteins along with some which were not so successful — were amalgamated into a continuous RNA band.

Finally, with the development of proteins capable of catalyzing the formation of DNA from RNA (reverse transcriptase), DNA took over the function of information repository, and RNA got relegated to the role of the intermediary it has today.

Figure 6.4   The Circus. Flowing in from the cosmos, information loops back onto itself to produce the circular information complex we call Life. The complex is represented in silhouette, hiding its maze of interconnections. In the Circus logogram shown in Figure *7.1*, the influx from the cosmos is symbolized by the *top* arrow, and the return flow to the cosmos under the law of information conservation, by the *bottom* arrow.

In that molecular world centered on DNA, loop was added on information loop, as the repository got longer and longer. By the development of adaptors of various three-dimensional forms—RNAs, proteins, sugars—the loops sprouted connecting branches, forging a huge complex. The complex became twisty and labyrinthine, but all through evolution it maintained its circular shape, for it grew by piling up information circles, more than by piling up information in each circle.

The complex of information loops cuts a spiffy silhouette (Fig. *6.4*). But what should one call it? What name is there that would be apt for such an incorporeal something, a circle of circles that rolls through time, grasping more and more information? There is none in our daily language for something so utterly unique, for sure. So I prowled in Greek and Latin dictionaries after a vocable with good roots. I came up with *circus* where our word "circle" comes from; "circle" stems from the Latin *circulus*, the diminutive of *circus*. Circus has a nice ring to it and a nice bounce befitting something so alive. But it is its Greek root that will clear away, I think, the hems and haws that naturally crop up with such a neologism. The root is *kirkos*, which means "ring" *and* "hawk." This homonym also has enough of Circe rubbed on it, so that—fair warning!—the user is at risk to get smitten with it. Anyway, certainly we will never meet a more rapacious hawk or a grander circus.

The show will go on for as long as the information flows in from the cosmos. But what comes in, must go out—that is the Law—and so the arrow in the logogram on page 116, pointing away from the Circus, rounds out the picture and reminds us that all the hoarded information is but a loan. When the loan comes due in 4 billion years (give or take a few hundred thousand, depending on the temperament of our sun), all those *circuli* will go back to where they came from. Nevertheless, to those who are inside the Circus, it will always seem the greatest show on Earth, though I can't speak for the One who is outside it.

Part Two

Information Flow Inside Cells

The Two Information Realms

The Flip Sides of Information: Life's Two Realms

Having discussed how, by means of information loops, a molecular organization might have come out of quantum chemistry, and how, by loop interlacing, such an organization might have pulled itself up in time by its own bootstraps to higher and higher levels, we now take up the information trail in the year 3,500,000,000 post-Circus.

By then that organization has cut its wisdom teeth. The information loops have formed a vast network and the structure materialized has gotten enormously complex. Nevertheless, the guyline of all that information—the linear nucleic-acid structure that serves as a repository—still shines through, and the ancestral division of information depository and temporal expression is as clear as ever.

We might say that information leads a double life: a life in a one-dimensional world where the information is perpetuated in the form of a string of nucleotides, and another in a three-dimensional world where it is acted out in the form of proteins and other structures of a short-lived sort. The information itself, the abstract entity, is the same in either world at a given time; the same entity just appears in different dimensions and uses different makeups. It is a little like making a line drawing and a sculpture of the same subject. Both express similar information, but one is rendered in two dimensions and the other in three. The constraints are different, though. Whereas the artist has leeway in his expressions and is the better for it, Evolution brings forth hers under the tyrannical rule of equation 3; all her expressions are mathematical transforms, and the only license she is permitted is to choose among random events she does not control.

But why this split of information into separate realms? The answer which most immediately comes to mind is that the split affords a division of function: one realm conserves the information and the other does the work. This is true, but the answer only scratches the surface and smacks of teleology. A twoness that goes through the entire biological sphere is bound to have a more fundamental reason.

So, before we turn to the nitty-gritty of the information flow in organisms, let us wing our way once more out to the origin of life to see the roots of the duality. These can be traced to a fundamental paradox: a conflict between conservation of information and transfer of information. This conflict arose at the very start of the green twig of the cosmic tree, with the first attempts at conserving information by means of organic molecules. That conservation was at loggerheads with transfer of information; such transfer necessarily entails a change in molecular information, a change in molecular structure. Thus, the paradox: if the molecules concerned are absolutely stable, they can't transfer information, and if they transfer information, they can't be absolutely stable.

The evolutionary solution of the dilemma was a Solomonic compromise. The molecular information was partitioned among two realms—one maximizing conservation and the other maximizing the transfer. The two realms were embodied by molecules of distinct structural properties: one kind linear, like DNA, a form that can be easily and faithfully reproduced, and that is stable enough to stay intact in between reproductions; and a second kind three-dimensional, like the proteins, of more temporal and mundane ilk, that can turn over and interact with other molecules. The first kind doesn't do anything; it can't move. And segregated from the other realm, it can be shielded from the ordinary hustle and bustle in the world (in higher organisms the DNA is secluded in the cell nucleus). The second kind can move; impelled by short-range electromagnetic forces, its structure can move between determinate positions and, in turn, set in motion other structures in determinate directions. And the actual mechanical work produced by this electromagnetic pushing and pulling is useful in natural selection; it provides a performance quantity by which the information can be judged in evolutionary competition.[1]

There are several terms commonly in use concerning the two realms: "genotype" and "genome," which apply to the conservatory realm, and "phenotype" and "soma," which apply to the three-dimensional realm. These are time-honored names coming from a century of work in genetics. Though not strictly coterminous, they are close enough to the above classification to be considered synonyms of it. However, an unfortunate misusage has crept in in recent years. In the aftermath of molecular biology's triumphal march waving DNA as its

1. Early on, the work performance of the protein fits is the sole basis for this evolutionary judgment (Chapter 6), but as the biological information grows, the judging of performance entails more and more information tiers and gets more and more complex. But it is always in the three-dimensional realm, not in the one-dimensional one, where the choices are made. Ever since Darwin, students of evolution have recognized that the phenotype, not the genotype, is the test object in the laboratory of evolution.

banner, it became standing practice to refer to DNA as "*the* information," reserving this term for that molecule alone. Perhaps that came about because this do-nothing molecule comes closer to the abstract essence of information than the three-dimensional expressions do, or perhaps, because this molecule appears to be the source of information when we look just at a little stretch of the flow, outside the circular context. Old habits die hard, but let's hope that this one proves the exception because such restricted usage of the term obscures the picture. The molecules in both biological realms carry information—an RNA, a protein, or a sugar is as much an informational molecule as DNA is. The quantities they carry individually are different, to be sure, but if we could weigh the total amounts of core information in the molecules of the two realms, they would about balance—the two realms are but the flip sides of the same information.

The Two Major Information Streams

Let us see now what happens between the two realms. We are not so much interested in what goes on inside the realms (that we leave to the books of physiology and biochemistry, which should be consulted), as what goes on between them. Our main concern lies with the information transfer between the realms, the flow whereby one realm gives rise to the other—or, to be more precise, we should say, whereby one realm contributes to the rise of the other, for the ultimate source of all biological information, we have seen, is the cosmos; the flows between the realms go in circles.

Our problem, then, is how to come to grips with the circles, the information Circus, which are given a logographic encore in Figure *7.1*. Its loops confuse the dickens out of us; they have no beginning or end. So, we shall resort to an old trick scientists never tire of: to dissecting a piece of the whole—in this case, a segment of the Circus, the stretch between DNA and protein.

Thus, we fix on DNA and use it as our point of departure. We see then in each cell the information emanating, as it were, from the DNA in two major streams: one stream duplicates the information in the DNA, and the other, which consists of thousands of streamlets, imparts that information to the three-dimensional world (Fig. *7.2*). The first stream preserves the information depository over the generations of a cell, and the second keeps the cell alive. These streams are the descendants of stretches of the primeval circles of information (Fig. *6.1*), now as mightily swollen as the rest of the Circus of which they form a part.

We shall deal in this chapter with both streams, as we find them in every cell today. In the following chapter we shall see their further branchings, as they give rise to organisms with multiple cells. Throughout, it will be good from time to time to take off our blinders and glance at the logogram, to be reminded that what we are looking at are not straight lines that start at DNA and end at protein

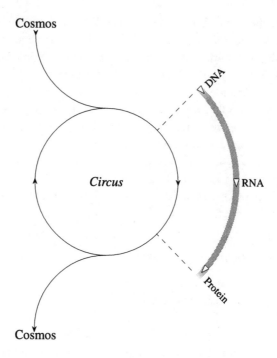

Figure 7.1 The Circus logogram and the little segment we focus on.
(The multitude of parallel circles in Figure 6.4 have been lumped together in the logogram.)

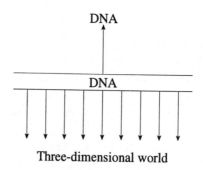

Figure 7.2 The two information fluxes from DNA.

(or, for that matter, anywhere else), but are stretches of closed loops where no one can say what comes first, the chicken or the egg.

The Quintessence of Conservation of Information: DNA

The secret of these information streams lies hidden somehow in the unique structure of the DNA. This double helical molecule made of nucleotide units is a crystal, but one of a most unusual sort: the units are all arrayed in one dimension. And as if this were not unusual enough, the array is not periodic. Ever since the discovery of this one-dimensional aperiodic structure by James Watson and Francis Crick in 1953, it has been sounded out with everything in the physicists' repertoire, hardly escaping a question about solid state that was the fad of the year: Is it a semiconductor? A superconductor? Has it anomalous magnetic properties? Does it carry solitons?[2] There were flurries of excitement, but in the end the answer was always, no. Rather than in a special solid-state property, the uniqueness of the DNA molecule lies in the complementarity of its two strands, in the chemistry of its nucleotide units, and in the aperiodicity of these units—a determinate (not a random) aperiodicity. The aperiodicity gives the molecule an immense capacity for storing information (there are 10^9 nucleotide units in our DNA), and the other two qualities enable it to reproduce that information economically. Storage and reproduction are inseparable facets of conservation of information.

We have dealt already with the storage. Let us now look at the reproduction of this immense information amount. The reproduction is an information transfer of a special kind: the entire information in the molecule is transmitted to the recipient structure, a structure which is not an information transform but the exact duplicate of the original—a total conservation of the information. All crystals can reproduce themselves, but the problem of reproducing an ordinary periodic crystal is trivial compared with that of reproducing an aperiodic one, especially one as long as DNA. From whichever angle we look at it, the task of getting total conservation here appears titanic: How is such a complex process achieved within the bounds of evolutionary information economy? How is the energy that goes into the organization of such a long molecule parceled out so that it is not blasted to pieces? What keeps the information—and to repeat, we are dealing with astronomic amounts—from becoming scrambled over generations of duplicates? These three questions are related. They chart the evolutionary search for extraordinary molecular talent.

Let us first look at the problem overall in terms of the principle of information economy enunciated in Chapter 6 to get our bearings. This principle

2.   Solitons are solitary waves that differ from ordinary waves; they are nonlinear and highly stable.

demands that, among the competing duplicating molecular processes that were once on the evolutionary scene, the process now before us was the one that cost the least information. To get a feel for the magnitude of the problem, consider the reproduction of a wax figure. The usual way for sculptors is to make a plaster mold and to use this as a negative for replication. This yields one replica per cycle. The procedure entails a minimum of four stages: two stages of molding and two of separation (Fig. *7.3 A*). At each stage, there is spatial organization of matter. In the mold stages molecules are orientated at the positive/negative interfaces, and in the separation stages big chunks of matter are orientated. All this obviously is information-costly, but no simpler method has occurred to anyone in the three thousand years of the art of sculpting.

Now, taking a stab at a more canny method, imagine a wax figure with a built-in negative (Fig. *7.3 B*). Here we can do without the plaster mold: from one structure, almost magically, as it were, two identical ones emerge—a duplication in the true sense of the Latin *duplicare*, a "doubling." The trick is to start with an original that is made of two complementary halves—a structure that has two potential negatives built into itself. So, each half can make a complementary half, can make a complementary half, can make a complementary half, . . . One could not ask for a more information-saving replication—we save the mold and the separation stage, to boot. And the end result is a perfect conservation: the original is conserved in the duplicates, one-half of the original in each duplicate.

That, in essence, is the mode of replication living beings use for their DNAs. Indeed, the DNA crystal, with its two antiparallel helices, is tailor-made for that. It has a built-in complementarity. Its two helical nucleotide chains have exactly the same unit sequence (and the same information content) but going in opposite direction to one another. So, when the structure splits in two, each helix can serve as a negative for its own reproduction by using short-range intermolecular forces (Fig. *7.4*)—a mechanism of breathtaking informational economy.

Behind all this elegance is the internal symmetry of the double helix. The cylindricity of this structure leaves a space exactly right for a hydrogen-bonded adenine-thymine rung or a guanine-cytosine rung, but not for an adenine-cytosine or thymine-guanine rung. And because of these spatial constraints, the nucleotide units of the two helices are uniquely complementary to one another. It would be hard to improve on the clarity of Watson and Crick's original announcement of this complementarity or "base pairing" (1953), which followed within a month their formulation of the DNA structure:

> If the actual order of the bases on one of the pair of the chains were given, one could write down the exact order of the bases on the other one, because of the specific pairing. Thus one chain is, as it were, the complement of the other, and it is this feature which suggests how the deoxyribonucleic acid might duplicate itself.

A

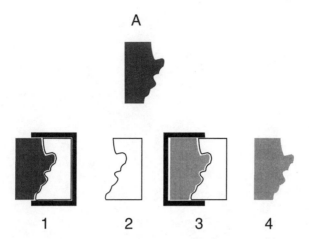

1   2   3   4

Figure 7.3 A To replicate this head shape by the ordinary method used by sculptors, a complementary template (*white*) is molded on the original (*darkly shaded*) (1), which is then separated (2), serves to mold the duplicate (3), and is finally separated from the latter (4). The procedure involves a minimum of four steps.

B

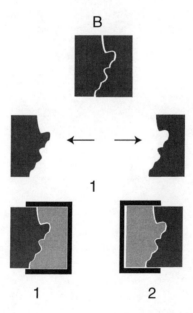

1

1   2

Figure 7.3 B In the duplication of this object consisting of two complementing halves, each half can serve as template for the other. The original (*darkly shaded*) can be duplicated in two steps (saving one step of molding and separation) if one half is conserved in each duplicate.

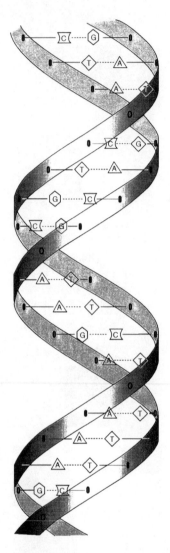

Figure 7.4   The two antiparallel strands of the DNA double helix. A, adenine; T, thymine; G, guanine; C, cytosine. Only an A-T or a G-C pair fits the space between the helical strands.

. . . Our model for deoxyribonucleic acid is, in effect, a *pair* of templates, each of which is complementary to the other. We imagine that prior to duplication the hydrogen bonds are broken, and the two chains unwind and separate. Each chain then acts as a template for the formation onto itself of a new companion chain, so that eventually we shall have *two* pairs of chains, where we only had one before. Moreover, the sequence of the pairs of bases will have been duplicated exactly.[3]

The Information Subsidy for DNA Replication and Error Correction

The duplication of the DNA molecule, thus, comes as close to a self-reproduction as anything ever could with such a giant molecule. But for all that elegance, information must necessarily be brought in also from outside; the two helices of the original don't come apart by themselves, nor do the nucleotide units of the nascent chain link up by themselves. Indeed, the extrinsic information here amounts to a tidy sum. Just consider how much information must go into the orderly unzippering and unwinding of the long double helix. The nucleotide pairs on the two strands are stably locked in place; if we just used brute force, we would have to well-nigh boil them to pry them apart. So, already the unzippering here takes a good amount of information. This amount is supplied by an enzyme. The smooth unwinding of the coils takes another portion that comes from a whole set of enzymes. Then, the linking of the nucleotide units demands its share of information. Each unit here carries a high-energy phosphate piggyback, defraying the entropy dues for the linkage, but the lion's share is supplied by an enzyme complex, the DNA polymerases.

Other extrinsic information must go in to correct errors in the copy process. Thanks to the high specificity of the base pairing, the copying is rather meticulous, but it is not above an occasional skipping of a base or a slipping in of an adenine to where a guanine should be. The frequency of such errors can be estimated as 1 incorrect base in 10^4 internucleotide linkages. Such errors would soon lead to the scrambling of the DNA information, were it not for corrective measures. The plain fact that the biological species around us are stable over many generations tells us that there must be error correction in the DNA duplication process. How does this correction come about?

Error correction is an organizing process that requires information. In the more familiar duplications—the typing of manuscripts or their setting into print—the rectifying information issues from a human being who compares the duplicate against a vocabulary standard, deletes the mismatches, and fills in the correct information pieces. More information goes into this than meets the eye, but such information processing is not beyond the capability of a state-of-the-art computer. One could affix a sensor and a deletion device (deletor) to a roving type-guide, and program the computer to send the ensemble chucking back and forth over the typing. A system of this sort could do the word-by-word copy-editing of this book.

3. The word "base" here refers to the nitrogenous bases—adenine (A), thymine (T), guanine (G), or cytosine (C)—the characteristic components of the nucleotide unit (after which the units are named). The other two components of each unit, a sugar and a phosphate, are invariant.

In the case of the DNA, the copyediting is done by one of the DNA polymerases (Polymerase I) which roves along a DNA template strand. The enzyme, in collaboration with the template strand, checks the correctness of each base pair formed before proceeding to ligate the next nucleotide. If a misfit occurs, this is excised and the right nucleotide is put in its place. The enzyme complex contains the information for detecting the nucleotide error, for excising the wrong nucleotide, and for replacing it. It is sensor, deletor, and type-guide, all in one, and the roving protein machine chucks back to the next correct nucleotide (often just the next neighbor). As a result, the errors are kept down to one per one billion nucleotide units copied. This is a performance no human book typesetter could hope to match; it amounts to printing one wrong letter in a book about two thousand times the size of this one!

Information Is Pumped into DNA as It Leaks Out

For all its peerless qualities, DNA is but mortal stuff: a composite of nitrogen rings, sugar, and phosphate—a frail structure that gets nicked, mutated, and broken up by thermal energy, radiation, and other things. Indeed, there are a number of notorious weak spots on the DNA structure: the junctures between the purine bases and sugars are prone to come apart as the molecule jostles at ordinary temperatures; the bonds between nucleotides can split; or the bases' nitrogen rings can decompose by action of ultraviolet, ionizing radiations, or alkylating agents in DNA's normal habitat. So, under constant attack at these and other weak spots on a long front, the molecule ceaselessly leaks information. The more solid crystals in the inorganic world also invariably leak some information, but that is a mere trickle by comparison. In our own DNA some ten thousand purine bases are lost every day by the thermal jostling alone!

Thus, to preserve its information, fresh information must be pumped into the DNA on a continuous basis and the thermodynamic price be paid. This means that the informational cost of this molecule is paid many times over during the lifetime of a cell. The attending energy supply is basically not different from that in the large-scale duplication process, except that it goes into local restoration jobs—regluing junctures, cutting out spoiled bases, putting new ones in their place—and each of these operations demands its share of extrinsic information. That information is provided by enzymes, like the DNA polymerases, which detect deteriorated nucleotides and implement their replacement.

But the lion's share of the information for the repair processes comes from the DNA itself, namely, from the strand which, just as in the "original" formation of the molecule, serves as template. And this points up again the elegance of the internal complementarity of the DNA molecule, the complementarity of its two halves. This peerless feature makes it possible to form the molecule and to restore it with the same outlay of information. One can hardly think of anything that could have served the evolutionary enterprise of conserving information better or be more telling about its economy.

The Spooling of the Golden Thread

Apart from the continuous information input by its repair enzymes, the functionality of the DNA molecule depends on a set of proteins with a largely mechanical role. A molecule as thin and long as DNA couldn't possibly be left to its own device and be of much use as a storage and readout tape. It would get twisted and turned by the thermals in a thousand ways and knotted up. Consider our own DNA, a thread 6 billion bases long. Stretched out in one piece, it would measure 1.8 meters—about two million times the diameter of the average nucleus. Just to fit it into that space, the thread obviously has to be tightly packed. Indeed, our cells do it one better: they keep the thread neatly wound— not much is left to chance when it comes to *that* weft!

At the heart of this fine fettle is a set of mini-proteins. These proteins—or *histones*, as they are called—are only 100 to 200 amino acids long. Still, so important and universal is their function that there is no living being on earth without histones. These are among the oldest proteins known, and they haven't changed much over more than a billion years (there is hardly a difference between our own histones and those of cows or peas). When one thinks of proteins, the image of enzymes usually comes to mind, and we have seen that these macromolecules are accomplished Maxwell demons. The little histones, though, don't quite yet make the demon grade, at least not insofar as local recognition of DNA is concerned. They have but some general *Fingerspitzengefühl* for DNA; they can hold on to the double helix because of the positive charge of their arginine and lysine amino acids; but they do so indiscriminately, at any DNA locale.

The combination of DNA and histones is called *chromatin* and its higher cadre, *chromosome*. With a little staining, those formations become visible in the light microscope, and the profiles of our own 46 chromosomes are not unfamiliar. However, only recently have these structures been put under a strong-enough loupe for us to understand what they really are and what their proteins do. These proteins seem to spool the DNA: eight protein subunits of four different histone kinds combine to form a bobbin—*nucleosome* in the anatomist's terminology—around which the DNA thread is rolled twice; the bobbins, in turn, form arrays that keep the DNA adroitly wound (Fig. *7.5*).

So, we can see now how so much DNA yardage gets orderly stowed into a cell's nuclear space. However, this brings up another issue, one concerning the readout of the DNA tape. The very density of the packing raises the question of how the transcription enzyme gets on to the tape. How does this bulky polymerase machine (about 500,000 daltons) bestraddle the thread if this is so tightly wound, and how does it slide down on it the length of a gene, with all

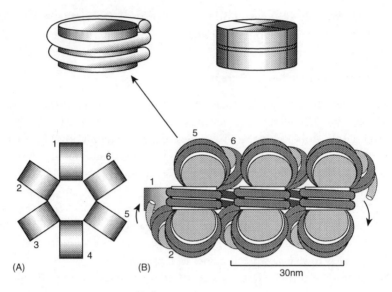

Figure 7.5   The DNA spooling apparatus. *Top:* A spool unit (*nucleosome*) consisting of eight histone subunits, around which the double stranded DNA is wound twice. *Bottom:* A model of the apparatus with the spools helically arrayed—six units per helical turn—in top view *(A)* and in side view *(B)*. (Bottom diagrams reproduced from Alberts et al., *Molecular Biology of the Cell,* by permission of Garland Publishing.)

those bobbins (half its own size) in the way? A plausible, though as yet unsettled, answer is that the eight-piece bobbin structure comes apart as the enzyme passes. Evolution has in store for us one of her rope and ring tricks here. We will get acquainted with some of them further on; alas, this one—easily the most dazzling of all—she plays close to her chest.

Cells Know More Than They Let On

I'd leave that question be, if it didn't go so much deeper than thread mechanics. In fact, it reaches into the *arcanum arcanorum* of biology: how the cells in a developing organism become different from one another—or put more cogently in information terms—how some portions of a cell's DNA information, often large blocks of it, are rather permanently shut out from the three-dimensional world. Consider what happens to our own embryonic cells. Early on in the womb those cells will express most any gene. But in the course of time they become more and more specialized in their gene activities: the cells destined to form our skin will transcribe one set of genes, while those constituting our muscles transcribe another set, and so on. But perhaps more telling here is what they do *not* transcribe. It's that untapped information that puts us on to the mechanisms of how they become different from one another, for they all possess the same genetic information. Every one of the hundred trillion cells that make up our adult organism (except those specializing in sex) contain the entire collection of bits, the same library of genetic instructions for making every piece of us, though they don't use all of it in their adult differentiated state.

Each cell type, thus, has some unemployed knowledge—our skin cells have unused knowledge about how to make bone, our nerve cells have unemployed knowledge about how to make muscle. But what is it that keeps those cells from employing all that knowledge? What stops a nerve cell from making the muscle-protein myosin or the skin-protein keratin? The answer must lie somewhere in the mechanism whereby information gets transferred from the DNA, for there is no longer any transfer from the genes encoding these proteins after maturation of the cell. Indeed, whole sets of genes then remain forever silent.

The complete answer is not yet in, but the histones seem to be involved somehow in this large-scale squelching of gene activities. The available experimental data suggest that these proteins prevent whole DNA segments from being transcribed in the differentiating cells. Perhaps the histone bobbins no longer allow these segments to be unwound—say, they tighten the DNA thread, and so put these DNA segments off limits for the transcription machine.

DNA's Various Information Tiers

At bottom is the question of how the cells tap the information from their DNA threads—the accessible portions of the threads. Before we launch into that, let us briefly flash back to the primary function of the DNA, the conservation of information. The full battery of conservation mechanisms we have seen—the wholesale copying and partial restorations (repairs) of the DNA—come into play every time a cell in our body divides. The yard-long string of DNA we carry in each of our body cells was originally put together when the parental egg and sperm cells that called us into being fused. That string then got duplicated trillions of times, and so each of our body cells carries inside an identical DNA copy, a duplicate of the original information depository.

This ancient information script in us—which has many pages in common with that in the other creatures on Earth, including the lowly bacteria—is written in thoroughgoing, prudently redundant detail. It contains everything our cells need to know. But how is this vast stockpile of information put to use, how is its information sent out to the three-dimensional world and where does it go? The depository uses the RNA as a go-between here. Thousands of different RNA molecules carry the information from the depository, breaking the information flow up into thousands of parallel lines. These lines fan out into a complex maze not too far from the depository, but over their initial portions, their first two stages, they are coherent (Fig. 7.6). These stages represent the well-known aphorism

DNA → RNA → Protein

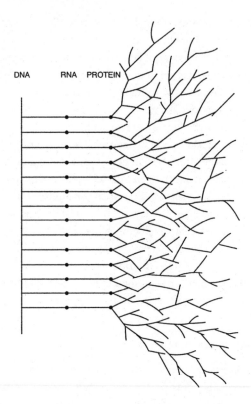

DNA   RNA   PROTEIN

Figure 7.6   The information flow from the DNA depository to the three-dimensional realm is divided up into many parallel streams which are coherent over the first two stages (DNA → RNA → Protein) and then fan out into a maze.

where the arrows indicate the flux of information between corresponding molecular structures. The arrows here imply the existence of a precise one-dimensional structural arrangement, a precise linear sequence of informational units chosen from a standard set—four units in the nucleic-acid structures and twenty in the protein structures.

Altogether, the flux along these stages and the maze beyond reflect the various informational hierarchies in the depository. These are the different information tiers of the DNA—the different layers of meaning ensconced in that master script. In the first tier the various DNA stretches encode the various RNA molecules—a straightforward encipherment, base for base. The decoding of this tier is called *transcription*. In the second tier the DNA stretches encode protein—one triplet of bases for one amino acid. This is the *genetic code,* and the decoding here is called *translation*. Then, beyond protein come the tiers for higher cellular and supracellular organizations.

At present, we know the codes only for the first two tiers, and so we may hope to coherently follow the information stream only from these two. And this we shall try to do in the remainder of this chapter.

The DNA Transform Pro Tempore

"Transcription" and "translation" are the words molecular biologists generally use when they deal with the first- and second-tier DNA decodings. This nomenclature—a lexicographic one—may sound odd to coding experts and information theorists, but it is apt. The RNAs are not unlike prints that are transcribed from a master plan—working blueprints for everyday use—and the second-tier decoding is very much like a language translation.

Chemically, RNA is much like DNA: a chain of nucleotides with a monotone phosphate-sugar backbone from which four kinds of bases stick out. Three of the bases, adenine (A), guanine (G), and cytosine (C) are the same as in DNA, but the fourth is uracil (U) instead of thymine (T). The sugar in the RNA backbone is ribose rather than the deoxyribose in DNA. Ribose has a hydroxyl (OH) group attached to its ring (where the deoxyribose has only a hydrogen atom), and this little extra makes RNA more prone to be split in the cell water. The molecule, therefore, can be turned over many times during the lifetime of a cell, and the information sent out from the DNA depository can be changed in a timely fashion and adapted to the circumstances prevailing in the cells. RNA is a DNA pro tempore, we might say.

Higher cells have several types of such pro tems; their RNA comes in several forms, nearly always single-stranded: messenger RNA (*mRNA*), transfer RNA (*tRNA*), ribosomal RNA (*rRNA*), and small nuclear RNA (*snRNA*)—all recipients of DNA information. For the time being, we will concern ourselves with mRNA. We shall meet some of the others in due course.

The First-Tier Code

mRNA is a simple information transform of DNA. In fact, it is so straightforward a transform, that the information equivalence is seen at once if one holds one structure against the other: the modular units of the two structures fit together, so that A is complementary to U, and G is complementary to C. Thus, the code of the first information tier is immediately transparent; it is but the rule of base pairing, and the entire book of words is just a list of four (Fig. *7.7*).

In the transmission of information from DNA to mRNA, only a fraction of the information in the first tier gets transferred at a time: the amount needed for making a protein, that is, the quantum contained in the corresponding exons, plus an extra amount contained in the intervening introns, which is not transmitted to the protein stage (Chapter 6). The amount in the exons is what opera-

```
A   U

G   C

C   G

T   A
```

Figure 7.7 The four-word lexicon of the first information tier. The four DNA nucleotide bases *A*denine, *G*uanine, *C*ytosine and *T*hymine are listed on the left, and the RNA bases they code for, on the right (*U*racil takes the place of *T*hymine in the RNA).

tionally constitutes a *gene,* and the mRNA embodying that information quantum gets shipped from the nucleus to the cytoplasm, where its information is read out by the protein-assembling machinery and disseminated in the three-dimensional cellular world.

The Genetic Code

A good deal of molecular hardware goes into the readout of the mRNA information, but let us gloss over this for the moment and consider only the logic of the information transfer—the logic of the molecular language translation. We have two classes of molecules here engaged in information transmission: on the one hand, a large DNA molecule, a long string of nucleotides; and on the other, a few thousand chains of amino acids. Somehow, that string of nucleotides of four different kinds must contain in coded form the specifications for the thousands of amino acid chains (which contain twenty different kinds of amino acids).

The simplest form of coding with such linear unit arrays is a cipher based on the sequence of digits. But how does a sequence of 4 digits—the nucleotides A, G, C, and T (or their RNA pro tems)—encipher the sequence of 20 amino acids in the protein? Or to use linguistic terms, how is the four-letter language of DNA and RNA translated into the twenty-letter language of protein? Put this way, the problem is a straightforward cryptographic one. For starters, we only need to calculate the number of possible permutations to see how long the code-words—the *codons*—would have to be in order to yield twenty translations. A three-letter codon would allow $4^3 = 64$ permutations, while the next lower quantum, a two-letter codon, would permit only $4^2 = 16$. Thus, a three-digit codon is the minimum that could provide for twenty different amino acids. Sixty-four translation possibilities may seem like an overkill when only twenty are needed, but that's the way the cookie crumbles with these digits—there is an inherent excess information in the logic structure of the transform here.

So, Lady Evolution had no choice but to adopt a three-letter codon. But did she have perhaps a choice in the reading format? Could the three-digit code-script be made more compact, for example, by an overlap of adjacent codons (Fig. *7.8*)? This is what a competent cryptographer would do if space were at a premium. He'd shift the reading frame serially, say, one digit at a time (this is how spies fit their messages onto pin heads and microdots). However, such compactness has its limitations; only some coding sequences could follow others without jumbling the message. One can live with such restrictions if the messages are short, but not if they are long, as in the case of the messages enciphering hundreds of amino acid sequences in protein. *Thus, a nonoverlapping code was a Hobson's Choice and Evolution adopted a three-letter code which is read from a fixed point in a straightforward manner, three bases at a time.*[4]

Now what of the extra forty or so coding possibilities, the intrinsic excess information in the second tier? Where did all that information end up, or let's say it outright, how well was it invested?—we have seen enough of Evolution's mores to come to expect no waste. Figure *7.9* shows how the assets were finally apportioned: 61 codons for one amino acid or another, and 3 codons for "stop" signals.

This allotment is as remarkable as it is revealing. Most of the extra information went into synonyms—only a fraction, into punctuation marks. Such richness of meaning warms the cockles of the poet's heart, though the lady's motives, I daresay, probably were more down-to-earth. Redundancy just may have been her best possible option. Consider the alternatives: most point mutations of the twenty coding triplets lead to nonsense or chain termination.

• • •

This genetic lexicon goes back a long, long way. Not an iota seems to have changed over two billion years; all living beings on earth, from bacteria to humans, use the same sixty-four-word code.[5] Nothing could proclaim the unity of life more strongly than that universality of information logic.

4.   This determinacy applies only to the mathematical logic of the genetic code, not to its molecular details. Indeed, all attempts to deduce the 64-word lexicon of the second tier on the basis of molecular complementarity or other chemical grounds have been unsuccessful so far.

5.   There are a few known deviations. Some protozoa, which branched off early in evolution, use the AGA and AGG codons for "stop," instead of "arginine." Mitochondria have their own DNA which uses a code with somewhat less information than its cellular counterpart.

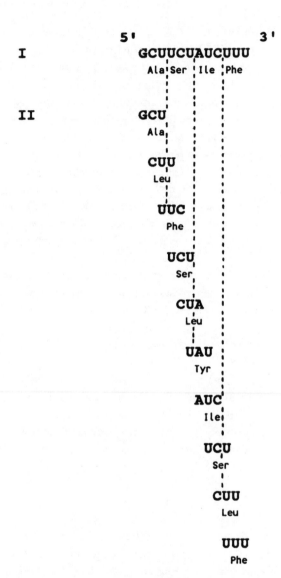

Figure 7.8 (*I*) The genetic code is a nonoverlapping code: the message is read from a fixed starting point from left to right (5' to 3'), three nucleotides at a time. Thus, the RNA codescript on top translates to the 4-unit polypeptide, Ala(Alanine)-Ser(Serine)-Ile(Isoleucine)-Phe(Phenylalanine). (*II*) An example of an overlapping code is shown for a comparison. The reading frame here shifts by one nucleotide in successive readings; the above codescript would translate to a polypeptide with six additional units, Alanine-Leucine-Phenylalanine-Serine-Leucine-Tyrosine-Isoleucine-Serine-Leucine-Phenylalanine. Such compact coding costs more information than the genetic code.

	U	C	A	G
U	UUU Phe UUC Phe UUA Leu UUG Leu	UCU Ser UCC Ser UCA Ser UCG Ser	UAU Tyr UAC Tyr UAA Stop UAG Stop	UGU Cys UGC Cys UGA Stop UGG Trp
C	CUU Leu CUC Leu CUA Leu CUG Leu	CCU Pro CCC Pro CCA Pro CCG Pro	CAU His CAC His CAA Gln CAG Gln	CGU Arg CGC Arg CGA Arg CGG Arg
A	AUU Ile AUC Ile AUA Ile AUG Met	ACU Thr ACC Thr ACA Thr ACG Thr	AAU Asn AAC Asn AAA Lys AAG Lys	AGU Ser AGC Ser AGA Arg AGG Arg
G	GUU Val GUC Val GUA Val GUG Val	GCU Ala GCC Ala GCA Ala GCG Ala	GAU Asp GAC Asp GAA Glu GAG Glu	GGU Gly GGC Gly GGA Gly GGG Gly

ALA	–	Alanine	LEU	–	Leucine
ARG	–	Arginine	LYS	–	Lysine
ASN	–	Asparagine	MET	–	Methionine
ASP	–	Aspartic acid	PHE	–	Phenylalanine
CYS	–	Cysteine	PRO	–	Proline
GLN	–	Glutamine	SER	–	Serine
GLU	–	Glutamic acid	THR	–	Threonine
GLY	–	Glycine	TRY	–	Tryptophan
HIS	–	Histidine	TYR	–	Tyrosine
ILEU	–	Isoleucine	VAL	–	Valine

Figure 7.9 The 64-word lexicon of the second information tier. The four nucleotide bases of messenger RNA are symbolized by their first letters, as in Figure 7.7, and the 20 amino acids, by three letters (the full names are tabulated above); "Stop" stands for chain termination. The codons read from left to right (5' → 3'). The third nucleotide of each codon corresponds to the wobble place in the interaction site of the adaptor; therefore, the coding determinacy of this codon is less than that of the other two. AUG codes for chain initiation apart from coding for methionine.

The genetic code was discovered in the 1960s, within ten years of the other landmark in biological science, the discovery of the structure of the DNA. The key to the coding logic was obtained between 1954 and 1958 through the efforts of George Gamow, and of Francis Crick and Sidney Brenner and colleagues. This was followed by a period of intense biochemical experimentation, and by 1965 nearly all the codewords had been worked out by Marshal Nirenberg, Severo Ochoa, Gobind Khorana, Philip Leder, and colleagues.

Transcription

We turn now to the hardware for the decoding of the two DNA tiers. In the first tier, the information flows between complementary molecular structures—from DNA nucleotide units to RNA nucleotide units. The DNA serves as a template for the RNA nucleotide assemblage much as it does in DNA replication, and an enzyme, the *RNA polymerase,* supplies much of the extrinsic information. This enzyme is a large protein ensemble (nearly 500,000 daltons), one of those information-heavy molecular machines mentioned before. This one carries off a number of complex jobs: (1) it separates the two DNA strands of the double helix, exposing the nucleotides that are to serve as template; (2) it finds the starting point of the DNA sequence to be transcribed; (3) it links the RNA nucleotides into a chain; and (4) it stops the transcription exactly where the template information for the gene transcript ends.

The operation begins with the polymerase machine swiftly roving along the DNA double helix in search of a place to start. Once that place is found, the machine shifts gears and clicks ahead, unwinding a stretch of double helix just in front, as it links up, one by one, the RNA nucleotides that pair with the exposed template strand. The RNA chain, thus, grows by one nucleotide at a time until the machine stops. The completed strand, which may be anywhere between 100 and 10,000 nucleotides long, then falls away and travels to the cytoplasm.

All these operations run without a hitch: the DNA double helix is rewound in the rear of the advancing polymerase machine at the same rate as it is unwound in front. Meanwhile, the RNA chain grows, forming a short-lived double helix with the DNA template strand; and as the chain grows, it unwinds from the template in step with the rewinding DNA double helix. Or, better, let the numbers tell the tale: the RNA-DNA double helix is about 12 base pairs long, just short of a full turn of the DNA double helix, and the RNA chain grows at the rate of 50 nucleotides per second, the same rate it comes out of the machine. All these lengths and speeds are finely honed. There is not much leeway here; let the RNA-DNA helix just be a bit longer or the RNA-chain extrusion a bit slower, and you have the Gordian knot for life!

Such a complex operation entails enormous amounts of information. Our most immediate question concerns the topological information: How does the RNA polymerase know where to start? Obviously, it must find the right DNA place, the beginning of a gene, or else it would transcribe nonsense. The beginning of the transcript is signaled by special DNA motifs. In many eukaryote genes the motif is the nucleotide sequence thymine-adenine-thymine-adenine—TATA—a motif located just upstream of the protein-coding sequences. It is motifs like these that the enzyme is looking for when it roves along the DNA, and once it finds them, it firmly hooks on to the DNA double helix to commence operation. The enzyme itself spans a 60-nucleotide length of double helix, and starts transcribing 25 nucleotides downstream of the motif (the transcript starts usually with an A or a G). The motif not only tells the enzyme where to start from, but also in what direction to go; it is oriented in the direction 3' to 5' on the template, pointing downstream, as it were (the direction is defined by the orientation of the phosphate-sugar backbone, the 3' and 5' carbons demarcating the chain ends). Because nucleotide chains are synthesized only in the 5' to 3' direction, only one of the DNA strands gets transcribed.

What makes the machine stop transcribing is less understood. But at least in the case of prokaryotes there is a good hint: a characteristic DNA motif, a palindrome of Gs and Cs which in the RNA transcript tend to self-pair and fold over in a hairpin loop. The polymerase machine there seems to pause and the tail end of the transcript, a string of Us, peels off the DNA template. The little machine now is free to search again for a starter place and begin another transcript.

Splicing

As in DNA replication, there is editing at the transcription stage. But the editorial job here is more compositional than corrective. The RNA polymerase does not match the feat of its counterpart in DNA replication who detects and corrects errors on the nascent strand. There are thus more errors in the transmission of information here: 1 error per 10^4 or 10^5 nucleotides transcribed. But we can live with that error rate. The mistakes are short-lived, not perpetuated as in DNA replication.

The first thing that happens—and this is even before any editing begins—is the capping of the nascent mRNA; a methylated G nucleotide is put on the 5' end of the RNA within a second of its coming out of the polymerase machine (this cap will play an important role in translation). Then, as the RNA strand gets to the 3' end, it is cut by an enzyme that recognizes the sequence AAUAAA, and yet another enzyme adds a string of some 150 to 200 As to the tail.

Only then starts the real editorial job: the pulling together of the information to be translated. The transcript contains the information of both the exons and the introns in the DNA. The introns, as we have seen in Chapter 6, are important in the evolution of the one-dimensional information depository, but they may have no immediate uses in the three-dimensional world; they don't ordinarily translate to protein. The editorial job, then, is to cut the introns out and to splice the exons together (Fig. *7.10*). This job is not unlike the editing of a movie, but the basic information for the editing here is in the movie itself, and the processing of that RNA information plus the cutting and splicing are all done by macromolecular machines, the *spliceosomes*.

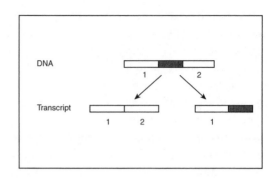

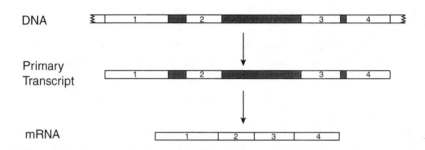

Figure 7.10 Condensation of DNA information by RNA splicing. The introns (*gray*) of the primary RNA transcript are excised and the exons (*numbered*) are spliced together. *Inset:* Some eukaryotic genes can be processed in alternate ways, depending on the presence of splice sites. Here the primary transcripts of the intron (*gray*) and exons *1* and *2* can be arranged in two ways. *Left*, the intron is removed, and the transcripts of exon *1* and *2* are spliced and translated into protein. *Right*, the intron transcript is retained, and the transcript of exon *2* is not spliced; the message of exon *1* and of the intron are translated into a different protein.

Spliceosomes are complexes of protein and small nuclear RNA molecules (*snRNA*) with only about 100 to 300 nucleotides, which form large assemblies that attach themselves to the primary transcript. These machines know where to cut and where to splice. If their cutting were off by even one nucleotide, there would be a frameshift and scrambling of the information. In fact, the editing and splicing proceeds from the 5' to the correct 3' end, without skipping a beat. For genes with multiple introns, such as those commonly found in eukaryotic cells, this implies a series of coordinated steps of recognition, cutting, and splicing.

The eukaryotic introns nearly always begin with GU and end with AG, and in vertebrates these sequences are part of a common starter and end motif. But where precisely the information for all the splicing operations comes from and how it flows is not known. Fortunately, there exists a simpler splicing entity at the lower phylogenetic rungs, which is made solely of RNA. This entity helps us come to grips with the splicing mechanism and lets us in on an age-old information cabala dating back to early evolutionary times.

The Genie that Dances to Its Own Tune

We have come across this entity before (Chapter 6), an RNA of an amazingly self-sufficient sort that manages to make RNA without the assistance from protein; it is template and Maxwell genie all in one. Now, this customer can pull off yet another trick: self-splicing. Behind this second feat is an oddly metamorphic string of nucleotide units which unlatches itself from the genie's structure. The string—some four-hundred RNA nucleotide units long—straddles two translatable RNA sequences, but it is not translated itself. Thus, we have two exons and one intron here, a short RNA script where an intron gets spliced out (Fig. *7.11*).

The chemistry of the splicing has been studied in detail in Thomas Cech's laboratory and, based on the results, we can sketch the information flow. The operation starts with the cutting of the RNA, the breaking of a bond at the intron's 5' end by a guanosine nucleotide of the medium—the nucleotide's hydroxyl (OH) group does the attack. Then, the bond at the intron's 3' end is broken; here the intron uses its own hydroxyl group for the attack, the group at its 5' end now dangling free. And finally, the two exons are bonded together (Fig. *7.11: 1–3*). All this requires catalytic information; there is no chance that a chemical cascade like that could occur spontaneously. So, the question is then, where does that information come from for those three reactions? Cech's experiments give a startling answer: the RNA itself.

Indeed, the answer originally caught everyone by surprise. In the 150 years of biochemistry, no one had ever seen an enzyme that wasn't a protein. But the experimental evidence left no room for doubt. This sort of RNA, it turned out, has the three-dimensional wherewithal for catalysis. Its intron string tends to fold up into a somewhat twisted hairpin structure (which gets stabilized by hydrogen bonds between complementary bases). As a Maxwell demon, this structure is not in the same class as the proteins, but its information power is enough for the catalysis at hand. It even seems to have some allosteric talent; it switches conformation as it gets unlatched. In its initial conformation, we may assume, this genie supplies the catalytic information for the first cut (the hydroxyl attack at the 5' end), and in its other conformation, it provides the information for the second cut and for the exon splicing (Fig. *7.11*). The only extrinsic element in the operation is the free guanosine; but this merely lends a hand and fuels the entropy flux for the first cutting job (the flux can be driven by the guanosine gradient).

So, all the basic information for the splicing operation here is built into this one RNA molecule. Such self-sufficiency is so staggering that no number of exclamation marks would give it its due. But is this RNA really a single entity? It certainly starts as one, but then—hocus pocus—it becomes two. In the single

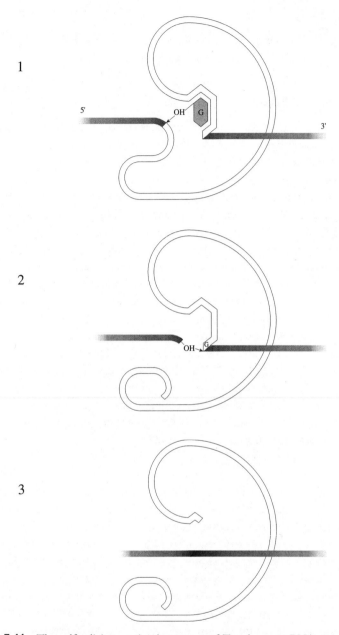

Figure 7.11 The self-splicing genie. A cartoon of Tetrahymena RNA, representing the exons and the intron (dark) in two conformations supplying the information for the sequence of events leading to the removal of the intron and the splicing of the exons: *(1)* the hydroxyl (OH) group of the free guanosine nucleotide (hexagon, G) in the medium breaks the phosphodiester bond at the intron's 5' end; *(2)* this frees a hydroxyl group there which, in turn, breaks the phosphodiester bond of a constitutive guanine (G) at the intron's 3' end and, then, *(3)* the two exons are bonded together. The two intron conformations in *1* and *2* are assumed to fit a guanine nucleotide and to juxtapose the reactive groups for bond breakage and reunion in the various steps. At each step one phosphate ester bond is directly exchanged for another.

state it is information repository or script, and in the dual state it acts out some of that script. This tricky molecule belongs to more than one world, as it were: by virtue of the linear order of its nucleotide units, it belongs to the one-dimensional information world; and by virtue of the three-dimensional configurations of its intron segment, it belongs to the action world.

A Punishment that Fits the Crime

This metamorphic RNA gives us an inkling of the sort of Maxwell genies that may have inhabited the primeval world of RNA. They are the ones who, we assume, brought off the first informational amalgamations, the splicing of the aboriginal R pieces (Chapter 6). And later, in the DNA world, RNAs of that same ilk (the snRNA in spliceosomes) took over the splicing task, in association with protein, allowing the recombination of exon and intron information. So, this RNA type has been instrumental in the combinatorial information enterprise all through biological evolution. It is truly a primary element of evolutionary advance. Endowed with this RNA, an organism could shuffle and recombine pieces of duplicate exon information, giving it a clear edge over an organism who could not; the shuffled pieces already had been successfully tried out to serve some function or other. On the other hand, an organism who lacked that combinatorial ability had to start the trials always from scratch. And even more developmental power fell to the one who could recruit intron pieces.

To appreciate this evolutionary advantage, we only have to look at some of the creatures around us. Many of them have lagged behind. Take bacteria, for example. These are about as old as we are, but they lack introns and splicing genies. However, if our notion of R-conservation is right, they must have possessed them in primordial times and lost them later on. If this meant trespassing on hallowed ground (conservation of information), it may not necessarily have been such a bad thing over the short evolutionary run. By ridding themselves of all that unused information ballast, those creatures gained genetic responsiveness to a changing environment; without that ballast, they could replicate their information depositories and themselves much faster than we can. Today's bacteria commonly reproduce themselves at the rate of one generation every twenty minutes—in one-and-a-half years they produce as many generations as we did since Cro-Magnon! Under the environmental selection pressures, they are able to change their genetic information through inherited mutations and produce variants in a relatively short time. However, that fast genetic adaptability was won at a steep price. The moment those creatures gave up the old information, progress was possible only by point mutations in translatable gene territory, and that only goes so far. Such mutations can bring forth new protein, but they necessarily wipe out the original information. On the other hand, those beings who kept the old pieces of information and eventually used them to generate new protein achieved virtually endless innovation without sacrificing old translatable information; they only had to develop new splice sites by mutations in intron territory.

So, the commandment, *make thy information larger,* is not one to be trifled with. Those who dare to disobey it for temporal gains get punished in biblical fashion: they and their offspring are forever damned to sit on the lower branches of the tree of life.

How to Bridge the Information Gap

After this side-glance at the hazy past, we continue our trek along the cellular information stream, from RNA to protein. First a summary of our sightings so far: the DNA information gets processed right after its transcription to RNA. As a result, the RNA, which starts as a faithful transcript of the DNA sequence, is shortened (Fig. *7.10*). Except for a few genes, like those coding for histones, the eukaryotic genes are split. Many have multiple introns, and some, like the gene for the collagen of our tendons, have as many as 50. Thus, the mRNA gets in the main much shortened by the processing; in some instances the condensation can amount to as much as 90 percent. The condensed, spliced version of the mRNA now is ready to go to the readout machinery, the *ribosomes,* where the mRNA information is transferred to protein.

At that stage we face a tricky transmission problem: How does so much information get across? There is precious little contact between mRNA and protein molecules—certainly not enough for transferring by electromagnetic interaction the huge amounts of information contained in a codon. The nucleotide triplets offer no complementary cavities that could serve for weak interactions with the amino acids, nor do they offer surfaces that would allow strong interactions of the template sort. However we try to align the two kinds of molecules or however much we turn them to face each other, there is always a wide gap.

To bridge that gap must have been a hard nut to crack for Evolution. The problem called for a filler that could serve as a transmission bridge, and such fillers are not easy to come by in the molecular world. The solution came in the form of a small three-dimensional RNA molecule, the tRNA, a molecule which at one end fits the RNA codon and at the other binds covalently the amino acid.

The gap problem is not unique to this stretch of the information stream. It comes up often, as we shall see, once the stream gets past the protein. In all these cases an interactive intermediary serves to establish information continuity. Such intermediaries are called *adaptors.* The name is apt. Like the adapters we use for connecting nonfitting electricity plugs, they connect nonfitting molecules. Most adaptors are proteins. However, here at this early stage in the stream and bespeaking an ancient evolutionary role, the adaptors are tRNAs.

The tRNA molecules are short nucleotide chains (about 80 nucleotides) forming stably folded hairpin structures twisted into short double helices. Each molecule has a codon-binding site and an amino acid–binding site. The codon-binding site offers a nucleotide sequence complementary to that of the mRNA codon. The site is located on a helical loop that runs antiparallel to the mRNA codon (after the fashion of the two strands in the DNA double helix), and site and codon can hydrogen-bond following the standard base-pair rules, but with some leeway: the third place of the codon tolerates a nonstandard partner; there the adaptor wobbles a bit. This sloppiness saves information. The system can get away with fewer RNA species to adapt 20 amino acids to 61 codons. In fact, it accomplishes this function with as few as 31 species instead of 61, as would be required if the interaction were wobble free. The wobble has still another benefit: it allows the tRNA molecules to come off the codon more rapidly after the adaptor job is done.

On the other interaction site of the tRNA adaptor, the transmission of information is more accurate. This accuracy is owed to the input of information from a set of twenty enzymes (*aminoacyl-tRNA synthetases*), one for each amino acid, which couple the amino acids to their appropriate tRNAs. Each enzyme has two specific cavities: one cavity fits one of the twenty amino acids, and the other fits one particular tRNA species. Thus, these enzymes can instruct the formation of twenty different tRNA–amino acid linkages, whereby each tRNA molecule gets specifically coupled to an amino acid. The error rates in these couplings are on the order of 1 in 3,000.

So, let's sum up. The gap between nucleic acid and protein is bridged by a set of adaptors. These adaptors plug a little sloppily into the codons, but tightly into the amino acids. The adaptors get the RNA message across with good fidelity: on the whole, each codon is translated into a specific amino acid. Despite code-script synonymity, the message is not ambiguous.

The untold want by life and land ne'er granted,
Now voyager sail thou forth to seek and find.
—Walt Whitman, *Leaves of Grass*

The Game of Seek and Find

Seek and Find

We now veer off the mainstream for a while to survey how all that enormous flux gets aportioned and channeled. The question immediately before us is, how do the messages from the thousands of genes get to the right destinations? Consider their conveyance from RNA to protein. The continuity of the information flux here depends on a series of specific molecular associations: the 20 different enzymes, the aminoacyl-tRNA synthetases, join up with their appropriate amino acids; these 20 molecular combinations, in turn, associate with their appropriate tRNA adaptors, and the resulting 31 amino acid–tRNA couples hitch up to the appropriate mRNA codons. But how do all these molecules find one another in the hodgepodge of the cell? How do they sort themselves out?

Let's take into consideration only the enzymes and amino acids, to narrow the issue. Each molecule here shares the cellular space with 40 different cognate species which are not materially segregated—there are no walls between them, nor a dynamic water structure that sweeps them apart. All those molecules are scattered about randomly by thermal energy, tossing and tumbling in the cell water, occasionally bouncing into one another. Millions of molecular encounters take place every instant in this jumble, yet only a few are of the fulfilling sort where the right enzyme meets the right amino acid. But those few matches make the endless roaming worthwhile.

The good matches here come about by serendipitous strike. Lots of strikes are needed to make such a game productive. Luck may not smile very often, but if an enzyme molecule lives long enough, it stands a fair chance to meet the object of its fancy—there are enough amino acid partners around. All a molecule really has to do is travel. And that, most everywhere in the cell water, has an affordable price. The random movement of the molecules costs no information; their ride on the thermals is free.

This is a peculiar game of seek and find. Like all sports of Evolution, it is a game of chance *and* information. The latter is the heartblood of the game, and each participant here contributes a share of information—exactly the amount needed to tell a partner apart from the rest of the crowd. Without that, there would be no finding. Information is at the core of any cognitive act, and in the case of the enzyme here, the basic information is sculptured into its molecular structure—a shape fitting the playmate. Indeed, the entire basic score of this matching play is imprinted in the structure of the molecular partners from the start, and so everything is preset for their union. The pair only need to get within electromagnetic range of one another . . . and the two become one.

The seek-and-find game then continues between the joined couple and the next playmate, the tRNA, and so on. Each successful encounter is marked by an electromagnetic interplay between the partners, an encounter where molecular configurations shift and information changes hand. This dynamic interplay provides the undergirding of the information flow from DNA to protein, that is, from DNA codon to amino acid—the sum total of the individual information transfers between the myriad of molecular partners constitutes the cellular information mainstream.

How to Keep Molecular Information Broadcasts under Wraps

At first sight, this zigzaggy way of sending information may seem odd or even slipshod compared to the way messages are sent by telephone or other solid-line communication systems. But the cell's style is really only more diffuse. The meandering messages from the codons get as surely to their destinations inside the cell as radio messages do from New York to London. If anything, the cellular information broadcasts offer more privacy; sender and receiver use a cipher, a personal code. Thus, despite its diffuseness, the flux of information from codon to amino acid, in effect, constitutes a communication line, and there can be hundreds of such lines prattling at the same time in the cell, without significant cross talk.

The diffuse method of sending signals probably dates back to the very dawn of life, when the first biomolecules emerged in the water pools on Earth. But why this coding? Why did Evolution not wall off the cellular space to obtain privacy for senders and receivers? That would surely not be expecting the impossible, considering that she produced so many kinds of polymers of which barriers or pipes could be fashioned. But her reasons go deeper than the matériel problem—they always do —and if we check out the information costs we see why.

Consider what it takes to connect a sender and a receiver of molecular information by means of a pipeline. First off, one needs to know the locations of sender and receiver (points *A* and *B*, Fig. *8.1*). Then, as the line is laid and

Figure 8.1   The high cost of channeled communication. To lay a pipeline from the sender of information *A* to the receiver *B*, one needs to know the positions of *A* and *B* and of each pipe, as the line is put together from *A* to *B*.

advanced from *A* to *B*, one needs to be apprised of the intermediate locations demarcated by the ends of each pipe, in order to adjust the positions. All this requires a good deal of information; indeed, all forms of partitioning the cellular water space, not just pipelines, have a high information tag. Add to that the difficulty of cramming thousands of partitions into a cell, and you get a sense of the full magnitude of the problem.

It should not be surprising, therefore, that thrifty Evolution chose the coded mode for message separation. That mode prevails not only in the mainstream inside the cells, but also, as we shall see, in the connections of the streams between cells. Only at the higher tiers do we find pipes and cell partitions attending to the function of intra- and intercellular communication. But those are johnnies-come-lately; the coded diffuse mode was there from the start.

The method of message separation here, in principle, is not different from the one we use when we want to keep our radio or telephone communications under wraps. We couch our messages in a special language that only the intended receiver understands; or in technical words, we use a codescript that only we and the receiver have the key to. In cellular communication, the codescript is in the form of a special molecular configuration, a spatial arrangement of atoms that is singular enough so as not to be confused with other arrangements of atoms that might occur elsewhere in the system. That three-dimensional singularity is the nub of biological coding.

The Virtues of Molecular Coherence

The keys to a codescript embody the information that makes the communication private or cryptic, as the case may be. There can be large amounts of information in such keys. Indeed, the keys which military agencies or spies use are often so loaded with information that they are totally out of proportion with the message itself. Just think of how much information is poured into a key for our run-of-the-mill spy: a wordlist or an alphabet is cooked up; a grammar may be invented if a fancy cryptographer gets in the act; and all this has to be explained to spy and master. Cells are not as profligate. They lay out only the barest minimum of information—they follow the principle of information economy—yet they get as much communication privacy as they need. In fact, by following the principle, they get an optimal coding which, apart from providing privacy, pemits them to overcome the noise in the communication lines, as we shall see.

The secret of that efficiency lies in the coherent modular construction of the cellular communication system. We have already had a glimpse of this remarkable construction encompassing two realms—a one-dimensional repository realm and a three-dimensional action realm—where information is built into molecular modules which bear a common lineal stamp. But if we detach ourselves for a moment from the static tectonics and focus on the flow of information, we see the whole menagerie in a functional light: the information dwelling in the repository realm sends out partial transforms of itself in one- and three-dimensional modular form to serve as signals in the action realm.

Thus, the modular entities in the two realms are *coherent,* and this by itself speaks volumes: the modular signal constructs inherently fulfill the condition of singularity for coding, namely, for sending information in a particular form that only the intended receiver has the key to. It is hard to find an analogy that even remotely comes close to this sort of coherence. The closest metaphor I can come up with is a grail—the set of engraved bowls once used in antiquity by the Greeks and Romans. The bowls fit neatly one above the other, the heraldic insignia going through the stack from top to bottom. In the biological grail, the fit between pieces is more intricate and the congruency not as straightforward; nevertheless, when all the modules and adaptors are lined up, the lineal coherence is unmistakable—the pieces hang together, bearing the ancestral stamp of singularity.

It is this through-and-through molecular coherence which makes the biological communication system so unique and so elegantly parsimonious. What other communication system is there where information is *securely* broadcasted with such economy of means—where the information carriers are signals and keys all in one!

The Transfiguration Called *Biological Specificity*

It will not have escaped the reader versed in biology that we are harping on the theme of *biological specificity.* Specificity, as the dictionary defines it, is the *quality or state of being peculiar.* The term biological specificity implies this, too, but its meaning is deeper and farther reaching. Looked at solely as a molecular property, that is, of molecules isolated from the living system as a whole, the meaning of biological specificity seems not so far off. It has, then, the sense of *chemical specificity*, a specificity where the state of being peculiar is a singular molecular form or complexion. But there is surely more than that to a specificity that ranges through all biological information strata—from the one-dimensional to the three-dimensional realm, from molecules to whole organisms.

The notion of biological specificity, thus, has collective undertones, and that is why it defies easy definition. When I looked up the word in textbooks of biology, only those dealing with zoology and botany give definitions, and these are largely couched in the traditional terms of Darwinian *speciation*, the specificity that applies to whole organisms. In the books covering the contemporary fields of molecular biology and cell biology, the word "specificity" comes up constantly, but no definition. It seems that the modern scientific developments have tied the tongues of the new biologists, who are not otherwise known for coyness. This is understandable. Alone the developments concerning the genealogy of macromolecules were enough to put the lid on definitions, not to speak of the conundrum of two realms.

However, having come thus far, we might as well take one more step and try a definition. But first let's have another look at the realms. We note at once that what dwells in two realms here is information, not energy or matter. This may not help us much with our definition problem, but it puts us at ease, for there is nothing unorthodox with *information* being in more than one realm. Quite the contrary, this is common fare in the cosmic kitchen; information takes up residence in all possible realms of the universe. And this has been so since the beginnings—to swap states of order of energy or matter is the oldest game in the universe.

In living beings that game has the same rules as everywhere else—it is always within the pale of physics—and the sport can go on and on, so long as there is a source of negative entropy to pay for it. What's out of the ordinary here, of course, is that it is played with macromolecules. We may rightly call the game *transfiguration*, because with these chunks of matter it's mainly a swapping of configurational states. Such molecular complexions arise under the rules of mathematical transformation. But they are further subject to another rule: they must serially fit with one another to allow transmission of their quanta of information. This rule is the boundary condition for electromagnetic interaction between molecules, which constrains the number of individual possible complexions to one or very few. It is the uniqueness of this serial fit, this collective state of being peculiar, which comes closest to characterizing the elusive quality of biological specificity.

The Maturing of Biological Communication

We now set the matter of molecular coherence aside and see about a developmental question: How did the communications in the mainstream become so clear—so free of ambiguities and errors? The question is germane because of the yawning gap between mRNA and protein. That gap, as we have seen, is filled in by a whole delegation of go-betweens, the tRNA adaptors and the amino acid enzymes. These macromolecules—there are 51 species of them—are the cognitive elements that make the transmission of information in the mainstream tight, the code stringent, and the translations clear.

This may not have always been the case. The first translation systems are unlikely to have had the luxury of protein assistance, but presumably had to make do with three-dimensional RNAs of less power of recognition. Indeed, it is not absolutely necessary to have the whole gap-filling delegation to get some protein. Polypeptides of modest size can be produced on RNA templates in the test tube, without the help of enzymes. There are then more ambiguities in translation, but what comes out is still protein. The amino acid enzymes, or even the tRNA-adaptor gear, may have developed gradually. Perhaps, early on in evolution, a faulty sort of information transmission—a translation full of ambiguity, mediated by just a few adaptor species—was all there was. The transmissions may have gotten better and better and the translational "meanings" sharper and sharper as the number of go-betweens increased and their information caliber improved.

Such a gradual seasoning would actually not be very different from what happens to our spoken languages. These start, as studies of primitive societies have revealed, with the simplest of sounds: grunts, pops, and clicks. Phonemes of this sort are information lightweights which offer little in the line of coding and, hence, are full of ambiguities. But, eventually, the utterances get richer, the information transforms become more coherent, and the meanings, sharper . . . and, one fine morning, one of the babblers declaims Hamlet!

How Biological Systems Break the Fetters of Quantum Chemistry

This much about translational ambiguities. What about errors? Their lack hits one still more in the eye than the lack of ambiguity. Even the tightest adaptor or the most canny enzyme demon cannot help some information from becoming lost in transfer. Inevitably, information gets degraded in transmission, and this is so regardless of how the information is carried—whether by electricity, radio waves, radar, light, or what have you. There is an elegant mathematical proof to that effect, one of the brainchildren of Shannon from the 1940s, when he drew together for the first time communication and statistical mechanics. Thermodynamic entropy is irreducibly associated with communication noise. Now, molecules dissolved in water are rather noisy information carriers, and the internally mobile macromolecules especially so; aside from the usual degradations due to hits of matter and radiant energy, they degrade information during their large-scale electromagnetic transmissions and their endless configurational changes.

For all these reasons, we would expect substantial informational losses in the flow down the mainstream and plenty of transmission errors. Nevertheless, few errors actually get through. Most are doctored out by special information loops in the sidearms of the stream, the circular portions of the flow we have been skirting so far. In our journey down the information stream, we shunted our boat, by a little sleight of hand, to the "mainstream" in the DNA-to-protein segment. Had we sailed down by the book, we might have been sucked in sideways and gone in dizzying circles. These are the loops that feed information back and control the main flow—braking, quickening, stabilizing, compensating, and correcting it everywhere. Without them, the molecular affairs in living beings would be as unruly as they are in ordinary liquid-state matter, and the communications as prone to error. The information doesn't just run in circles there; as anywhere else in the Circus, information flows in from the outside (Fig. *7.1*). This is what nurses the Circus, compensates for the informational degradation here, and ultimately brings about the high fidelity of transmission.

Take, for example, the information transfer at the DNA replication stage with its prodigiously low error rate of one in one billion. Such fidelity of transmission could by no manner of means have been achieved in a world of molecules that follows the laws of statistical mechanics pure and simply. As in any chemical equilibrium system the whole here is, roughly speaking, the sum of its quantum statistical parts. There is complete time symmetry in that quantum world; things are exactly the same when they run backward in time as when they run forward. And the kinetics are linear, that is, the rate changes are proportional to the number of molecules (chemical concentrations). Now, the circling information changes that vernacular state of things—it breaks the shackles of quantum chemistry, as it were—and the associated influx of information holds the transmission errors down. Looped together here in a feedback circle are the DNA polymerizing enzyme (polymerase I) and its substrate: after a nucleotide gets linked to the nascent DNA strand through the information flowing from the enzyme, information flows back to the enzyme from that nucleotide's base and its partner's base on the template strand; and with this feedback the enzyme either moves on to instruct the next linkage if the bases matched or, as often as not, it moves back to cut the nucleotide out in case of a mismatch. The enzyme here contains all the hardware for the control: the sensor element for the base match, the pieces that propel the enzyme, and the cutting pieces.

Not all feedback systems of the mainstream achieve that kind of fidelity; though, by human engineering standards, they are still very good. And whenever you find an error level that is well below the purely statistical mechanical one—be that in communications inside or between cells—it is a safe bet that the trick is turned by an information circle.

The Creative Kaleidoscope

The conglomerate of loops of the basic cellular information stream constitutes a sophisticated compensatory filter network—an "active" filter in engineering terms—that keeps the error rate down in the segment from DNA to protein. But what of the errors that take place upstream of this segment, the alterations in base sequence arising from instabilities in the primary DNA strand itself or the one-in-one-billion error that slips through the filter at the stage of DNA replication? Those errors usually go under the name of mutations. They are, nevertheless, errors—they are deviations from the original. But those sorts of errors are not cut out. On the contrary, once they are past the DNA replication stage, they are not recognized as errors but are treated by the filter stages downstream as if they were the real McCoy.

Thus, in effect, this mutated DNA information uses the normal amplifier and filter loops to get itself accurately transmitted at the various stages of the stream, and so has as good a chance to be represented in protein as the original information has. Thus, we find three types of unitary mutations in DNA and protein: substitutions of one base or one amino acid unit for another, deletions of these units, and insertion of extra units—exactly the kind of deviations from the original one would expect a priori from such a modular arrangement.

The DNA mutations are largely caused by random hits of other molecules or by random hits of radiation. The most effective hits are those by molecules that have a reactive group or by radiations that have enough punch to break a covalent bond, namely, radiations of wave lengths shorter than those of visible light, like the ultraviolet, X rays, and cosmic rays. In bacterial DNA, the DNA which has been studied most thoroughly, mutations arise at an average rate of 1 in 10^9–10^{10} bases for each replication. Thus, a gene of 1,200 bases coding for an average-sized protein of 400 amino acids would undergo 1 mutation in about 10^6–10^7 cell generations. Most of these mutations are useless or harmful; the mutated proteins function less well than the original ones, or not at all. However, once in a while a mutation will produce a better protein, and such deviations are the source of a slow but steady evolutionary advance. They allow the otherwise conservative system to gradually accumulate new favorable traits.

At the aforesaid mutation rate, the inheritable changes are necessarily rare, at least so in organisms that engage in sex. To be passed on from one generation to another, the mutations here must occur in a germ line cell, a sperm or an egg cell; on the average, there may not be much more than one mutation in a germ line every 200,000 years. Such slow rates are beneficial, if not necessary, adaptations. Really fast rates would fly in the face of the evolutionary strategy of conserving information; if a gene would change each time a cell divides, the offspring would no longer resemble the parents.[1]

A slow mutation rate, on the other hand, is a constant source of genetic variability that natural selection lays hold of at the level of its three-dimensional

---

1.   In fact, this is what can happen to viruses: their genome gets visibly scrambled in a short time, and this is carried to an extreme in the AIDS virus.

information transforms (phenotype). Even so, mutations are chancy business. To be at the mercy of such random events is like putting one's head in the lion's mouth. With each mutation the organism risks its life and even its progeny. Higher organisms manage to lower the risk. They make two copies of some genes: one copy for survival and the other for speculation. Most of the time the mutations will be useless and natural selection sees to it that the second copy is lost. But when that rare event, a *functional* variant gene, occurs, both the variant and the old gene will be preserved.

The mutable gene material constitutes a sort of kaleidoscope. You shake a kaleidoscope a little, and its pieces form ever fresh patterns. In the genetic kaleidoscope the pieces are nucleotides and the mutations do the shaking. But there is an important difference: the genetic kaleidoscope has loops that implement information gain. So, whereas in the toy kaleidoscope the total amount of information never changes (the pieces are merely rearranged), in the genetic one the amount increases with time and, pari passu, the number of mutations increases. Thus, in this circinate and stochastic way, we assume the kaleidoscope grew over the 3 billion years it has been on earth, its nucleotide pieces waxing from a handful to 10^9. And who knows how many pieces are still in the offing over the remaining period of grace!

The Paradox of the Progressive Reactionary

But how much progress can one expect from a genetic kaleidoscope that is hemmed in by feedback loops? Is the aforementioned mutation rate high enough to account for the great variety of living beings on the globe? We are not concerned here so much about the early biomolecular events, as about what happened later during the development of whole organisms. Early on, before the advent of noise- and error-suppressing feedback circuits, the mutation rate must have been quite fast—the kaleidoscope was then a whirling wheel of change. But suppressor circuits have been around for at least a billion years, for much of the time when the animal and plant variety emerged. So, our concern is justified. Indeed, we meet here with one of the oldest questions of biology: How is novelty generated in evolution, how are species developed? Here is the problem in the words of the one who started much of the scientific thinking about it:

> whilst this planet has gone cycling on according to the fixed law of gravity, from so simple a beginning endless forms most beautiful and most wonderful have been, and are being evolved.

Those are Darwin's closing words in his *Origin of Species*. Nowadays we can put a number to those "endless forms": 20 to 60 million species—a conservative estimate, though such estimations have shown a tendency of going up ever since people put their minds to cataloging living beings. A hundred years before Darwin, the Jesuit priest Anthansius Kircher estimated that 310 animal species went with Noah into the Ark. Well, in this day and age the assessment has risen to tens of millions—some 100,000 species of protozoa, 500,000 species of worms, 100,000 species of mollusks, 30,000 species of fish, 10,000 species of birds, and so on, have been pigeonholed.

The intervening century and a half since Darwin has seen many attempts at coming to grips with his problem of "endless forms." Literally no stone was left unturned to get to the bottom of how those forms came about; the search for clues went on in paleontology, taxonomy, embryology, genetics, and molecular biology. Gradually over the years, the focus of the inquiry changed from whole organisms to molecules, but at every level loomed this perplexing problem: the biological system changed its outward form, yet it clung to the old and tried; it varied and diversified, yet it conserved and stood fast.

And the problem still looms today. But there are some rays of light; equipped with the keys to the codescript, the present generation of biologists can flip through the pages of the DNA script in search of telltale signs of stepped-up mutations. So, we now can ask two sober questions: Do mutations get an occasional reprieve from the noise- and error-suppressing information accoutrements? And more daringly, is there, besides the DNA kaleidoscope, another one that is less throttled? The available evidence, we shall see, gives a qualified nod to these questions; and when we are through with stripping Darwin's statement to the pallid terms of information, we may hope to be a little closer to understanding why that quirky twig of the cosmic tree, this reactionary of reactionaries, is so perplexingly progressive.

What If the Cellular Information Stream Ran Backward?

Before we start our inquiry, let's step back for a moment and give the information mainstream another glance. In the endless cellular endeavor of making protein, the information fluxes in thousands of parallel communication lines all run in the same direction, DNA to protein. This, we might say, is the law and order amid the bustle in the cell—it is what gives that flutter and flurry of molecules a direction. But let us wing here above the thick of things and ponder what might happen if that information flow would run backward. We will keep to the first and second information tier—having seen how costly it is to get information to flow across the chasm between mRNA and protein, it isn't hard to resist a siren song from beyond the two tiers. So, suppose that there were such a reverse flow from mRNA to DNA, the stretch with naturally fitting molecules, where all that may be needed to make things go is an enzyme of the sort the retroviruses use to transcribe backward. This is an unconventional thought, to be sure, but it can be useful to play the devil's advocate from time to time.

We see right off that such a movement of information to and fro, if left unchecked, would create havoc. It would destroy the information that gives life—the macromolecular information so carefully hoarded over aeons would soon be degraded by transmission errors. Just consider what the mutations originating in the mRNA would do: these would be transcribed to complementary DNA, some eventually being incorporated by recombination into the parent DNA; the genomic information would be scrambled.

Clearly, an unrestrained two-directional information flow is incompatible with conservation of biological information. But what if the reverse flow were somehow kept in check and became viable only on certain occasions? Such a *selective* flow could allow a subset of processed mutant information, some RNA mutations, to get into the DNA repository, opening undreamed-of possibilities for genomic control. That flow—the lifting of its restraints—might conceivably be geared to environmental conditions. It is known that transcription, the regular forward one, can be regulated by environmental factors. In bacteria, for instance, transcription is switched on when the cells get stressed by lack of food in their environment. Thus, suppose reverse transcription were habitually repressed and the repression were lifted when a certain food molecule necessary for cellular growth (replication) is missing from the medium. Mutations arising in the mRNA would then be backtranscribed to DNA during the stress period.

Now, suppose further that among these particular mutations there happened to be one (or a subset) that permitted the cells to use a substitute foodstuff for growth, a molecule that is available in their environment. That mutation would then stand a better-than-average chance to be incorporated into the genome of the cells' progeny during DNA replication. So, in general, mutations that neutralize the condition of stress—mutations that are advantageous to the cells in this restricted sense—would occur more frequently than those which are not.

This model relies essentially on Darwinian selection, and so we can place its mutations in a canonical pigeonhole, that of *adaptive mutations*. This gives the mechanism a semblance of orthodoxy, though it hardly changes its innards. But more important, from the information point of view, is the evolutionary question whether the mechanism is economical enough to have been affordable a billion years ago. One aspect of the mechanism, the counterflow itself, easily passes muster; the reverse transcriptases and recombinant linker enzymes have been around since early biomolecular times. But what of the control aspects? It is mainly to that end that our question is directed, for by furthering "advantageous" mutations, the mechanism affords the cell a degree of control, and cybernetic soft- and hardware are not precisely bargain-basement items.

However, if we take a closer look, we see that the control here—if we can call it control—is of a most primitive sort. It doesn't even come near to being cybernetic; all the basic processes in this model—the RNA mutations and the interactions with the environmental factors—are random. And a mechanism operating so close to thermodynamic equilibrium has a small information tag. The smidgen of control over mutations it affords the cell is owed to the temporal coincidence between reverse transcription and cell growth; the growth is stochastically linked to the effect of the mutation, the phenotypic expression of the mutant information, and this gives rise to the temporary synchronism, the transient window for the selecting of mutations.

You get what you pay for, as the saying goes, and the cell here gets a small-time control. It never has the mutations completely in hand. The cell is no more a master of the events than we are when we try to fix a radio by banging it a bit here and rattling it a bit there, and the aberrant pieces happen to fall into the proper places. But the point is, when it works, it works.

So, let us sum up: the cell in this model has a subset of its informational macromolecules subjected to natural selection during the span of time DNA replication and reverse transcription coincide, and (without feedback loops) significantly improves its chance for advantageous ("adaptive") mutations.

The Elusive Second Wheel of Change

That all still sounds innocuous enough but, as with all heresies, beneath the surface lurk the quicksands. That tack takes us, as intended, to the origin of evolutionary variability but, with only the slightest push, we head smack into an old spiny question: Can the genome be directed by the phenotype? And this is only a step away from an even spinier question: Can the genome learn from experience? These were questions of hot dispute, a controversy which raged throughout the nineteenth century into the twentieth.

The problem is still here today, but we can now lean on the DNA script to deal with it more levelheadedly. The situation has changed, indeed, though only since about fifteen years, when one began to compare the nucleotide sequences of genes. And the knowledge gained makes the possibility that the phenotype may have influence over genetic mutability no longer seem a priori so remote as it must have seemed only a few years before to anyone who was not a vitalist. To transcribe in reverse from RNA to DNA, it now appears, is not something only retroviruses do. A whole family of sequences in the mammalian genomic DNA (the *Alu* family and the genes encoding the *small nuclear RNAs*) show telltale signs of having passed at one time through reverse transcription and reintegration. And as for ourselves, certain duplicates of structural genes in our chromosomes, a subclass of the so-called pseudogenes, all bear the marks of RNAs that have been transcribed back into DNA (a string of As at the 3' ends). Such repeated gene sequences and pseudogenes have been found in the germlines and early embryos of various mammalian species, and so the thought comes naturally that perhaps potentially many sequences in the genome of higher organisms can go from RNA to DNA.

These findings give us enough to stand on to make a stab at the question of the second kaleidoscope we raised in the preceding pages. Stated more fully, the question is this: Is there, besides the mutation generator at the DNA level, another one at the RNA level? The first is the classic forward-flow mechanism. It is reined in so tightly by corrective information feedback that the genetic variations produced are few and far between (the error rate is $\leq 10^{-9}$). Now, the second generator, the one in question, would tick faster; there are no fancy feedback brakes at the RNA level (the error rate in transcription is $10^{-3}-10^{-4}$, that of uncorrected nucleic-acid synthesis). And what makes this mechanism even more appealing, the variations produced would be amplified by alternate RNA splicing and operate on many gene copies—50,000 copies of mRNA per gene are not unusual.

Such an RNA kaleidoscope could in a matter of months generate a degree of variation for which the DNA kaleidoscope would need hundreds of thousands of years. There is no problem of information scrambling in the repository here; this RNA information is devoid of splicing signals and signals of transcriptional control, and so would not unduly interfere with the expression or regulation of the information in the DNA. The variant information generated by this kaleidoscope, because it comes from mRNA (or structural RNA), has a good chance to be functionally expressed. But only a fraction of that information would find its way into the DNA repository by the retrograde route, because the backflow of information is governed by stochastic events. So, despite the lack of compensatory feedback, that second wheel of change doesn't whirl free. Judging by the number of pseudogenes and *Alu* sequences found in our genome, such reverse-processed mutations may have occurred with a frequency of one every 100–200 million years for each gene in the germline, accounting for some 20 percent of today's genome.

The dual kaleidoscope mechanism may be the answer to the riddle of how the evolving living system can be both conservative and progressive. The system's cells, in effect, would have two channels for introducing novel information into the genome: a noisy channel—a holdover from the RNA age—which feeds processed information backward into the information repository, ensuring novelty, and a relatively noise-free channel where the thermodynamic communication entropy is offset, ensuring stability.

Can Organisms Promote Useful Mutations?

The genetic data above make a good case for reverse transcription. But what about adaptive mutations? Is there any evidence for cells preferentially producing or retaining mutations that are advantageous to them? Not so long ago the answer would have been a ringing "no," but now there's something in the wind.

Before we consider the new clues, let us briefly review why adaptive mutations—or for that matter, any selective mechanism of mutation—were deemed so implausible. A series of experiments carried out on bacteria in the 1940s and 1950s, in particular the penetrating work of Salvatore Luria and Max Delbrück, had left the lesson that mutations are random events. Some regions of the genome eventually proved to be more susceptible than others, especially the regions with certain bases (methylated cytosine)—the "hot spots" discovered by Seymour Benzer—but, by and large, mutations seemed to follow the statistics of random-hit kinetics. And as to adaptive mutations, the evidence then available from experiments on bacterial resistance to antibiotics and viruses spoke against them. In a series of ingenious experiments in which siblings of bacteria exposed to antibiotics or bacteriophage were analyzed, Joshua Lederberg and colleagues had shown that the mutations which made the cells resistant to the antibiotics or the viruses would have occurred regardless of whether the cells had been in contact with these agents or not; the mutations occurred without the cells or their ancestors ever having been exposed to the pressures of selection. Clearly, these particular mutations had arisen spontaneously before the cells could have had any indication of the advantage that the mutations offered in the prevailing environmental conditions.

However, it now turns out that the mutational process is not always so entirely separate from its phenotypic consequences as in those experiments. In 1988 John Cairns and his colleagues broke the ice with the discovery that there are mutations that occur more often when they are beneficial to the bacteria than when they are not. Cairns worked with strains that were incapable of using the sugar lactose as an energy source (*Lac⁻* mutants) and were living in a nutritionally depleted environment. Such *Lac⁻* strains tended to become capable of using lactose when they were fed on that sugar as their only energy source, and this acquired capacity—a reversion to *Lac⁺*—was inheritable and occurred more often than when the strains were raised on another energy source, like glucose. One of the strains (frameshift mutants *LacZ*) reverted with such alacrity that it was possible to show that most (at least 90%) of those cells that reverted in lactose medium would not have done so in glucose medium.

Here, then, was a clear-cut case where mutations occurred more frequently when they were useful. Does this mean that the cells can actually choose which mutations will occur? Has their phenotype gained a hold over the genotype? We might have been tempted to jump to conclusions, had the stochastic model we considered before not forewarned us that the cells need not be in control here, at least not in the usual deterministic or cybernetic sense, to get preferentially useful mutations. A random trial-and-error mechanism based on backflow of genetic information, namely, a mechanism allowing reverse transcription of mutant mRNA information which *happens* to code for useful protein, may suffice. Cairns proposes a mechanism of that sort. Other kinds of mechanisms may be envisioned, including more orthodox ones, but it would be fruitless to enlarge upon our cogitative schemes at this early stage of the experimental game.

We would dearly love to forge ahead with models, of course. Models are the lifeblood of science; they drive the experimental endeavor and prepare the mind to see the next fact. However, things are not precisely plain-vanilla here. The mutational mechanism still needs to be experimentally explored in greater depth and the path for informational backflow to be charted. Here we are in territory where one needs not just to fill in an existing map, but where one has to figure out the whole geography from scratch. So, we'd better bridle our speculations, lest we run the risk that someone like LeCarré's Smiley asks us one day: "What dreams did you cherish that had so little of the world in them?"

The Solid-State Apparatus at the Headgates of the Mainstream

We now bring our journey down the mainstream to an end and return to our point of departure, to what one might call the headgates of the stream. There various side arms converge that feed in crucial regulatory information: the instructions that control the outflow from the thousands of genes. Those instructions are relayed by an array of protein molecules flanking the DNA double helix. The array constitutes a formidable information-processing apparatus that is as close to solid state as biological things ever get to be.

The flow of information from a gene begins, we have seen, with the RNA polymerase machine attaching itself to a site on the DNA double helix and then chugging down to transcribe one of the strands for the length of a gene. But how does the polymerase find that particular DNA site? What information guides it to that place and makes it start? These are the basic questions of gene control, and here is where the solid-state apparatus comes in.

In brief outline, that solid-state apparatus consists of interlocking protein arrays that process cellular information and channel it into the polymerase machine, telling it when and where to operate. Each gene has its own information-processing array of proteins and each array receives information via three discrete lines. One line issues from certain DNA districts and carries the basic positional instructions. The other two lines—one comes from elsewhere inside the cell and the other, from outside—supplement those instructions.

We will dig a little into this information-processing apparatus to get acquainted with its gear, and from that vantage point, we will endeavor to tackle the software aspects, starting with those concerning the two lines coming from inside the cell. The third line, the cell-extrinsic one, we leave to the next chapter.

Three Protein Designs for Tapping Positional Information from DNA

The first indications that the DNA somehow supplied positional information for its transcription came from work with bacterial genes, showing that, besides the base sequences encoding protein, there were sequences—short DNA stretches—

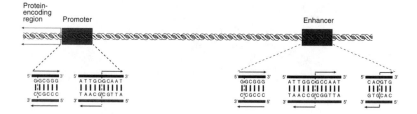

Figure 8.2 A stretch of DNA displaying the promoter and enhancer regions and a portion of the protein-encoding region of a gene. The promoter and enhancer sequences of eukaryotes typically contain several motifs, which are docking sites for regulatory proteins. In this particular example the promoter exhibits two motifs and the enhancer, three. Two of the motifs here are palindromes; their halves read the same in the 5' to 3' as in the 3' to 5' direction (the dashed vertical lines indicate one of the axes' symmetry).

with a regulatory function. These regulatory elements, called *promoters*, seemed to dictate where on the DNA the RNA polymerase will start transcribing. The promoter base sequences lie adjacent to the protein-encoding ones. More clues eventually came from studies on viral genes, revealing still other regulatory elements on the DNA. These elements, called *enhancers*, can lie far away from the protein-encoding sequences they control (Fig. *8.2*). Together, the promoters and enhancers provide the information that determines where on the DNA the transcription machinery will operate and how fast.

The promoters and enhancers contain distinctive base sequence motifs. One such motif we already have come across: the TATA sequence, a motif present in the promoters of many eukaryote genes. Another motif, to cite just one more example, is the sequence CCAAT, which is commonly found 50 to 100 base pairs upstream from the start site of eukaryote genes. Such motifs are somehow recognized by the proteins of the headgate apparatus. These proteins, which abound in the sap of the cell nucleus, bind to the DNA regions which carry the motifs. In general, proteins are not very good at binding nucleic acids, but these proteins are special. They have protruding domains which plug into the double helix, so to speak; the domains are so contoured as to make enough contact with the DNA for multiple atomic interactions. The domains often fit the major groove that spirals down the double helix much like the thread of a screw. The bottom of the groove is paved with the nitrogen and oxygen atoms from the edges of the bases, which allow focal interactions (including hydrogen bondings) with amino-acid side chains of the protein domains.

Three structural designs of protein have been identified that can plug into the DNA double helix. Two of them are protein domains made of tightly coiled polypeptide chains that can slip into the major DNA groove. In a third design, a

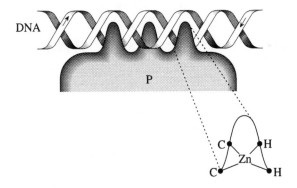

Figure 8.3   The zinc-finger—a modular protein design for plugging into the DNA double helix. Stabilized by the tetrahedral coordination with a zinc ion (*Zn*), a stretch of polypeptide chain, between a pair of cysteines (*C*) and histidines (*H*), loops out from the protein (*P*) like a finger pointing into the major groove of the double helix. In the illustrated example, three fingers point into the groove alternately in front and behind (touching guanine clusters on the noncoding strand). Some multiplex proteins have twelve such modular fingers or more. (After A. Klug and D. Rhodes, *Trends in Biochem. Sci.* 12:464–469, 1987, by permission from Elsevier Science).

modular one, the domain unit consists of a short polypeptide chain (30 amino acids) folded around a zinc atom, which sticks out from the body of the protein like a finger (Fig. *8.3*). Typically, several such "zinc fingers" plug into the DNA, and the length of DNA involved in the interaction depends on the number of these modules.[2]

What Do Proteins See Where They Plug into the DNA?

These are just the first peeks into the protein hardware involved in gene regulation. The field is only a few years old and, no doubt, we shall soon learn more. But not to lose ourselves in the forest of structure, let us pry a little into the cognitive act of these proteins, their ability as Maxwell demons, for any of these proteins evidently offers a *Fingerspitzengefühl* of a much higher order than that of the histones involved in the DNA spooling. How do these proteins draw information from the DNA—how do they finger it? What precisely does a polypeptide helix that slips into the DNA groove sense there? Does it give heed to the base sequence or to some three-dimensional aspect of the groove? This question goes to the fundamentals. Take the CCAAT sequence, for example. This motif certainly has something to do with the recognition, for if it is experimentally

---

2.   The two designs based on coiled polypeptide chains are α-helical, that is, helices wound in clockwise direction. These designs were brought to light by Mark Ptashne, Steven McKnight, and Paul Sigler. The "zinc-finger" design was discovered by Aaron Klug and colleagues.

scrambled, the cognate protein will not attach; or if the motif is transplanted elsewhere on the DNA, the protein will go there. But this does not tell us whether the protein recognizes the sequence as such or a wider geometric aspect that goes with the sequence.

So far in our dealings with the biological information flow we were spared the worry of that question because we purposely stayed with the mainstream, the "straight" run from DNA to protein. There, all that matters is the one-dimensional aspect of the DNA information store. The linear sequence of the DNA bases contains the essential information for all the information transforms in that run and the genetic code rules it all. But things get more complex at the higher information tiers, and as soon as we venture out of the main stream to the side arms, we are in higher-tier territory. So here, at the DNA-protein junctures, we cannot escape the question whether the DNA structure in all its three dimensions, rather than just one, holds the key to the information transfer.

The molecules at these junctures, like all organic molecules, are nearsighted to an extreme. Their information horizon, their scope for electromagnetic interaction, as we have seen in Chapter 5, is just 4 Å; hence, the question of how those proteins can recognize anything in the DNA groove—how their amino-acid side chains and the DNA bases, structures which ordinarily don't fit together so well, get enough interaction. Are the focal atomic interactions of the sort we have just seen enough for the information transfer that goes on at this gene control level? Or is the information transfer geared to some broader three-dimensional aspect of the DNA double helix? These are issues which are not yet within our ken. There is no question that the base sequences in the promoter and enhancer motifs are important—there are genetic-engineering experiments galore to attest to that. But these experiments don't tell one whether the information transferred emanates directly from these one-dimensional sequences or from the three-dimensional structure of which the sequences form a part. The base sequence and the local three-dimensional DNA structure conceivably might be linked in a systematic manner. There are actually some hints to that effect coming from X-ray crystallography of short synthetic DNAs. For example, in the common B form of DNA, the exact tilt of the bases and the helical-twist angle between base pairs depends on which nucleotides are next to each other in the sequence—and some gene-regulatory proteins, which don't make contact with the base (e.g., the repressor protein in bacteriophages), do discriminate between those three-dimensional differences.

Thus, there may be more to the regulatory motifs of DNA than just their base sequences. The recent advances in biotechnology have made our reading of the sequences in a stretch of DNA as simple as ABC, and so our research and knowledge tend to be skewed toward the one-dimensional. But, after all, it is not what *we* see, but what the nearsighted proteins see, that counts in *this* game of life.

How proteins manage to draw off information from the DNA is a key question in genetic physiology. The experimental doors here have just been opened, and X-ray data of the interacting molecular structures are beginning to come in. So far, the headlines in molecular biology have been made by those who concentrated on the one-dimensional aspects of the information repository. Such single-mindedness has paid high dividends; the brisk advances of the past thirty years, indeed virtually all we know about the "straight" run of the Circus, are owed to it. But as we sidle up to the circinate parts, we may have to take off the blinders, and the cognitive aspect of the proteins that plug into the DNA is certainly one to watch. Perhaps the time now has come for those who see things the three-dimensional way to break the news. We badly need to widen our horizon to understand what happens at the strategic Circus junctures where the information flow turns onto itself, and those relatively simple protein demons may well hold the key.

If the Mountain Will Not Come to Mahomet . . .

Let us now check out the other side of the protein apparatus at the DNA headgates, the side that relays the positional information to the transcription machine. At this interface things are apt to be more familiar. A good deal of experimental genetic engineering with headgate proteins has been done and the indications are that some of these regulatory proteins come endowed with two distinct domains: one is designed to engage the DNA, and the other to engage the transcription enzyme. So, these regulatory proteins seem to bridge the gap for transfer of information between DNA and transcription machine. They are adaptors in the same sense as the tRNA molecules or the tRNA synthetase molecules are in the information transfer between mRNA and protein; one interface of the adaptor receives information from DNA, namely, topological information, and the other transfers information to the transcription machinery, the RNA polymerase itself or a helper protein.

Thus, we may envision that the transmission of information to the transcription machine is implemented by the same sort of weak electrostatic interactions we have seen operating at other protein-protein transmission sites. The message coming from a DNA region containing a promoter motif here need not be very complex. "Bind me"—may be all this motif is saying to the regulatory protein ambling in the nuclear sap. And as the protein obliges, it is given the exact Cartesian coordinates for a start site of transcription. The roving RNA polymerase then gets nabbed there as the "bind me" theme is replayed on the other face of the regulatory protein and the positional information passed on. All it may take, then, is a start command, say, a conformational change on that face, and the polymerase machine will go chug-chug down the DNA. And it will keep going until it hits a termination motif.

So much about information coming from promoters. What about information from enhancers which may lie hundreds or thousands of bases away from the start site of transcription? How does that faraway information get to the polymerase machine? Well, if the mountain will not come to Mahomet, Mahomet must go to the mountain. The DNA helix is accommodating in this regard. It has enough conformational freedom at some spots to loop out and bend over like a flexible spring, bringing the enhancer with its attached protein to the vicinity of the starter region of the gene (Fig. *8.4, a*).

This way, an enhancer-bound protein may interface with a promoter-bound protein, and the information from both may be passed on in concert to the polymerase machine. Such cooperateness makes for a secure and responsive control. There is room for quite a few proteins at eukaryote promoter and enhancer sites; so, the local information flow and processing for this control can get complex (Fig. *8.4, c*). Indeed, in higher organisms, scores of regulatory proteins may be involved, and such large-scale cooperations are commonplace, if not de rigueur, in gene control.

The Multiplex Adaptor

It is too early to say how the various cooperating proteins on the eukaryotic DNA precisely interact with one another. However, a plausible picture of gene regulation is now slowly emerging, a model where each gene has its own peculiar set of proteins regulating transcription. Some of those proteins may be shared by different genes—which represents economy in both regulatory and protein-encoding DNA sequences—but the combination of the various proteins is probably unique for a gene or a class of genes. It is that combination—the set as a whole, rather than the single protein—that would provide the key for the control here.

Using proteins as keys is run-of-the-mill in living systems, but why should such keys be made of multiple proteins in the case of gene control? Why not single proteins? The evolutionary reasons may be traced to the lack of complementarity between DNA and protein surfaces. Evolution, we have seen, found a way out of that fix by employing special proteins with zinc fingers or other protrusions that can slip into the DNA groove, permitting interactions at small spots. However, the information that gets through such polka-dot contacts doesn't amount to much. The solution was to multiply the contact spots and use several proteins to tap and sum their information. The tapping is done by the domains which plug into the DNA, and the summing by the cooperative action of the proteins which plug into each other. All put together, the interlocking proteins form a mosaic that links various regulatory sites on the DNA to one another both laterally, within a promoter or an enhancer region, and vertically, between a promoter and an enhancer region (Fig. *8.4, c*).

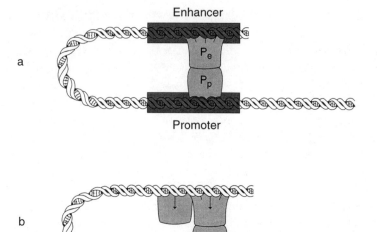

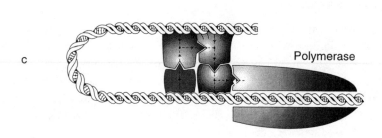

Figure 8.4 The relay apparatus controlling gene transcription. A model. (*a*) The looping of the DNA double helix allows a regulatory protein bound to the enhancer region (P_e) to come in contact with a regulatory protein bound to the promoter region (P_p) of a gene. (*a* to *c*): The diagrams show the progressive assembly of four regulatory proteins. (*c*) The complete set of proteins can process the information from four double-helical districts and relay it to the RNA polymerase which engages at the set's output end. The dashed lines and arrows indicate how the information flows into the mosaic and is transferred by allosteric proteins (*dovetailing*) to the polymerase—one of several possible flow schemes. For simplicity, the set is depicted as comprising only four regulatory proteins; in real life there may be more.

The operational principle of the model may now be succinctly stated: *the regulatory-protein mosaic is an adaptor made of separate pieces, a **multiplex adaptor**. On one end, the multiplex plugs into the DNA and, on the other, into the RNA polymerase. Information about the location of the start of transcriptional units flows into the multiple plug-ins at the various promoter and enhancer sites, and is relayed from protein to protein in the adaptor mosaic, to finally converge on the RNA polymerase.*

This gives the why and wherefore of the solid-state apparatus of the genes. But our bare-bones description of its internal information flow hardly does justice to the power and elegance of this apparatus. By breaking up the transcription-controlling information flow into various segments and distributing it over several mosaic pieces, Evolution, with one masterstroke, achieved both specificity of control and economy of information. Her gambit was to make the information flow from DNA to RNA polymerase contingent on the fit between the interlocking mosaic pieces. Each of the internal interfaces of the mosaic, in mathematical-logic terms, represents a decision point or a logic gate where, depending on surface matching, the message is either passed on or not. And as this digital information processing is repeated at several internal interfaces, a high degree of gene specificity results, even if the interactions at the individual external interfaces with the DNA are not so specific. The second achievement, information economy, simply pans out from the same gambit, since some of the mosaic pieces could be used again and again in diverse combinations.

This multiplex adaptor model is based on experimental studies that are in mid-progress in the laboratories of Steven McKnight, Mark Ptashne, Keith Yamamoto, and others. The evidence on hand is still far from being complete, and so some overhaul of the model may be necessary in the next few years. But the multiplex principle, I hope, will stand.

Supramolecular Organization Out of Molecular Hum

To consolidate ideas, let us briefly go back over the hardware at the headgates of the cellular information mainstream. All the hardware pieces for the readout of information, the RNA polymerase, the readout head—to resort once more to the forgiving tape recorder as a trope—and the regulatory proteins, the elements which control the readout head, are mobile. The tape itself, the DNA strand, is fixed. The regulatory proteins amble about in the sap of the cell nucleus until they bump into appropriate docking sites on the tape. Then they become fixed to the tape and, in due course, a group of regulatory proteins assembles, forming a mosaic which, as postulated in the model, relays instructions to the readout head. The head slides randomly along the tape until it comes upon a completed mosaic. There it engages and would receive instructions to commence operations.

The various pieces here, while on their own, just roam about aimlessly, playing seek-and-find, the habitual sport of biological molecules. They can see a docking station only when they happen to get within 4-Å range of it. Then they hitch up to leech their drop of information. Thus, piece by piece, a set of proteins eventually assembles on the DNA, forming a solid-state apparatus which marshals a fair amount of information.

So, all around the DNA in the nucleus of the cell is the hum of a myriad of molecules, but only those molecules that are suitably shaped will hook up. The solid-state organization here is born of the occasional lucky bounce that brings a dovetailing molecule to the docking place. A few molecules wouldn't stand a chance in that statistical game. But the chance gets better and better as more molecules get in the fray, and the number of molecular copies that are actually produced in cells virtually ensures a winning hand. So, every *active* gene would have a regulatory apparatus in place.

Where does all that information come from which pulls such a complex organization out of chaos? We have seen that, above thermodynamic equilibrium, some types of chaos can bring forth organizations. But things are different here from that swirling and seething world of deterministic chaos, where a molecular organization, a dynamic pattern, suddenly materializes on a swing away from equilibrium. Huge numbers of molecules synchronize their movement there to form cohesive patterns, the "dissipative structures." In the present case, in contrast, we have no large-scale molecular synchrony. We have coherence, of course, but when the molecules cohere, they do so one pair at a time. But the contrast goes deeper than that. Whereas the organizations arising in deterministic chaos are *fundamentally* random and, hence, unpredictable, the organization here is highly predictable. The probability may not be quite as high as the apple falling down from Newton's tree, but it is a pretty safe bet that all like cells in an organism will have identical solid-state apparatuses at a given active gene.

This determinacy, like that of all things we call "anatomy" in living beings, comes from the interlocking of the constitutive molecules. The interlocking somehow pulls the organization together. The dovetailing faces of the molecules are the pivots of the organization, the nucleating centers, we might say. The free molecules are constantly being shuffled by thermal energy and, like the pieces of a mosaic, they eventually fall into place at their dovetailing faces. There and then, at close range, electromagnetic forces are brought to bear on the molecules, pulling and pushing them to stable interlocking positions.

So, it is from the dovetailing faces of the regulatory protein molecules, the domains specialized for interaction, where the bulk of the information flows that shapes this biological structure. If we pick up the information trail from these faces and backtrack on the mainstream, we should find the source: the DNA sequences that encode the interactive domains. The linear structure of these stretches of DNA, thus, holds the core information that shapes both the domains and the multimolecular assemblies.

Aside from this, there is another intrinsic information source for the multi-molecular assemblies: the DNA double helix. This double-stranded structure, namely, its docking districts, serve as a matrix of sorts. This is not as tight a matrix as the single-stranded DNA is for the assembly of mRNA in the mainstream. It is more like a frame than a mold. But it is good enough to immobilize the regulatory proteins and give them their bearings in space. How precisely this information transfer is achieved remains to be worked out at the atomic scale.

Matrices for assembling supramolecular structures will become frequent sights in the following chapters, as we move up to higher information tiers. The simple frame matrix of DNA will give way there to matrices of lipid, planar structures where the constituent fatty-acid molecules are neatly arrayed in two layers—the ubiquitous membranes inside and around cells. Those lipid matrices are the products of a special protein machinery; they are proxies of the DNA, we might say. They give the structured components to be assembled a spatial orientation and force them to move in a plane, instead of three dimensions as they do in the cell water. This is good for the economy; by constraining the molecular movement to two dimensions, membrane matrices achieve organization at low information cost. But all that pales against the savings the cell realizes in the assembly of the multiplex. Indeed, the living system gets away with murder here: it uses its DNA both as matrix and generatrix. Such self-contained parsimony should come as no surprise. It goes back to early times, when that was all Evolution could afford. It is but her coeval bent—her usual way of making a virtue of necessity.

A Ledger for Supramolecular Information

The solid-state headgate apparatus is a good example of what biomolecules can attain by playing their game of seek and find. Such a rich organization, of course, doesn't come out without its dose of information and much of it comes from the DNA. Now that the story is told, let's take a brief look at the information ledger of that molecular and supramolecular organization. First, there is the portion of information flowing from the linear structure of the DNA—this supplies the core information for the assembly of the individual regulatory proteins and for their subsequent conglomeration (the information built into their dovetailing domains). Then, there is the portion flowing from the double-stranded DNA—the matrix information for the multiplex assembly. And finally, there is the portion flowing in from the environment. Although we have not yet talked about it in the present context, we know from Chapter 4 that the last is the portion no organization in the universe can do without; this is the information input for the obligatory entropy quid pro quo, the information which living systems extract from the molecules in their environment and from the photons showering the earth. That extrinsic information is as necessary for the molecular assemblies and compositions as the information we called the core, and it amounts to a tidy sum.

In our musing about information, we naturally gravitate to the "core." We instinctively feel that this huge store accumulated over aeons makes the difference between living and nonliving things. This is what is missing from the organizations in mineral crystals and turbulent liquids, however spectacular those may be. It is the center which directs that esoteric twig of cosmic evolution and steers it through the sea of time. There are a thousand and one waves of entropy lapping at the twig that would tear it apart, but they meet their match in the information potential in that core.

Part Three

Information Flow Between Cells

Rules of the Intercellular Communication Game

The End of the Cell Era: The Merger of Circuses

We pause here a moment in our information tracking, for a perspective of time. Let us take our bearings and place the main events in the biological information flow encountered so far on our cosmic chart (Fig. 3.3). The flow started about 4 billion years ago. It began with a trickle, as the first self-reproducing molecular systems emerged out of the chemistry of the cooling Earth, and grew gradually to the stream found in all cells today.

The first 3 billion years or so of this snail-paced plot—and this includes the time it took for the basic Circus to be engendered—were dominated by the little individualistic actors, the cells. We may call this long span the *Age of the Cell.* Shielded from the environment and from one another, each cell then ran its own Circus—each one a fully autonomous unit fending for itself. But that primal golden age eventually came to an end, as some cells gave up a little of their individuality and banded together. This too happened gradually; little by little, those cells linked up and put their Circuses in tune—no big jumps here, no sensational twists. The One who penned this extravaganza doesn't need to titillate— He has a captive audience.

Thus, the spotlight shifted from the selfish to the social cells. Not that the stage then suddenly teemed with selfless beings. The new units, the cooperative cell ensembles, were every bit as self-seeking and self-serving as the units of yesteryear. And they were incomparably more effective in the evolutionary struggle—the best orchestrated ensembles clambered to stardom. So the pace of the plot quickened and over the following billion years ever better coordinated and organized ensembles made the scene, filling every nook and cranny on the surface

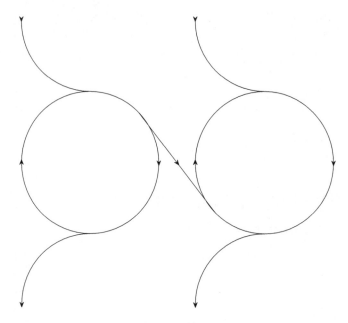

Figure 9.1   A symbolic representation of Circus linkage.

of the earth. A new era, the *Age of the Organism*, was on its way.[1] Some 10–30 million organismic species are now on the planet, and the age is still young.

The longest part of this history is the Age of the Cell. It took some two and a half billion years to get from the first cells to the first multicellular organisms, about twice as long as to get from the first multicellular organisms to us. That alone speaks volumes. Evidently, the toughest part was to get those solipsistic cellular beings to cooperate. After the first cooperative successes, the millions of organisms followed like an avalanche.

Now, the formation of such cellular partnerships implies informational linkage, a merging of the Circuses, which is represented symbolically in Figure *9.1*. It is this *informational* cooperation that was so hard to achieve. Indeed, this went against the whole cellular grain, and nothing could prepare us for the difficulties that initially lay in the way. For one, there was the cell membrane impeding informational linkage. The primary function of this ancient coat was

1.   The familiar term "organism," as applied to living systems, is not as old as it is well-worn. The naturalist George-Louis Leclerc, Count de Buffon, introduced it in the mid-1700s in formulating his zoological unity principle, a thesis proposing that animals are formed from organs, according to a hierarchical plan. The concept of the cell as the organismic unit was introduced a century later by Mathias Schleiden. (Before Leclerc, the term "organism" had a meaning related to general organization rather than to organ; it was used for nonbiological things that were ordered. In this sense, the word already appears in Evelyn's *Sylva* in the 1600s.)

precisely to achieve isolation—to keep the cell's own information in and alien information out. It was under this protective shield that the biological enterprise had prospered. But with the coming of the new age, that shield now stood in the way of the Circus merger, and ways and means had to be developed to penetrate or sidestep it. But as if this weren't problem enough, there loomed another one: the cellular ciphers and codes. Each individual Circus, indeed, every line of communication in a cell, was guarded by molecular locks; and, in the aggregate, these constituted an even more formidable barrier against information merger than the membrane did.

Thus, the transition to multicellular organisms entailed the breaching of two cellular defenses: an outer perimeter and an inner guard. In this chapter we will see how Evolution fared in this long battle; we will be apprised of her amazing breaching strategies and her solutions to the problem of the ciphers and codes.

The Innards of the Cell Membrane

The clues to the membrane-breaching strategies can be found in the structure of that remarkable coat of the cells. It is an extremely thin veneer (70–90 angstroms), yet a splendid insulator, owing to its fatty nature. Its basic matrix consists of just two layers of fatty molecules. Each molecule has a tail made of fatty acids and a head made of phosphate or some other electrically charged (polar) group. The molecules in the double layer are neatly arrayed: their tails face one another and their heads face the water (Fig. *9.2*).

That two-layered arrangement is what comes naturally to these molecules when they are in water; a double layer is thermodynamically the most efficient way for them to be deployed. There is nothing out of the ordinary about this deployment. Other fatty molecules with phosphate heads (phospholipids)—soap molecules, for example—will form bilayers too, and they do so spontaneously without help from cells or other props. In fact, it was the work with such soap films by the physicist Irving Langmuir, in the first quarter of this century, that paved the way for an understanding of the cell membrane structure, well before there was any physical or chemical knowledge about cell membranes to speak of. The fatty acids, like other hydrocarbon chains, are nonpolar, that is, they carry no net electrical charge or dipoles. Because of that, they do not mix with strongly charged molecules like water; they are *hydrophobic*. The head groups, however, are polar and are attracted to the water (*hydrophilic*). Thus, when these molecules are put in water, they tend to orient themselves, so that their heads will be in contact with the water and their tails will face one another—a two-layered configuration. Such thin bilayer films form any time we make a soap bubble. It's the cheapest organization with such molecules; it costs little information.

The lipid bilayer is a rather tight structure. Its molecules are closely packed—10^6 lipid molecules to the square micrometer—leaving scant room between

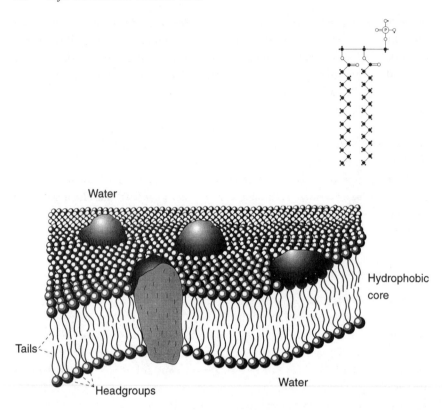

Figure 9.2 A diagramatic view of the cell membrane, showing the lipid matrix, an ordered double layer of lipid molecules, with protein molecules immersed or tacked onto it. *Inset*: the chemical structure of a phospholipid, the major bilayer constituent. The two tails are hydrocarbon chains; the head group is phosphate.

fatty-acid tails. Thus, the polar water and its polar solutes are kept out of the nonpolar membrane core. This is what makes the cell membrane tight; despite its thinness—it has no more than one ten-thousandth the thickness of a hair—it is a good diffusion barrier against water and matter dissolved in it (it slows the movement of water and its solutes by a factor of 10^6–10^9).

Their dense packing notwithstanding, the molecules have some freedom of movement. They can rotate about their long axes and wiggle their tails. They also can move sideways in each layer. In fact, there is quite a bit of turmoil in the structure: the lipid molecules change places with one another laterally in a layer, every tenth of a microsecond. The lipid bilayer, thus, is rather fluid. It is in a peculiar state, half liquid, half crystal—a state physical chemists call a *liquid crystal*.

However, this fluidity applies only in the lateral sense, that is, within the plane of the membrane. Perpendicular to that plane, the bilayer structure is rather rigid, as behooves a crystal. The electric field of the water on either side of the membrane has the lipid polar heads in a firm grip, not allowing the molecules to flip spontaneously. There is also strong impediment for them to slide alongside each other perpendicularly to the plane. And all this, plus the low water permeability of the lipid layer, makes the cell membrane such a good information insulator.

But there is more to cell membranes than just lipids. There is a fair amount of protein mixed in—about 50–70 percent by weight in animal cells. In a bird's-eye view of the membrane, the proteins stand out as big hunks. Some lie on the surface of the bilayer, others are deeply embedded in it (Fig. 9.2). The word "embedding," however, should not give us the impression that these hunks are immobile. Perhaps, we'd better think of them as submerged and floating in the largely fluid lipid—somewhat like icebergs in the sea. These proteins can move sideways in that semiliquid film with speeds of the order of 10 micrometers per minute; so, a typical protein can make it in about half an hour from one end to another in a mammalian cell. Add to this some allosteric talent, enabling them to move internally, and you have all the necessary ingredients for a shunt of the information insulator at the cell surface.

Indeed, the membrane proteins are uniquely suited for getting information through that insulator; they can transmit along and across the bilayer. The transmission along the membrane, a diffusional information transfer between protein molecules at different membrane spots, typically takes on the order of seconds; and the transmission across by allosteric movement, about one-millionth of that.

These are all features that must have been naturally selected for during that long span leading up to the organismic age. But splendid as these features are for the information-transmission function, they would hardly be enough if the membrane didn't stay cohesive all the time. There is quite some swing and flurry in that film—the lipid molecules change places, the protein molecules twitch and slither sideways through the lipid, the lipid and protein molecules rotate— yet the membrane stays tight at all times. Thus, there must also have been selection for intermolecular cohesiveness, namely, for the proper protein/lipid mix where the two mobile molecular moieties always stay in tight contact and thermodynamic harmony with one another.

It took aeons to perfect that mix. Lipids may once have been the principal or even the sole constituents of cell membranes, but eventually, and little by little, proteins were blended in. These are birds of a different feather from the proteins inside the cells; they are adapted to float in lipid, instead of water—their polar amino-acid units are tucked in. The principle of their structural design is rooted in thermodynamic harmony: the polar amino acids are kept away from contact with the fatty-acid tails in the bilayer, while the contact between the nonpolar amino acids and those tails is maximized. This principle was uncovered by Jonathan Singer in 1972. As in the case of the proteins inside the cells, many of the membrane proteins consist of coiled polypeptide chains—right-handed coils (α-helices). In a variety of them the chains cross the entire bilayer, once or several

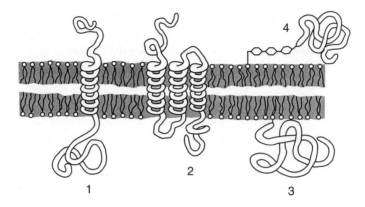

Figure 9.3   Membrane proteins. The diagrams on the left show transmembrane proteins crossing the bilayer (*1*) as a single α helix or (*2*) as a series of such helices. The diagrams on the right show more loosely associated proteins; one protein (*3*) is covalently bonded to a fatty acid in the bilayer, the other (*4*) is attached by way of a sugar to a phospholipid in the outer half of the bilayer.

times (Fig. *9.3*), and so have long nonpolar stretches maintaining harmony with the hydrophobic membrane core; their short, predominantly polar stretches face the polar heads of the bilayer and the water. These polypeptide chains are inserted into the membrane right after they come out of the ribosomes; a special protein apparatus in the endoplasmic reticulum literally funnels them into the lipid bilayer, slipping them past the energy barriers.

In another protein type in membranes the polypeptide chains are not coiled up but stretched out, forming a sheet (*ß-sheet*). This type follows the above design principle too but, because the chains are straighter, the transmembrane runs are not so long. Finally, there is a class of proteins that are more loosely tacked onto the membrane. Here the polypeptide chains are mainly in the water flanking the membrane on one or the other side, and contact between polar residues and water is maximized. These chains are covalently attached to particular lipid molecules in the bilayer—some are bonded directly to fatty acid chains, others are bonded by way of a sugar group (Fig. *9.3:3,4*).

Strategies for Breaching the Membrane Barrier

Now to our question of how this formidable cellular bulwark was breached in evolution. From our perch overlooking the phylogenetic tree, we can see four breaching strategies: one strategy uses lipids to penetrate the lipid membrane barrier, and the other three use proteins that blend in with it; all four take advantage of its liquid crystalline state.

The various tactical plans are sketched in Figure *9.4*. The first strategy is the simplest and perhaps the oldest one. The cell that does the breaching employs a

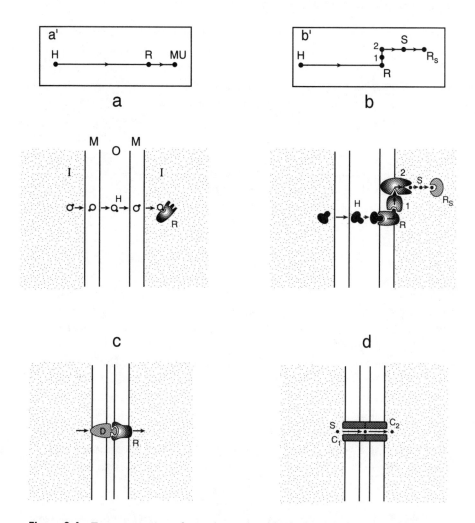

Figure 9.4 Four strategies of membrane breaching for Circus connection. *Top:* Hormonal strategies: *a*, a small lipid molecule (steroid) penetrates the membrane lipid, carrying the connecting information—the *message*—into the target-cell interior, where it is decoded by R, a receptor protein; *b*, a membrane-impermeant molecule, a protein, carries the message to the target-cell membrane, where it is decoded and relayed by a set of proteins (R, *1*, and *2*) to the cell interior—S is the secondary (retransmitted) signal and R_s, its cytoplasmic receptor. The corresponding communication-line schematics, including the target-Circus stage the lines tie into—*Mu*, the DNA-headgate multiplex or T, the R_s-target—are shown in *a'*, *b'*. (*M*) cell membrane; (*O*) the water outside the cells; (*I*) cell interior. *Bottom:* Direct cell-interaction strategies: *c*, a pair of membrane proteins constitute a (fixed) information-transmission chain, with D as the information donor and R as the receiver; *d*, a pair of membrane proteins (C_1, C_2) form an intercellular tunnel for cytoplasmic signal molecules (S).

small fatty molecule that will appear disarmingly familiar to the other cell: a look-alike of the cholesterol residing in its own membrane. That fatty look-alike—a *steroid*—is sent out as a messenger, reaches its target by way of the extracellular medium, and slips quietly through the target cell's bilayer. Once inside the cell, it latches to a protein which plugs into the multiplex at the DNA headgates, completing the most straightforward of all Circus connections (Fig. *9.4 a*).

The second strategy is a variation on the messenger theme. The messenger here is a protein. Like the lipid, this messenger gets to the target cell via the extracellular medium, but it uses a proxy for the membrane breaching: a protein of the target cell membrane. As the messenger arrives at the target cell, it docks at that protein, donating its piece of information. The information flow then continues inward, via protein relays in the membrane, eventually merging with the target Circus by means of a secondary cytoplasmic signal (Fig. *9.4 b*). The messengers between the cells are called *hormones*—a collective term, including molecules of lipid, protein, and other ilk.

The next two strategies are for cells that are in contact with one another. Here the breaching of the membrane is accomplished by pairs of dovetailing transmembrane proteins that stick out enough to interact across the narrow gap between the adjacent membranes. In one of the strategies—the third in our account—the two proteins constitute a fixed transmission chain: one protein donates information and the other receives it—a binary molecular interaction tying the Circuses together (Fig. *9.4 c*). And finally, in the fourth strategy, the two proteins form an intercellular tunnel for cytoplasmic signal molecules. These proteins belong to a sort that constitute pipelike structures in cell membranes, structures with a continuous hydrophilic core. When the two proteins come in contact with one another at their projecting extracellular ends, they bond, forming a leakproof water conduit through which cytoplasmic molecules of up to about 1,000 molecular weight can go directly from cell to cell (Fig. *9.4 d*).

Thus, all in all, the membrane breaching problem met with four elegant solutions. In all four, the membrane's basic function, its primordial role of serving as a barrier between cell interior and exterior, was preserved. There were necessarily some losses of individuality by the cells, but in all cases the individuality of the information repositories, the genetic individuality, was preserved.

The Bare Bones of Intercellular Communication

That much about penetrating the outer perimeter. Now to the problem of how to get through the inner guard—the messenger information somehow must get past the battery of safe information locks, the codes and ciphers of the target Circus. These were big obstacles—if the cell membranes were walls, then the cellular information codes were bulwarks. However, the solution of this dilemma, unlike the one of membrane breaching, did not call for new strategies.

All through evolution cells had made a fine art of encoding their internal communications by the use of protein hardware shaped like locks and keys. So in going from single-cell unit to multicellular organization, they only had to hang on to an old vocation. It was largely a matter of sprucing up old proteins and using them as passkeys. Indeed, in due time, a variety of such keys materialized that were suitable for the Circus linkage. Or perhaps we should call the passkeys adaptors, for they bridge the gap between nonfitting molecules.

Call them what you wish—the case at bottom is they are the molecules that led the march from cell to us. Today's social cells are literally teeming with such molecules ministering to the inter-Circus game. The messenger molecules and their adaptors form information-transmission chains. Some of these chains, like those involving steroid hormones, have just two members: a messenger and a receptor (Fig. *9.4 a'*). This simple two-piece, mobile molecular bridge may have been the first inter-Circus linkage to emerge—its dime-a-dozen lipid piece looks just like something that might have been affordable when cells came of age. The chains involving protein hormones have higher information tags. Not only do they have posh messengers, but also complex relays. Such chains can mount up to quite a few members, especially at their cytoplasmic ends; Figure *9.4 b'* is but a minimal diagram.

These, then, are the bare bones of intercellular communication: *chains of dovetailing molecules, which splice into the Circuses by their ends.* They provide the informational continuity between Circuses, a linkage so smooth that it is hard to say where one Circus starts and the other ends—a perfect tandem (Fig. *9.1*).

We fixed on the Circuses here rather than on the cells. Had we done it the other way round, our view would have narrowed to what goes on outside the cells, namely, in the water space delineated by their membranes. We would have seen then but one link of the chain—only the messenger in the case of hormonal communications. That link, it is true, provides the firing spark for all the activities down the line, but it amounts to but a fraction of the total information that connects two Circuses—there is retransmission of information all along, and each retransmission stage contributes its own information share. The intercellular communications go in crescendos, we might say; they may start with the whisper of a little steroid lipid but swell to a full chorus of proteins in the end.

The Intercellular Communication Channels

The inter-Circus chains are the essential links for coordinated cell interaction, the foundations of the entire multicellular organismic enterprise. But to understand that enterprise, we also need to understand the pertinent software—learning about the molecular hardware, important as it is, only goes so far. So, we will set aside for awhile our molecular account and consider the logic of intercellular communication.

To start, we will try to define the basic elements in this intercellular network in terms of information theory. That theory, specifically its branch, communication theory, specifies five stages for any communication system, regardless of its form: an *information source*, an *encoder*, a *communication channel*, a *decoder* (or *receiver*), and a *user*. The information source generates the information to be transmitted; the encoder transforms the information into a suitable message form for transmission over the communication channel; and the decoder performs the inverse operation of the encoder, or approximately so, for the user at the other end of the channel.

This elementary series may be expanded by any number of substages, depending on the complexity of the network involved. For our purposes it will be useful to add a substage after the encoder, a *transmitter*, for we find such a go-between in all communications beamed out of the DNA. This stage—in effect, a second encoder—converts the information in the message into yet another form and, more often than not, there are more than one transmitter reissuing information down the line. Let's call the output of such transmitters *signals*, for short.

Given the above definitions, the equivalent stages in intercellular communication may be readily identified, though now and then we get into a jam because the biological vocabulary was largely coined before the informational logic of the communicating molecular chains was understood. However, there is no problem of terminology that a lexicon won't fix. Try this word list, for instance, for the channels of communication entailing hormones of the lipid (steroid) variety:

DNA	=	information source
transcription machinery	=	encoder
messenger RNA	=	message
translation machinery/ hormone synthesis/ release apparatus	=	transmitter
hormone	=	signal
DNA-headgate multiplex	=	decoder or receiver
DNA polymerase	=	user

Except for one ungainly double entendre, the translations run smoothly. (Several transmitter stages here have been lumped into one, for simplicity.)

Things get more tricky with communications entailing hormones of the protein sort. Then, there may be transmitters both in the membrane and in the cytoplasm of the target cell—a long series of retransmissions—and even more than one user, when the communication lines fork. In biology those retransmitters are collectively called *receptors*, the name I will use throughout the book, though occasionally I will fall back on the communication-theory term, to remind us that they are not end stages. Communication channels with multiple transmitters are not so easily tracked down in cells. The seasoned sleuth here looks first for the user of the message—if he finds a user, he knows he has come to the end of a line. He then backtracks along the signal chain and, though he may not always strike a rich vein, at least he can hope to come up with the basic elements of the channel. Many an intercellular communication channel was bagged precisely that way.

A communication channel is defined as much by the elements at its ends (the encoder and decoder) as it is by the medium that carries the signals. So, a microphone, an earphone, and a couple of wires connecting the two make a channel for telephone communication; or the same devices (with different encoders) and a pair of antennas, instead of the wires, form a channel for radio communication. The microphone and its encoder convert the voice message into pulses of electrical current (pulses of electrons) or into pulses of photons (electromagnetic waves), as the case may be. The earphone and its decoders, in turn, reconvert these signals into sound waves and decode them for the user. Similarly, we may say, a sailor waving a set of flags from the deck of a ship and an observer on another constitute a channel; and, by the same token, an Indian moving a mackinaw over a fire and a cohort observing these smoke signals on a mountain constitute one. Here, too, the signals are carried by photons, though the decoders (built into the eyes and brains of the observers) are immensely more sophisticated than those in any of our telephone or radio channels. Or, to cite an instance where the medium is less monotonous, the shipwrecked mariner sending a letter in a bottle and the finder across the sea constitute a channel where signals are carried by matter and photons.

Thus, the medium for information channels can be either energy or matter, and any tool that will do the coding job is fair game in communication engineering. The examples above give a taste of that. In particular, the last, the one with the variegated channel medium, gets us set for intercellular communication where the channel medium, as a rule, alternates between macroscopic matter and energy field, namely, organic or inorganic molecules and electromagnetic energy; the molecules carry the information over the long hauls through liquid medium, and the electromagnetic energy carries it over the short hops (3–4 angstroms) between the molecular carriers (with the electrons occasionally lending a helping hand).

The Imperturbable Biological Signal Stuff

Communication channels, in one form or other, are behind all organizations in the universe. For human organizations this has been obvious since ancient times. The old Greeks held communication in such high regard that they exalted an earthborn courier to Olympus, to officiate as a messenger of the gods. This was Hermes, Lord of the Communication Realm. I don't aspire to rewrite Greek mythology but, looking at his exploits through the eyes of physics, I daresay Hermes is somebody who would be the delight of any travel agent. He isn't choosy about how he travels; he rides on photons, electrons, molecules, or in a plain old bottle—he has the entire gamut of particles at his beck and call. At the Circus entry points (represented by the inflow arrows in Figs. *7.1* and *9.1*), he betakes himself to photons, but elsewhere in the Circus, and especially in his inter-Circus journeys, he turns mostly to macroscopic things and uses particles and energy fields only in passing, as it were.

This predilection for macroscopic vessels of information is evident in all inter-Circus pathways. Take, for example, a hormonal pathway with multiple transmit-ters, such as the one in Figure *9.4 b*. The message here meanders in one particular molecular form for seconds or minutes through the membrane or cytoplasm; then, in a flash, it leaps to the next molecular form, to stay there a good while, and so on. The inter-Circus information dwells mostly in solid matter. Only for a fleeting instant is it in electromagnetic-energy form. Though, this instant is the moment of truth in cell communication, the moment when the information gets transferred from one molecule to another—when one molecule makes the atoms of the next molecule in the communication chain deploy themselves in an analog pattern. Conservatively estimated, the average time spent in matter here is ten million times longer, and this is rather typical for intercellular communication channels.

By using molecules for the long hauls and electromagnetic energy for the minute (intermolecular) gaps, the cells reap large benefits. They are spared the troubles of signal distortion and interference, which plague our radar, radio, and television channels. The electromagnetic signals in those systems are distorted by nonlinearities in the encoders and decoders—they are bent out of shape, so to speak. They may be unbent again, though this runs into money. Mathematically speaking, a signal distortion amounts to a fixed operation that can be undone by the inverse operation. This precisely is what much of the signal-processing gear is for that clutters up the encoder and decoder boxes of our technological commu-nication channels. That gear is usually integrated into the circuitries there, cor-recting for distortions; without it we would hardly recognize a TV picture that is beamed from a station close to home, let alone one from planet Mars. However, such signal rehabilitation has a steep information price.

Cells are spared all that. Their massive molecular signals don't get as easily rumpled as the tenuous photon signals are. Not that these molecular signals are inherently stable. In fact, many of them get disarrayed or even dismantled in transmission. But the important thing is that the information they carry, that is, the portion of that information allocated to transfer, is not altered; a billion-year search for the right signal stuff has seen to that. That search ensured that the allocated information is immune to the attack of the proteolytic and other enzymes inside or outside cells. Moreover, it ensured that the cellular communication channels are normally rather free of interference from extraneous signals and electromagnetic fields. No human manufacturer of communication equipment offers such a warranty—just try out your car radio under an electric power line!

Cells have not much to fear on that account. Electromagnetic fields of strengths we might call "normal" do not significantly interfere with the molecular signals and their flow. Nor do thousand-meter-long waves, like those of radio; or meter-long ones, like those of TV (Fig. 9.5). Only at much shorter wavelengths do the signal molecules become appreciably disturbed and, as we get to wavelengths of one-tenth of a micrometer or smaller, the signal molecules begin to dance on the razor's edge. However, there is only one type of particle with a wavelength of that sort to pose real danger: the photons in the ultraviolet. And these get largely filtered out by the ozone and water in the atmosphere.

So, all in all, the intercellular communication channels, like their evolutionary predecessors inside the cells, are made of rather imperturbable stuff. The information transmitted through these channels is largely free of external interferences. It is as solidly bottled up as the message sent by our stranded mariner across the sea.

"Nature's Sole Mistake"

But treading on the heels of that splendid evolutionary achievement comes a sour note: the warranty against signal interference, which had been in effect for aeons, has expired. That warranty came with a conditional clause that I had abbreviated above to one word: *normally.* But let me now fully write it out: "the biological communication channels are interference-free in a normal environment, meaning an environment of the sort where the signals originally had been (naturally) selected in—an environment shielded by a layer of ozone."

Now, that shield, which has been protecting the Planet since primordial times, has been beaten and battered for the past hundred years by a creature which calls itself *Homo sapiens* ("wise man," in optimistic Latin). For the first time in a billion years, ultraviolet photons with wavelengths of atomic dimensions are getting through in considerable number. These high-quantum particles can make mincemeat of the organic signal molecules, putting the entire communication enterprise in danger.

However, things are not beyond repair—not yet. In the communication scheme of the universe, the crack amounts only to a tiny distortion; and if *Homo*

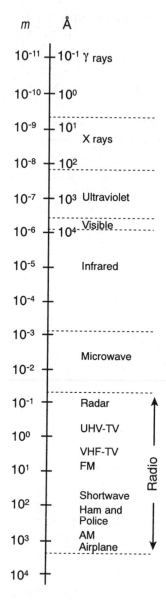

Figure 9.5 The electromagnetic spectrum. The range from radio waves to γ rays is shown on a scale where each successive division represents a tenfold increase in wavelength. The units are in meters (m) and angstroms (Å). The visible range lies between 4,100 Å (violet) and 6,800 Å (red). Atomic radii are of the order of 1 Å (10^0), and the sizes of the signal molecules in intercellular communication are about 10 Å (10^1).

lives up to his name, that distortion can be undone. Though beware! That tellurian also has been given another name: "Nature's sole mistake"—and this by two shrewd observers of the human condition, Gilbert and Sullivan. So, if you ask what the long-range prospects of the organismic communication enterprise are . . . well, I am reminded of what Chairman Mao reportedly said when asked what he thought of the French Revolution: "It's too soon to tell."

How to Get a Message Through a Noisy Channel: The Second Theorem

The point of every communication channel is to get the message through, though this can by no means be taken for granted. Information tends to be degraded along the way—that's the Second Law—and eventually will merge with the random events, the *noise*, in the channel. And in intercellular communication, there is a lot of noise, because the signal flow is largely left to the random walk of molecules. Thus, everything would seem to militate against the information making headway against the noise, and one would have all the reasons to be surprised that there is a message received at all. Yet, somehow, despite the odds, the message gets through. To see how, we will take a brief digression into information theory.

We already have seen how information is measured (equation *1*). If we apply this measure here and bring in time, we get a useful gauge for a communication channel: the time it takes for an information quantum to get from transmitter to receiver; or, in the case of a rather noisy channel, as in intercellular communication, the rate at which that channel can carry the information, without gross error. This upper limit is called *channel capacity*. We all have experienced that limit for ourselves at one time or another; but as for a lively description, it is hard to top that of a master penman, like Russell Baker. In his column in the *New York Times* he tells what happened one evening when he carried the surveying of the daily news too far and he tried to take in "McNeil/Lehrer, Tom Brokaw, Dan Rather, Peter Jennings, CNN, ESPN, Time, Newsweek, U.S. News and World Reportportportportpor . . ."

Now, there is a mathematical theorem which covers just this sort of situation—a theorem by Claude Shannon that deals with the problem of noise and the limit of information transmission. Random noise or, to be precise, the energy density of the noise, sets a limit to information transmission. Noise doesn't limit the amount of information per se—Russell Baker conceivably could have taken in all that gift of gab, given enough time—but it limits the rate, the number of bits of information that can go through a channel per unit time. The fact that the noise effect depends on time doesn't make it less disturbing—the channel gets trammeled from fore to aft. Indeed, there is no more efficient way of jamming a channel than flooding it with random noise (the experts in communication warfare have this down pat). But if time is not of essence, there is always a way to get your message across: repeat it. By sending the same information over and over, one eventually gets the better of the noise. Though slow, this method has the virtue of frugality; it's the one to go for if one doesn't have the where-withal to process out the noise—the poor man's choice—and Evolution used it to the hilt for her basic communication endeavors.

Hidden between the lines is the statistical notion of transmission accuracy. And this is what took us to Shannon's theorem—his Second Theorem. This the-orem proves that the random noise in a channel does not, by itself, set a limit to the transmission accuracy; for any channel (with finite memory), *a capacity can be defined*—and this is the most surprising outcome of the theorem—*where the probability of error per bit can be made arbitrarily small for any binary transmis-sion rate, by proper encoding and decoding*. Conversely, the probability of error cannot be made arbitrarily small when the rate of transmission exceeds the chan-nel capacity.

In plain language the second theorem says roughly this: *if one cannot elimi-nate noise, a message can be transmitted without significant error and without loss of transmission rate if the message has been properly encoded at the source*. This came as a surprise to everyone when the notion made its debut fifty years ago. Just think what the theorem promised: an error-free transmission in a noisy channel, without lowering the transmission rate, and all you have to do is come up with precisely the right code! It's a good thing that one cannot argue with a theorem, or else there would be quite a few skeptics. Well, the promise has since been fulfilled by scores of applications in communication engineering, and the theorem sits high in science's pantheon.

As for biological communication, this theorem holds the key to the question of how the messages get through those noisy molecular channels, as will become clear in the following pages. It turns out that Lady Evolution knew the theorem all along, and our knowledge of it will help us unweave a little her multi-cellular organismic fabric and afford us a rare glimpse into her master strategy.

The Virtues of Redundant Small Talk

First off, let us consider the basic strategy Evolution uses to solve the noise ques-tion. The problem is how to get, by thermal teeter-totter, a molecular signal through a communication channel—how to beat the statistical odds. To do so, the cells make signal molecules in multiple copies and send one copy after another through their channels. This, as we said, is the poor man's method, a straightforward tack: where one molecule singly must surely fail, one molecule in a hundred thousand or a hundred million has a sporting chance.

We see intuitively that the dispatching of more and more copies of a signal will eventually beat the odds. One actually can calculate those odds with the aid of probability theory. Although this theory had a somewhat frivolous begin-ning—it originally was started to get an advantage at the gambling table—it eventually acquired solid-rock foundations thanks to a set of axioms by the mathematician Kolmogoroff. This is the same set that the whole edifice of statistical mechanics in physical chemistry rests on. Thus, by bringing to bear probability theory on the movement of the signal molecules, we can find out whether an intercellular channel is up to snuff. All we need to know is the channel space (the signal dilution volume) and the average net velocity of the signal molecules (which, over the short range, is given by the molecules' diffusion rates in the lipid bilayer or in the cytoplasm and, over the long range, by their convection rates in the blood and other body liquids). Indeed, the answer turns out to be positive in all cases tested so far; the number of molecular copies is enough to virtually ensure a signal throughput.

A rather modest number of signal copies goes a long way because the probability of a throughput failure decreases exponentially with the number of copies. Here is how this works: assume that the individual signal molecule in a channel has the chance of one in a thousand to get through; then, the chance that a thousand copies would be lost altogether is $1/e$, about 0.37; with ten thousand copies that chance falls to about 5×10^{-5} and with one hundred thousand copies it dwindles to 4×10^{-44}.

There is no simpler formula for defeating communication noise than that copy method; *the recurrency of the same information, the redundancy, decreases the probability of transmission error*. It is the same formula we use when we try to communicate with someone across a noisy street or send an SOS from the middle of the ocean.

But why did cells turn to such an artless method? There is nothing sophisticated about redundancy, for sure. The answer can be given in four words: beggars can't be choosers. It was all that cells could afford at the start of their information endeavor. Later on when they began to cooperate, it is true, they had amassed considerable information capital and had mastered the art of extracting signals out of extrinsic signal noise. But their original (internal) communication scheme goes back to the dawn of the Cell Age when the luxury of processing out random noise was unaffordable. So, throughout the ages they stuck to their basic guns—they modified and recombined old information and produced signals in large editions. To copy or change information costs a lot less than to create new information, and copying one's own is no crime in Evolution's bailiwick.

However, there is a penance for being so uninventive: delay—it takes then a good while to get a message across. Although the tautology decreases the transmission errors in a channel, the transmission rate decreases too. Thus, the messages have to be kept short if one wants to say something in a reasonable time. Redundancy obviously is not practical for long speeches, but it is OK for short utterances. And this is precisely what cells go in for. They keep the communiqués between them to the bare bones and save their long speeches, the transmission of the lengthy strings of DNA and RNA information, for their internal business transactions. Indeed, compared to the large hunks of information we have seen circulating inside the cells, the information flowing between them isn't much— intercellular communication is small talk, we might say.

What Is a Code?

If we could eavesdrop on an intercellular communication channel, we would hear the repetitious prattle, a monotone signal staccato, drumming against the background noise. However, all that drumming would only be beating the air if the signals were not properly encoded. And that is the nub, for only by optimal encoding can clear communications be established in the presence of so much noise. And when we say "optimal encoding," we mean an enciphering that meets the conditions of Shannon's Second Theorem.

That theorem, which appeared in 1948, is one of the deep insights into the mysteries of information. Indeed, it was a bolt from the blue that put communication theorists and engineers onto the unsuspected powers of encoding. Until then, coding had been the purview of workers in cryptography, a rather esoteric field with military undertones. The theorem changed all that. Mathematicians and engineers now went in search of the land of promise: the formulas for signal encoding that would grant them clear communication, despite the noise. The rest is legend. The informational web that now knits us together on this planet, regardless of distance or rumble, attests to that.

However, little did the theorist and engineers know that Evolution once again had beaten them by aeons. She came up with tip-top coding formulas for knitting molecular communication webs long before human beings were around, even long before multicellular organisms were. Her search for the formulas must have started as she fashioned the first cybernetic circuits with the few organic chemicals then at hand, culminating, in single-cell times, with the genetic code, and in multicell times, with the codes for intercellular communication.

Now, what precisely is a communication code and how is it generated? That word pops up everywhere in biology, but rarely is an effort made to get to the bottom of the question of how a code comes into being. But first we need a definition of the word. The definition part is readily taken care of: *a code is a set of rules governing the order of symbols in communication*. This defines a code, regardless of the nature of the symbols, be they alphabetic letters, voice sounds, dots and dashes, DNA bases, amino acids, nerve impulses, or what have you (Fig. *9.6*). Codes are generally expressed as binary relations or as geometric correspondences between a domain and a counterdomain; one speaks of *mapping* in the latter case. Thus, in the International Morse Code 52, symbols consisting of sequences of dots and dashes map on 52 symbols of the alphabet, numbers, and punctuation marks; or in the genetic code, 61 of the possible

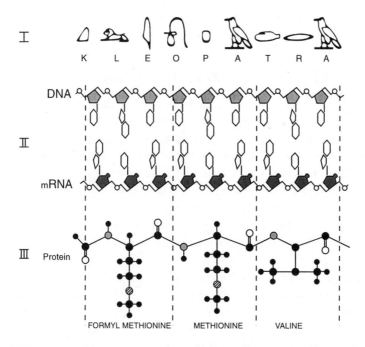

I

K L E O P A T R A

DNA

II

mRNA

III Protein

FORMYL METHIONINE   METHIONINE   VALINE

Figure 9.6 Three examples of encodings. (*I*) A word in Old-Egyptian. This set of symbols was inscribed in a stone of a temple on an island of the Nile River. Its translation follows a simple pair rule—a code broken by Jean-François Champollion at the turn of the past century. The symbols are phonetic; their equivalents in our alphabet are shown below (this was the first word that Champollion translated; the Kleopatra here is not the notorious one, but an earlier queen). (*II*) A syllable in genetic language. These symbols are inscribed on DNA. Here too, the transliteration (to RNA language) follows a simple pair rule, namely, Watson's and Crick's base-pair rule, the first-tier code (Fig. 7.7). (*III*) The final translation into protein language proceeds from a fixed point, three RNA symbols at a time, and the symbol transfiguration is governed by the genetic code (Fig. 7. 9). The DNA, RNA, and protein languages are written here in symbols of chemical structure: the bases adenine (*A*), guanine (*G*), cytosine (*C*), and thymine (*T*) or uracil (*U*), and the amino acids methionine and valine; the bases are shown on their sugar (deoxyribose, *pentagons*; ribose, *pentagons with heads*) and phosphate backbones. The first RNA symbol-triplet (AUG) codes for a methionine with a formyl group; in that form the amino acid cannot hook up to a polypeptide chain, and so this AUG marks the start of the reading frame.

symbol triplets of the RNA domain map on a set of 20 symbols of the polypeptide counterdomain.

In intercellular communication the domains and counterdomains are the signal molecules and their receptors, and the code is like the base-pair rules of the first-tier code of the DNA (Fig. 7.7), a simple rule between pairs of molecules of matching surfaces. The signal and receptor molecules here are specifically embossed so that transmission of information can occur only between a pair of closely matching partners; each signal domain maps on its respective receptor domain.

Why There Are No Double-Entendres in Biological Communications

Thus, the basic information for the encoding in intercellular communication (a high-class encoding complying with the Second Theorem) is all concentrated at the interacting molecular surfaces. And this information is what makes the communications unambiguous. We can now define an *unambiguous communication:* a communication in which each incoming message or signal at a receiver (or retransmitter) stage is encoded in only one way, and no two different incoming messages or signals are ever encoded the same way; or, stated in terms of mapping, a communication in which there is a strict one-to-one mapping of domains, so that for every element in the signal domain there is only one element in the counterdomain (Fig. *9.7 A*).

Such ironhanded mapping is the hallmark of all good communication systems, not only the biomolecular ones. A telegraph, a good teletypewriter, a fax link, a TV downlink to earth from a satellite, all work with just such a stringent set of rules, and the hardware that executes those rules is the encoder. In general, an encoder is a device—a deterministic device—that converts the message in one set of symbols into a new message, usually in a different set of symbols. Take an old-fashioned teletypewriter, for instance. This is a system where the mechanics of the conversions are seen at a glance. Here, a typewritten English message is converted into a pattern of holes punched into a tape, then into a sequence of electrical pulses on a telegraph wire, and back into English by another teletypewriter. In each conversion—transmission or retransmission in the corresponding intercellular communication scheme—the sequence of the symbols is ordered according to a set of logical rules, the *code* for each transmission segment.

The teletype codes are comparable in quality to the codes in intercellular communication. There is enough information in each teletype symbol to make the messages unambiguous. In fact, we can readily estimate the upper limit of the amount of information required for that; all we need is the number of coding symbols. These are the letters from *a* to *z* and the spaces—altogether 27 symbols in "telegraph English" (which has no punctuation, capital letters, or paragraphs). For an upper-limit estimate, we assume all 27 symbols to occur with equal probability. So, by virtue of equation *1*, we get log 27 = 4.76 bits—the amount of information per letter needed for unambiguous communication.[2]

Our verbal communications, our common spoken languages, are by and large not so stringently encoded; the mapping is divergent instead of one-to-one, as in

2.   The amount of information actually needed is less than that upper limit. If, instead of assuming equiprobabililty, one calculates the actual probabilities of the individual symbols in an ordinary English text, the amount of information turns out to be about 1 bit per letter.

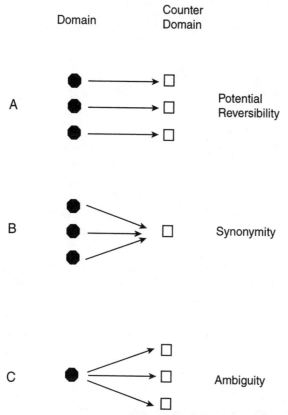

Figure 9.7   The three coding modes. (*A*) A one-to-one mapping where each element of the domain has a corresponding element in the counterdomain. This type of coding precludes both synonymity and ambiguity, and is potentially reversible. (*B*) A convergent mapping where more than one element of the domain maps on one element of the counterdomain, permitting synonymity. (*C*) A divergent mapping resulting in communications with ambiguities. In biological communication the information flow is one-directional (because of the presence of irreversible protein demons). Thus, in a convergent mapping, as in the case of the genetic code, there is synonymity, but no ambiguity. The arrows on the diagram indicate the direction of the information flow.

the teletypewriter above. Take, for example, the word "communication." It can mean transmission of information, passageway, social intercourse or impartation, and so forth. There are whole dictionaries of synonyms to keep tab of such divergent mappings. Communication engineers have tried to improve on those language ambiguities by constraining the order of the language symbols. Some of their recent translation machines do a credible job. Though, lest you think that they always score home runs, here is an authentic story about an English/Spanish machine. The machine was asked to translate a few passages from

Matthew. It did quite well until it got to 26:41: "the spirit is willing but the flesh is weak." The translation read: "el aguardiente es agradable pero la carne es insípida" ("the liquor is agreeable but the meat is stale").

We can live with such double entendres, at least with some, but our cells cannot. We can—and sometimes are the richer for it—because we usually can determine the meaning through the context, a process which requires complex computations by our brains. But our cells outside the brain are not wired to perform such computational tasks. And they don't need to, because their language is stringently encoded; the mapping of domains and counterdomains in intercellular communication is one-to-one, and hence admits neither synonymity nor ambiguity.

But what of the encodings that lack the simple chemical logic of *direct* complementarity between domains? The most notorious case of this sort is the master code in the mainstream of intracellular communication, the genetic code, which rules the order of symbols in the translation of the DNA sequences to protein. The table in Figure *7.9* tells us at a glance that a given amino acid may have more than one coding triplet: UUA, UUG, CUU, CUC, CUA, CUG, for instance, are all synonyms for leucine. A code of this sort is said to be "degenerate." That is OK despite the epithet, so long as the information flow goes in the convergent direction (Fig. *9.7 B*), as it normally does. The counterdomain here consists of only one element, and so a given triplet codes for no more than one amino acid. Thus, there is synonymity, but no ambiguity in the communications ruled by the genetic code.

The Question of the Origin of Natural Codes

But how did all these optimized and stringent codes of biological communication come about? Here we are up against one of the great enigmas of biology, which thus far has defied all efforts of penetration. Darwinian evolutionary theory offers no help here. Its principle of natural selection or survival of the fittest, as even staunch evolutionists do concede, entails a tautology: it identifies the fittest with the survivors, and so boils down to no more than a "survival of the survivors." Modern molecular biological versions of the principle, where the survivors are less sweepingly described in terms of preservation (and recombination) of genetic information, do somewhat better.[3] Still they are good only for

3.   Attempts of this general sort go back to long before molecular biology, even before Mendelian gene theory. Already in the nineteenth century, Moritz Wagner, a contemporary of Darwin, advanced the notion of an evolution by reproductive isolation without selection pressures; this notion (in Mendelian terms) amounts to an evolution by genetic drift, a mechanism where the gene pool of geographically isolated individuals differentiates from the gene pool of the main population. Such endeavors intensified as Mendelian theory eventually sank in, and knowledge about the heredity of traits was gained through experiments on population genetics and embryology, which lately have led to a number of neo-Darwinian interpretations setting forth *ad hoc* restrictions to Natural Selection.

treading in place, and that won't change until we have at least an operational definition, some measure of the fitness.

Outside biology, the question how codes get to be hardly comes up. It is usually uncalled for; the origins are obvious in the case of man-made codes. Take our telegraph code, for instance, or the codes bankers or military men use. Those are concocted more or less arbitrarily, arising almost fully fledged from the cryptographer's mind. But no one but the most die-hard creationist would believe that the codes of biological communication—the natural codes—have such one-fell-swoop beginnings. The question of the origins of natural codes has haunted two generations of information theorists interested in human languages and, regarding biomolecular communication, the question has loomed large ever since the genetic code was found.

Indeed, one cannot help being drawn to something that speaks so eloquently for the unity of life and the early origin of its master code, like the fact that all living beings use the same four-letter alphabet for their genetic information, the same twenty-letter alphabet for their proteins, and the same code to translate from one language to the other. Each of the 64 possible triplets of the 4 RNA bases here is a "codon" that assigns one of the 20 amino acids in the polypeptide chain. Such a code and the hardware that executes it are surely too complex to have started with one swoop.

Was there once a simpler code with fewer letters? The primordial polypeptide chains may have had fewer amino acids, and so a one- or two-letter code might have been sufficient to specify them. It seems, however, unlikely that the codons ever had less than three letters; there is no simple way for going from a one- or two-letter code to a three-letter one without wiping the entire preceding system out at the transitions. But it is possible that originally only one or two of the letters in the triplets were meaningful; the early translation machinery might have used only the first and second of the three bases in the primordial codons, and so the meaning—that is, the mapping—might have gradually sharpened to the one we find today.

But this, at best, only shifts the basic issue. Even if the picture of a gradual ripening of mapping is true, the fundamental question remains: How did the precursor code, the primordial genetic code, come into being? And the same question is echoed in intercellular communication, though, so far, we managed to skirt the issue by dealing with that communication in piecemeal fashion; we considered the transmissions along a communication channel stretch by stretch, rather than over the channel as a whole. However, as soon as we take the holistic view, the various retransmission stages fall by the wayside and all we have is a source encoder and a (final) receiver, and the signals and receptors in the various stretches become just so many pieces of hardware (or adaptors in the molecular-biology sense) between these two. Thus, the question of how the hard- and software here came into being is no less a mystery than that of the genetic code.

The Why and Wherefore of Encoding

To search for clues, we will take a look at the codes from the thermodynamics point of view. We resort once more to the expedient of trailing the stream of cellular information as if it had a central source. As we start from the DNA, we see that after two stages of linear encodings—the codes for DNA transcription and translation—the flow fans out over the three-dimensional world.

Such a streaming from the one-dimensional simple to the three-dimensional complex is something we can readily catch on to. It appeals to the mind. In our reasonings in mathematics or physics we usually go from the one-dimensional to the n-dimensional—by and large, things in our brain seem to fare well that way. However, as we all know too well, information will not necessarily flow just because there is a store at hand. "Store," "fountain," "source," are figures of speech for an *information potential*. But there is a world of difference between an information potential and an energy potential. Energy potentials are readily tapped. To get a flow from a water reservoir, all you need is a pipe that goes downhill. To siphon information from an information reservoir is not so easy; to harvest that kind of sap takes more of the same kind—in fact, so much that sometimes one wonders whether it's worth the trouble. Think of the amounts of information one must invest to wangle something out of a mathematics book. But that's the way things are under the Second Law: tapping information requires wits, and there are no two ways about it. A book by itself means nothing, unless one has the information to decipher it. What good would an ancient script of Egypt do without the Rosetta Stone, or a CD without a CD player?

If we look a little deeper, we see why so much information goes into enciphering or deciphering. The reasons lie in the interplay between information and entropy. These two entities must be conserved under the law, so that any interplay between them is always a quid pro quo. Take the largest reservoir of all, the DNA. The number of combinatorial possibilities for linearly ordering the symbols there, the total *possible* number of genetic messages, as we have seen in Chapter 1, is unimaginably large. If there were no constraints for the ordering of the symbols in transmission and every possible combination occurred with the same frequency, the entropy would be maximal; but so would be the transmission errors—we would have gibberish, instead of communication, inside or between our cells. On the other hand, if all combinatorial potential were frozen, the transmission errors would be minimal; but then we would have a sequence consisting of only one symbol—a message that says nothing. So, if the message is to be meaningful, the system must hold to a middle course between information and entropy, a course where it sacrifices some message variety in order to get the *best possible* message fidelity.

Those are the subtler registers of Shannon's Second Theorem, its deeper thermodynamic meaning which can be shown from first principles, freeing us from the usual trappings of communication engineering.[4] The optimization of encoding is a Solomonic compromise in the eternal battle between information and entropy. It means steering a course somewhere halfway between maximal information (minimal entropy) and maximal entropy. And this is the course which Evolution must have steered all along to achieve her master code for transmitting rather error-free messages from the DNA "source." The formula for optimal encoding here happens to be one for a linear order in which 61 symbols of the source domain get assigned to 20 symbols of the receiver domain. But the basic strategy for arriving at the other codes must have been the same.

This may be true even for the elusive higher codes governing the three-dimensional biological sphere. Though such codes are mathematically more complex, the binary relations between the pertinent domains can be worked out if the domain structures are precisely known. In principle, any three-dimensional situation can be reduced to a one-dimensional one. Indeed, we do this all the time: we describe complex three-dimensional objects—machines, people, land-scapes—in one-dimensional language; or we transform a symphony of a hundred instruments into a one-dimensional sequence of digitalized pulses and press the whole caboodle into a compact disk.

The best places for attacking the enigma of the origin of codes would seem to be in intercellular communication. The communication lines served by steroid hormones are ruled by rather straightforward logic and their adaptors are few. A number of those channels are accessible to chemical and thermodynamical probings, and we can readily see reflections of second-theorem encoding in their hardware: the embossments of signal and receptor molecules are finely honed to a fit that is neither too tight nor too loose. Too tight would mean maximal bonding between the signal and receptor surfaces, a solid-state amalgamation that would preclude (repeated) message transmission; while, on the other extreme, too loose a fit would mean sheer noise—again, no message. So, here too, a Solomonic compromise between information and entropy is reached; the complementarity between molecular surfaces is optimized for an interaction somewhere halfway between maximal information and maximal entropy, as represented by the amalgamating tight fit and the noisy loose fit, respectively.

The question is then, who played Solomon? How did the evolving molecular system get to that thermodynamic middle ground? Our best hopes for an answer may lie in *Game Theory*, a rather esoteric branch of mathematics concerned with the analysis of games of wits and chance.

4.   The body of information theory, originally formulated to address questions of communication engineering, has since been derived from first principles. That treatment—largely through the work of the mathematician A. Khinchin—goes beyond the usual trappings of engineering, reaching farther into the depths of information theory, including Shannon's theorem.

What Can We Learn from Game Theory?

The Theory of Games is the brainchild of the mathematician John von Neumann. It started with a paper about how to win at Matching Pennies. This is a simple game between two opponents, where each player puts down a penny, either head or tail up, unknown to the other; then the players uncover their coins and player *A* takes all if both coins show the same side, or else *B* takes all. In that paper von Neumann announced to the Mathematical Society of Göttingen that he had found a winning formula. The preoccupation with such a lowly game may not strike you as the most appropriate topic for a scientific society, especially one so august that it once had Gauss and Planck among its members. Nevertheless, thus started the Theory of Games, a rich vein in the human scientific endeavor. In fact, eventually the theory was to reach deep into the interplay between information and chance, breaking the seal of the arcanum of competition.

Today, the theory can be applied to a host of situations where opponents compete with one another for a given pot. Matching Pennies is a game entailing only a single decision, but the theory eventually seasoned to comprehend games where the competitors have to make several decisions, as in bridge and poker, or long strings of them, as in chess or in the deadly games of war and economics.

By and large, those are games where one side wins what the other one loses— *zero-sum games*. And for such games, the theory shows that there is a single "stable" course of action, a best possible strategy or *saddle point*, as it is called. That saddle-point strategy assures a player the maximum advantage, regardless of what his opponents may do. The individual moves cannot be predicted—that is in the nature of chance events—but if the players play long enough, the one using the saddle-point strategy has the best chance of winning.

The existence of such an optimal strategy would not be so surprising, perhaps, in the case of strictly deterministic games. Tick-tack-toe, for example, when perfectly played must result in a tie. But the stunning result of game theory—a result rigorously backed up by one of von Neumann's theorems (the "Central Theorem")—is that there also exists such an optimal strategy for the games that are not strictly determined. Even then (e.g., in chess), the player who adopts the saddle-point strategy will, in the long run, turn probability to his advantage: he acquires the power to make an undeterminate game determinate. One could hardly think of a more eloquent demonstration of the power of information.

All this is thermodynamically aboveboard, to be sure. The saddle-point player pays for his empowerment with information. This is the amount embodied by his strategy—a Gargantuan amount in the case of a chess game. This information flows into the game at each decision point, and this is why the game gets more and more determinate.

Different games have different saddle-point strategies, but all such strategies have one thing in common: *they specify a course of action where, in each move, the player maximizes the minimum gains and minimizes the maximal losses ("max-imin" and "minimax," for short).*[5] They are play-it-safe strategies, where the user always keeps a wary eye on the minimax. Not much excitement here, no derring-do, but it is the sure course that in the long run will give you the advantage because, modest as they may be, you are assured of some winnings when you keep the maximal losses you may suffer in each move to a minimum.

So, in those games of wits and chance, bliss lies in the saddle point. The longer you play, the surer you can be of winning. Now, what of the longest game of wits and chance on Earth, the one that Evolution has been playing for aeons? That is a zero-sum game, like the rest of them; its winnings—information—and losses—entropy—must balance out overall under the conservation law. Thus, however complex or over our heads the numerical computations may be, there ought to be a stable optimal course of action which von Neumann's Central Theorem promises us: the saddle-point strategy. Is that the strategy Evolution has been following, the secret formula for the middle course between informa-tion and entropy she has been steering for three-and-a-half billion years?

This takes us back to the question of the origins of biological encoding, which we left standing in the wings. We have seen that the codes constrain the order of symbols in the intracellular and intercellular communications in a way that secures the best possible message fidelity; they are optimized codes which strike a compromise somewhere halfway between maximal information and max-imal entropy. Does that middle course represent a situation where the minimum information gains are maximized and the maximum losses (entropy) are mini-mized? We have no answers yet, to be sure. But if one has stared at a problem for so long, even the glimmer of a wink is welcome and just the framing of the fun-damental question is something to write home about. The latter, the formula-tion of a problem in basic terms, is the necessary prelude to a solution—example after example in the history of science attests to that.

Now, can we, with the aid of game theory, solve the enigma of the origin of the codes and wing ourselves over the quicksand of evolution? Can we find the *physi-cal* constraints for the minimax/maximin course Evolution may have pursued? The answers will come only after much mathematical effort, but game theory offers an appealing framework for such an inquiry. And with some luck, there may be short-cuts. Indeed, a possible one comes to mind: the analysis of synonymity in cellular languages. Those synonyms are unlikely all to have the same probability for error-free transmission of information, and so an analysis of their efficiency for second-theorem encoding may streamline the effort (a good prospect also for understanding the why and wherefore of the degeneracy of the genetic code).

---

5. The mathematical term "saddle-point" is based on an analogy with the shape of a saddle. It corresponds to the intersection of two curves: one curve is represented by the rider's seat of the saddle, whose low point is the "maximin"; and the other curve is represented by the bottom side of the saddle, the curve that straddles the horse's back, whose high point is the "minimax."

Even so, the cracking of the numeric saddle-point solutions of this game of games will be a long road—the solutions for chess, for which whole batteries of computers are needed, will look like child's play against that. No one can say what lies at the end of an untrodden road, but the stakes here could not be higher: a scientific theory of evolution, namely, an explanation where the century-old tautology of the fittest and survivors is replaced by a matrix of physical constraints. Such a theory may, for the first time in biology, hold up a mirror to the past, as well as to the future.

The Master Demons of Intercellular Communication

Let us now take stock of the things we have wangled out thus far about intercellular communication: (1) *the communication channels are rather free of external perturbations;* (2) *the channels have a high level of intrinsic noise, that is, thermal noise;* (3) *the signals are repetitive, and so prevail over the noise;* (4) *the signals contain relatively little information and, thanks to that, they don't easily saturate the capacity of the channels;* and (5) *the communications are encoded for the best possible message fidelity, a second-theorem encoding, leading us to suspect the existence of a saddle-point strategy in the evolutionary game.*

What still remains to be dealt with are the transmitters of the signals—the "receptors" in the biologists' lingo. That term doesn't fully do these protein entities justice. They are certainly more than just receivers of information; they also have an output. At one site of their structure, they receive information, picking a molecule out of a molecular universe, and at another site, they let go an information quantum of their own. They are Maxwell demons, complete with the intrinsic (structural) information for cognition and the wherewithal for tapping the extrinsic information needed for resetting their structure after the cognitive act. They tap that information from an energy source, like phosphate; but some of them also can draw it directly from the electrical potential at the cell membrane.

If we follow the information spoor in the channels of communication between cells, we find a demon of this general sort at every stage where messages are received and retransmitted. With each cognition cycle such a demon hauls in some extrinsic information. Not much in a single run, but he does so over and over. Thus, going at a rate of a hundred or a thousand cycles per minute, a horde of demons brings in a tidy sum; and this is what enables cells to pay the high price of a sophistically encoded communication. These demons are the uncontested stars of the inter-Circus sport—they carry the better part of the thermodynamic burden of communication. Without these Maxwell entities, instead of communication, all we would have is noise.

Let us check briefly on the performance of these genies to see how well they measure up to that promise. First, their directionality. To get a message through a communication channel, the information flowing through it must have a direction. In the channels of our technology, the direction is given by irreversible (one-way) transmitter stations. This is also true for the intercellular channels, though there may be an occasional reversible demon too along the line. But it's the irreversible ones who bring home the bacon—they are the ones who get the message through. So, locally, over some stretches of an intercellular communication channel, the flow may go to and fro, but thanks to the irreversible transmitter demons—and there is always at least one of that sort in any channel—there is, on balance, one direction.

Second, their cross-linking talent. Intercellular communication channels form networks; the information doesn't just flow straight in one line from a source to a user, but it may merge with lines coming from other sources or be forked out to different users. Indeed, such merging is commonplace in cell membranes where the messages from different sources often converge upon a "receptor" which funnels the combined information into the cell circuitry. This receptor demon has two information inputs—two cognitive sites—plus an information output that is coupled to the inputs. He can handle two incoming signals at once and perform the function of a T junction whose inputs face the membrane interior and output faces the cytoplasm. This is an active T junction; the demon processes the incoming information. He can integrate the two signals and when the sum of the two is of the right quantum, he lets loose a signal of his own—he cross-correlates. This goes well beyond the power of the demon in Maxwell's gedankenexperiment.

Third, their amplifier capacity. Highly capable as they are, these demons can't do the impossible: go against the Second Law. That law decrees that information be degraded in transmission. Thus, the signal will tend to fade along the communication line and merge with the ubiquitous thermal fluctuations. This is no different from our telephone and radio transmissions, and no less a nuisance— the fizzling out of signals is forever trying the patience of communication engineers. They have come up with a number of good remedies, but most of these remedies would be too profligate for the frugal evolutionary workshop. Indeed, the intercellular communication networks employ the most economical means to overcome the noise: tautology. Their information sources produce thousands of copies of molecular signals and pour them into the front ends of their communication channels. Thus, even with some fading, the information may be significant at the user end. This essentially is the way steroid-hormone systems work. However, the method is not always quite so austere. In the case of protein-hormone systems, the message gets boosted along the way; a fresh set of signals get emitted at each intermediate transmitter stage in the membrane or inside the cell. This amplifier action, at the very least, manages to keep up with the transmission losses in the communication channel; but, as often as not, there is actual net amplification, and so even a weak message can come through or a worn-out one be restored.

The Game of Seek and Find in the Membrane

These transmitter demons, thus, are responsible for what ultimately counts in communication: getting the message through. And thanks to the ones inhabiting the cell membranes, the ancient game of seek and find inside the cells is also played in between the cells. Like the rest of the membrane proteins, these demons are in thermodynamic harmony with the lipid bilayer; they have their polar amino acids tucked in and make contact with the lipid core largely by means of their nonpolar amino acids. Thus, they can move sideways, making the entire membrane plane their playground.

The basic rules of the game in that playfield are the same as inside the cells: the participants swim randomly and use their wits to catch a matching partner. But the odds are different. The probability for catching the partner is higher inside the membrane than inside the cell, because all players move in the same plane. In terms of equation *1*, this higher probability means that the players somehow receive more information—that is the only way that the probability for encounter between such nearsighted (4-angstrom) players could be raised. Now, as for the source of that extra information, we don't have to look far. No need for high-energy phosphate here, or for other extrinsic donors. The source is right inside the membrane—inherent in its structure, in the orderly alignment and orientation of its lipid molecules. This lipid lattice constrains the participants in the game to move in two dimensions.

So if you wondered what that neat crystalline array was good for, here is the answer: *the crystal framework lifts the signals and demons up from the helter-skelter of three dimensions to the relative tidiness of two. It is a two-dimensional sliding track.* And if you consider all this now from the cells' economy point of view, you see that with such a track the cells get away with relatively few signal copies; they still rely on signal redundancy to ensure a happy ending of the game, but less so than in their diffuse interior.

Among the players in this shallow field, the cross-correlator demon occupies a special strategic position. He does a triple act: he receives and integrates the signal information coming along the sliding track and channels it into the cell interior. His two inputs are designed to handle relatively small information pieces and his output, to emit, upon integration, a more sizeable piece. This is subtle tactics for getting information through the membrane; it allows a molecular transmission of high fidelity, something not easy to achieve in such a thin film. The difficulty here is how to get optimal encoding, without making the signal molecules too large for that narrow crawlspace. Molecular signals necessarily have a mass and the optimally encoded ones, with their well-honed cavities and protuberances, have a significant mass, which slows them on the sliding track.

The cross-correlator demon was Evolution's answer to that dilemma, one of her crafty middle-course solutions. His dual cognitive talent and integrative ability allowed her to apportion the quantum required for the encoding among two signal molecules—to split the signal embossment in two, as it were.

The membranes of multicellular organisms are busy concourses of information. A typical cell in our organism receives hundreds or thousands of messages from diverse sources in a day. All that information goes along and through its membrane—many molecular signals are crisscrossing there. These molecules are relatively small and nimble and, thanks to the cross-correlator demon, they can be so small without detriment to the fidelity of transmission in the intercellular network. Now, add to this the (allosteric) speed of the demon—his triple act proceeds with acoustical velocity—and you get the full physiological and evolutionary sense of his strategic position in the intercellular signal traffic. Indeed, it is hard to imagine how the multicellular organismic enterprise could have prospered without him. Without this master demon, the traffic jams in the membrane would be more than flesh and blood could bear.

The Communications Between Cells Are Lean in Information But Rich in Meaning

Having familiarized ourselves with the players and the rules of the intercellular game, we will look now into the physiology of the communication. To start, let us see what sort of communiqués cells send to one another and what physiological effects these have.

The communiqués, the messages themselves, contain little information. They are just the briefest of statements, like "yes" or "no," or short commands, like "do" or "don't." Such terseness suits the cellular habit of tautology and is good for the economy. But really, no long statements are called for. As each cell brings a complete information system, its own Circus, to the intercourse, an aye, a chirp, or a bark can elicit complex actions. Thus, a little steroid will set in motion the transcription and translation apparatus of entire sets of genes, an even tinier nitric-oxide molecule will trigger an act of memory, a small peptide will start the whole replication machinery rolling in a cell, or a small amino acid will put an end to a communication-fest in our brains.

Thus, there is more information in intercellular communication than meets the eye. The information transmitted, the message between two cells, is only a small fraction of all the information involved in the communication process. And this is what I meant before when I said that the communications between cells go in crescendos.

In this regard, the intercellular communications are cast of the same mold as our language-communications. What we speak represents only a part of the whole information going into our conversations. The rest is in the context. Take, for instance, some of our common utterances: "How are you?" "Glad to meet you" "Congratulations!" These carry very little information by themselves, but in the general context, that is, including the pertinent information existing in the brains of the individuals engaged in communication, these utterances acquire full meaning. All our languages—spoken, written, or gestures—show this feature: a few sounds, a few alphabetic symbols, the barest outline of a charcoal drawing, a body movement, will evoke in us a rich repertory of information—a Lawrence Olivier could convey the deepest grief by a curling of three fingers, or hurl a knife with a look!

In all these cases a minimal signal elicits a large response in a prepared system, one that contains practically all the information that gets into play. The meaning arises out of the context, by which we mean the total situation in a communication. But what precisely constitutes the "context"? Though often used, this is one of those words for which we have more feeling than real knowledge. Our dictionaries give no clear-cut definition; nor should we expect them to, since so little is known about the programmatic software involved in language-communications, and even less about the hardware.

However, we are a little better off in intercellular communication; we know the molecular hardware and some of the software in a few (basic) cases. The context here is defined by the portions of the Circus information that are engaged in intercourse. Thus, *the context is the sum of all participating subsets of information of the target-cell Circus plus the information in the message (the smallest subset)*. The participating subsets, in effect, constitute a compartment—and we will call it an *active Circus compartment*. So, each message evokes its active compartment. I say, "evoke," because what the message does has the earmarks of an act of memory; the core information is in the Circus before the message arrives—it is always there, only waiting to be recalled. This recall capacity runs in the blood of all living beings, and the underlying memory principle probably is used over and over at various organismic levels, (including perhaps those high up in the brain which are so difficult to see through). To ferret the principle out, we will poke a little into the design of the lower Circus layers that hold the compartment reins.

How to Pull the Strings of a Cell

To begin with, we take a look at how these compartments change with time. Different intercellular messages evoke different active compartments, though these compartments may overlap. Thus, the portions of the target-cell Circus that become active in a cellular intercourse may vary from moment to moment, but the *potential* active compartment for any given message is invariant. This active compartmentalization is the product of a well-orchestrated control where both the Circus and the message play a part. The immediate targets of this control are the internal (intracellular) Circus connections, the molecular couplings in and between the Circus loops *within* a cell. In nonsocial cells this connectivity is all-pervasive; it ties all components of the cell into a whole, as we have seen. In multicellular organisms, the intracellular connectivity is reined in, though it always lies dormant under the veneer of controls.

The controlling information comes from special loops which are integral parts of the Circus of the target cell. These particular loops exert negative control over the intracellular Circus connectivity—they are negatively biased. The bias depends on the interplay with other members of that loops-within-loops fraternity. The whole interplay is elegantly balanced, so that in status quo conditions the bias is critically poised below its negative maximum. And it is at this point, then, where the intercellular messages come in. These messages can tip the balance, effectively setting the bias at its maximum and, thus, let loose a veritable information avalanche.

The control here involves loops that often are quite complex (we will see some piece-by-piece snapshots later), but the physiological bottom line is this: the controls that hold the Circus connectivity in rein are not always turned on or not fully turned on; but when they do get fully turned on by the actions of intercellular messages, they effectively switch off certain intra-Circus connections. These controllable connections are in strategic Circus locations, circumscribing latent active Circus compartments. Thus, the special control loops here function like circuit breakers, disconnecting a compartment from the rest of the Circus. Each species of message activates a specific set of loops, and so walls off a Circus portion, the active compartment.

This little nosing into Circus organization afforded us a glimpse of what goes on behind the scenes when a piece of information gets evoked. With full intent, we did our prying at a low hierarchic level—only a bit higher, and the view would have been too rococo. But conservationism is in the nature of the beast, and so the basic organizational principle of evocation here—the releasing of an information flow from a negatively controlled informational compartment—may well apply also to higher organismic levels.

Telecontrol

Consider now this controlled informational compartmentalization from the point of view of the cell that emits the message. By means of its message, this emitter cell pulls the strings of the target cell's connectivity, subjecting a Circus compartment to its command. The sway it holds over the target cell is not complete, but even a small piece of a life Circus is a universe.

This is a peculiar game of one-upmanship which is played between neighboring cells of an organism, but quite often also between cells that are far away from each other—in our own organism, some hundred thousand cell diameters away. By means of such telecontrol, the emitter cell impresses its own rhythm on a target cell. A cell's Circus already has an animation of its own; it is wired to be responsive to quite a number of environmental stimuli, and the basic wiring dates back to the Age of the Cell. But there is a further quickening now owing to the Circus intercourse, the organismic cell interactions. Bombarded with different messages from many parts of the organism, a cell is usually under multiple telecontrol. So, in the course of a day, a cell's active Circus compartments are continuously changing with the intercellular ebb and flow. Imagine, for a moment, we had it in our power to have these compartments lit up. We would then see the sequence of shifting compartmental patterns, and that flickering *laterna magica* would show us more about what animates an organism than a thousand mirrors of analytical plodding.

Now, this telecontrol of Circuses is itself part of an information loop, a loop of a higher order than those inside a Circus. It is a circle that runs between whole cells, namely, the emitter cells and the target cells, tying them together into a community. So, we have one of those loops-within-loops here: the inhibitory (negatively biased) control loop of compartmentalization, the Circus loop itself, and the inter-Circus loop to which the message belongs. And so round and round it goes, everything hanging together—as in a baroque canon or a good fugue—by an endlessly recursive theme.

t e n

The Broadcasting of Cellular Information

The Begetting of Social Graces

Toward the end of the second billennium of the evolutionary calendar, the principal players of the intercellular game were in place. These belonged to the Maxwellian fraternity of transmitters who had served the cells so well during their monistic age. By then, several members of that fraternity had reached peak form and were ready to strike out for greener pastures. In partnership with messenger and signal molecules, these protein demons eventually would stretch a communication network between the cells, with tentacles spreading out everywhere, vastly expanding their original internal networks.

This expansion was responsible for that first on Earth: social intercourse between cells. The newly acquired intercellular channels of communication allowed the cells to coordinate their activities and to undertake joint operations. Like all things evolutionary, the begetting of sociability was a gradual affair— each successful extension of the communication network gave being to a social grace, a style of cell coordination. So, the third billennium saw the genesis of a wide spectrum of cell coordinations, going from simple mass action (a mass buffering of cellular chemical activities) to cybernetic intercellular control.

In this part of the book we will peer into the information flow behind those joint cell operations. We will dissect some of the channels of communication here and get down to their molecular nuts and bolts. We begin with two channels of old pedigree, channels of hormonal communication.

The "Emanations from Organs"

First a word about the history of hormones. There is a widespread misconception about who discovered hormones. Our Western world generally ascribes the discovery of these substances to Théophile deBordeu, a physician who lived in France in the eighteenth century. This, indeed, was the first glimmering the West had of hormones, but the Chinese alchemists were on to them long before. They used steroids for medical purposes already in the tenth century and had purified some of them to near homogeneity by the sixteenth.

The West actually was quite slow on the draw. It was only in 1775 when deBordeu made the observation that grabbed the attention of that part of the world: breast growth, beard growth, voice pitch, and other human sex characteristics are related to the ovaries and testicles. DeBordeu pleaded his case well (he was also a lawyer), but really avant-garde were his concluding remarks: "Each organ of the body gives off emanations."

What precisely "emanated" still took awhile to be found out. In fact, it took several decades until those emanations were perceived to be of a chemical nature, not to speak of their cellular function. That had to wait until well into the nineteenth century for Mathias Schleiden's ground-breaking theory that organs are made of cell units (1838). Only after the cell theory had time to sink in—and it took a long time to do so—did the light come at the end of the tunnel: the emanations were substances emitted by the cells—or, in the physiologist Charles Brown-Séquard's words at the time (1891), "an active principle discharged by the cells into the circulation."

By the turn of the century, enough had become known about that active principle to think of it as a "chemical messenger" for short-range, as well as for long-range, intercellular communication. Thus, in 1904 Thomas R. Elliot, addressing the short-range aspect, proposed the notion that the nerves that control the heartbeat might give off adrenaline.[1]

And then came the name for that game: the physiologist Ernest Starling, lecturing to the Royal Society of London a year thereafter, introduced the word "hormone." Starling had consulted two Oxford linguists, William Hardy and William Vesey, who composed the word from the Greek *Ormon* (ὁρμάω), I excite. This seemed an apt neologism at the time, and it certainly was one as far as adrenaline was concerned. But in the byzantine world of organismic information flow, things are often not what they seem. It eventually transpired that that's not all that intercellular messengers do; some will inhibit. Alas, it was too late. But think how much more felicitous things might have been with one vowel changed: *hermone*, after Hermes, the messenger of the gods.

---

1.   The experimental demonstration that adrenaline was an intercellular messenger came seventeen years later when Otto Loewi showed that an adrenaline-like substance was released in the heart when its sympathetic nerves were stimulated, while an acetylcholine-like substance was given off upon parasympathetic (vagus) nerve stimulation (1921). This became the cornerstone for the present-day notion of chemical transmission in the nervous system.

The Pervasive Messengers

The messengers between the *cells* come in various sizes—in higher organisms they range from 10^2 to 10^4 daltons (Table 1). The steroids are among the smallest. We start with these because they execute one of the simplest Circus linkages. Indeed, so straightforward is that linkage that one can see the whole communication channel at a glance.

Table 1   Common Hormones in Vertebrates

Chemical Class	Mol Weight	Hormone	Chemical Class	Mol Weight	Hormone
Amino-acid derivatives	10^2	Acetylcholine Noradrenaline Dopamine Serotonin Glutamic acid GABA Adrenalin Thyroxin	Peptides	10^3	Substance P Endorphines Enkephalines Oxytocin Antidiuretic Somatostatin Adrenocorticotropin Angiotensin Insulin Epidermal growth factor (EGF) Glucagon
Fatty-acid derivatives	10^2	Prostaglandins			Parathyrin Secretin Gastrin
Sterol lipids (Steroids)	10^2	Cortisol (Cortisone) Aldosterone Testosterone Estrogen Progesterone Vitamin D_3	Proteins	10^4	Platelet-derived growth factor (PDGF) Thyrotropin Follicle stimulating (FSH) Luteinizing (LH) Gonadotropin Prolactin

Steroids

Class	Hormone	Source
Corticosteroids	Cortisol Aldosterone	Adrenal cortex Adrenal cortex
Sex steroids	Estrogen Progesterone Testosterone	Ovaries Ovaries Testes
Vitamin D derivative	Calcitrol	Liver, Kidney

Steroids are made from cholesterol, a highly conserved lipid component of cell membranes. They are produced in a few strategic places of the body—the adrenal cortex, the ovaries, the testes, the liver, and the kidneys—and from there are sent out to the four winds. Their cellular effects are extensive, nearly global. Take testosterone, for instance. It acts on tissues as diverse as hair, skin, muscle, gland, and nerve. Cortisol, the hormone from the adrenal cortex, has even a wider range of action—there is hardly a cell in our body which is not under its sway.

Such pervasiveness is typical of this entire lipid class of hormones. The reasons can be stated in a few words: steroid molecules are stable and hydrophobic, and there are enough of them around to infiltrate the cells by sheer mass action. Their stability permits them to survive the long ride in the blood and other body fluids; and their hydrophobicity enables them to permeate the lipid bilayer of the cell membranes—a quality inherited from the cholesterol, the membrane lipid they stem from. Thus, when a steroid molecule arrives at a target cell, it merely does what comes naturally to a cholesterol descendant: it slides into the lipid bilayer.

And from there it is only a small hop to the cell interior. All that is needed is a little push, and that is bound to come when plenty of steroid molecules are around; the machinery that produces these hormones in the emitter cells is geared to ensure that abundance. The cells of the adrenal cortex, for example, make enough copies of cortisol to maintain a concentration of some 10^{-9} moles per liter in our blood; whereas in the target cell there is hardly a free cortisol molecule to be found—as soon as one gets in, it is sopped up by the receptor proteins there. So, the way of a steroid molecule into the cells is a smooth and easy downhill slide.

Tango à Trois

However, getting into the cell is only half the story. The information carried by the steroid molecules, the message, must be channeled to its molecular target. But have these tiny molecules got what it takes to get the message through a noisy molecular communication channel? Do they hold enough information for an optimal encoding, as mandated by the Second Theorem? This is an issue we could hardly skirt with such peewees—a steroid weighs only 300 daltons, no more than a molecule of table sugar, and the various steroid types are not very different from one another. They all are made of six carbon rings and differ only in their bonding patterns and side groups (Fig. *10.1*). But an inspection of the steroid communication channel as a whole is immediately reassuring; the molecular stage next in line, the "receptor" protein, is an able Maxwell demon with an eye for precisely such bonding patterns and side groups. The "eye," the *steroid-binding domain*, is located on the demon's carboxylterminal side. Besides, there are two other cognate spots on his sizeable (800 amino-acid) body (Fig. *10.2*): a zinc-finger module that can engage the DNA double helix, and a cognate domain on his aminoterminal side, which we will get to in a minute.

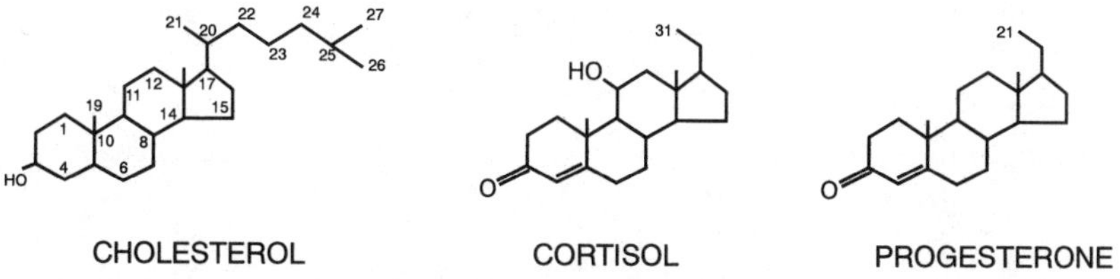

CHOLESTEROL   CORTISOL   PROGESTERONE

Figure 10.1   The steroid hormones are small lipid molecules derived from cholesterol. They all have the cholesterol nucleus, but fewer carbon atoms on the side
chain. The two examples shown, a hormone secreted by the adrenal cortex (cortisol)
and a sex hormone secreted by the corpus luteum, belong to the glucocorticoid and
prostagen families which have 21 carbon atoms (see Table 1).

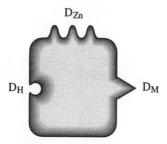

D_{Zn}

D_H D_M

Figure 10.2   The steroid demon, aka steroid receptor. D_H, hormone domain; D_{Zn},
zinc-finger (DNA-binding) domain; D_M, multiplex ("gene-activating") domain.

These triple Maxwellian features are shared by quite a number of demons
who transmit hormonal information to the headgates of the DNA—a large family of demons, including, besides the gamut of steroid receptors, the receptors of
thyroxin and retinoic acid. All are metamorphic proteins who can adopt two
conformations, one where the DNA-binding domain is operational and another
where it is not, and the binding of the respective hormones is what switches
them to that active conformation. Besides, in the case of the steroid demons,
another protein enters the act. This is a binding partner more on a par in size
than the tiny lipid, a partner which latches onto the demon when the steroid is
not around. And this big partner blocks the demon's interaction with the DNA
double helix (Fig. *10.3*). So, trading partners, the demon goes through a cycle
and transmits the hormone message to the headgates of the DNA.

Here is the complete libretto:

Act One. Little partner steps up in the cytoplasm // is swept up by the demon
// big partner slides off // demon and little one tango to the nucleus and on
into its sap // demon hitches up to the DNA headgates. *Act Two* (the demon

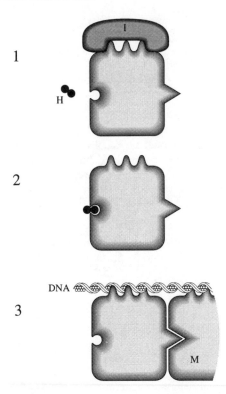

Figure 10.3   Three scenes from the Tango à Trois. *1*, A hormone (*H*) appears, while the demon is paired with his inhibitory partner (*I*). *2*, The hormone has engaged the demon, and the inhibitory partner has come off. *3*, The demon has plugged into the DNA double helix and into his neighbor in the multiplex (*M*).

> has finished his business at the headgates). Little one comes off // demon disengages from the headgates // big partner steps up in the nuclear sap // clings to demon // the two tango back to the cytoplasm.

And this little dance goes on every time a steroid gets into a cell. It is what gives continuity to the hormonal information flow inside the cell and allows the demon and his protein partner to be recycled.

The basic choreography here is fully written into the participating molecules, into their encoded domains of interaction. So all that is needed for the dance to run its course is enough participants. Indeed, the cell interior is teeming with partner molecules for the demon. And, as for the demon himself, there are typically some 1,000 to 10,000 copies of him inside a target cell, which is enough to statistically ensure the above tableau. The dance's direction (overall from cytoplasm to nucleus in Act One) is informationwise on the house, so to speak; the high concentration of DNA in the nucleus (100 grams per liter in our nuclei) will draw in virtually any demon-hormone couple from the (DNA-free) cytoplasm.

There is, of course, the obligatory spoonful of information that must be fed in from outside—the quantum needed to cock the demon for each round. That quantum comes, as with other demons we have seen, from downgrading phosphate energy and is infused at the end of Act One. So, it is in the entr'actes when this demon makes his bows to a higher law.

The Transfer of Steroid Hormone Information

Now, to the main business at hand, the transmittal of the hormonal message to the DNA headgates. Overall, the information transfer goes in two steps: hormone to demon and demon to headgates. The first step relies on redundancy and optimal (second-theorem) encoding to get the message through. The second step calls for more of the same, but rests less on tautology. It can do with less because there is a structural framework at the end—the DNA double helix with its entourage of proteins—which offers the demon topological cues to home in on its target on the DNA. This framework is as good as Cartesian coordinates.

And here is where the third cognate domain, the one located at the amino end, comes in. By dint of the information embossed in this site, the demon seeks out a particular protein in DNA's entourage, the regulatory protein for transcription of the target gene. His zinc-finger domain alone is not selective enough to find that gene; it is too promiscuous in its interaction with the DNA double helix and engages the major groove in numerous places. The amino-end domain narrows down those choices and sees to it that the hormone-carrying demon ends up at a particular gene (or a particular set of genes) and passes on the hormone message to the gene transcription machinery in that particular DNA locale.

The DNA Multiplex and Remote Gene Control

The precise mechanisms of that information transfer still escape us; not enough chemical and structural data are on hand to trace a complete picture of what happens at a DNA target locale (a promoter/enhancer region). So, meanwhile, we'll bell the cat and use our DNA multiplex model to envisage the information flow.

Thus, let me trot out again that model (Fig. *8.4*): the multiplex, a self-assembling mosaic of protein pieces (the "gene-regulatory proteins"), controls the information flow from the DNA reservoir; this solid-state apparatus merges topological information from the DNA double helix with regulatory information coming from various parts *inside* the cell and relays the integrated information to the transcription machinery, the RNA polymerase.

Now, here we add one more piece to the mosaic: the steroid demon. We make this receptor protein an integral part of the multiplex, a part which (in its hormone-bound form) fits with the rest of the protein pieces, completing the

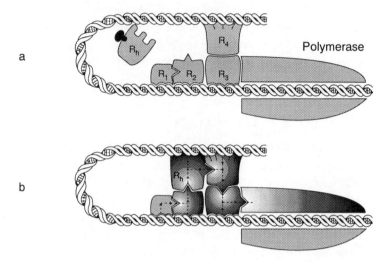

Figure 10.4   The remote-control multiplex model. The relay apparatus controlling gene transcription (see Fig. *8.4*) is endowed with an extra regulatory protein in pluricellular organisms, a receptor (R_h) that loops hormonal information into the integrated circuit board. The hormone-bound receptor, like other components of the multiplex, gets to its place by diffusion in the cell water. The diagram in *a* depicts the receptor (which has bound the hormone) on its way to the multiplex; the rest of the regulatory proteins are already in place here. In *b* the receptor has plugged in; one of its domains (zinc-finger) has engaged a specific enhancer region of the DNA, while another domain (aminoterminal) has engaged regulatory proteins R_2 and R_4. This completes the integrated circuit board of the multiplex which now can process DNA topological information and regulatory information, including hormonal information; the dashed lines and arrows indicate the topological and regulatory-protein information flows.

circuit board (Fig. *10.4*). When this protein joins the regulatory fray, the multiplex relays yet one more quantum of information to the transcription machine, a message from a distant cell. And this is the quantum that kicks off the machine.

The steroid demon functions as a switch in the circuit board here, a switch of remote control. This switch has a pivotal action in gene control and, because the demon is so widely present in tissues, that action is felt in all walks of organismic life. Nothing could show this more dramatically than a clinical example, a disease in which the cognitive function of the demon for testosterone, the male sex hormone, is altered by a mutation. The afflicted individuals develop the secondary sexual characteristics of females—breasts, female external genitalia, high voice, no beard, maternal instincts, and so on—although genetically they are male (and typically even have testes, though in the wrong places). This condition is known in medicine as the "feminization syndrome." It is a disease of the genetic switchboard, and no amount of testosterone will cure that.

The Rise of a Demon

How did such complex Maxwellian entities—proteins with three cognate sites—come into being? It may seem a bit cheeky to ask this a billion years after their inception, but there is a glimmer of clues in today's DNA which give us a betting chance. To begin with, we need to consider here the conjunction of two events: the development of the demon and the development of the hormone. How did this conjunction come about? It is hard to imagine that these events would have been in step; the probability is abysmally small. But separate developments, with a crossing of evolutionary paths, would seem to be within the realm of possibilities.

Let us ponder some mechanisms of evolutionary convergence. We will lean here on an idea that the biochemist Gordon Tomkins advanced a while ago. He suggested that the secondary signals in hormonal communication, before they served an intercellular function, may have been products of the metabolic cellular household and that their receptors may have derived from related household enzymes. Tomkins was guided by the chemistry of cyclic AMP, which serves as a signal of retransmission for a class of hormone messages inside cells, but his insight may well also apply to primary intercellular messages, like steroids.

The parties to the synthesis of steroids—cholesterol, ATP, molecular oxygen, and the enzyme cytochrome P-450 (and its kin, the enzymes in fatty-acid metabolism and cellular respiration)—are ancient players in cells. It is conceivable, therefore, that steroid-like molecules and protein domains capable of recognizing them had been in cells before the time of multicellular organisms. The information for the hormones and the partner demons would have been standing by, as it were. Thus, at the organismic dawn, the information in the cholesterol household products could have been used almost "as is," though the information in the enzyme genes would have needed some further tinkering to get their steroid-cognate stretches close to DNA stretches that encode gene-regulatory proteins (multiplex components from single-cell times). And here is where today's DNA script gives us something further to chew on: the DNA sequences for the demons of the various steroid hormones are similar and, what is more, the sequences for the demons of thyroxine and retinoic acid—hormones which are neither structurally nor biosynthetically related to the steroids—are rather similar.

So, the thought comes naturally that these demons originated from a shuffling of exons, which brought the information for the various cognate domains together. The transposition of an old enzyme exon next to the gene for an old zinc-finger protein, thus, might have given birth to a steroid demon which then took the place that a zinc-finger protein of the multiplex (a forerunner without a steroid domain) held before. And a few such encores, in which the genes for other old zinc-finger proteins were part of the blend, might have brought a whole set of multiplexes under the hormone's control. Or, even more economically, such a far-flung (but discriminating) control might have been achieved through duplication of DNA enhancer pieces with structural motifs where zinc fingers can plug into, and the subsequent transposition of these pieces to the vicinity of other genes. Such an evolutionary strategy would demand no great precision of the DNA translocations because the flexibility of the double-stranded DNA, its ability to make loops, gives the enhancer and promoter proteins of the multiplex box considerable leeway for alignment.

In keeping with the principle of information economy, the same old informational pieces which once proved useful—be that in the form of an enzyme domain, a gene-regulatory protein, or a DNA enhancer region—would be used here over and over again in different combinations, giving rise to novel compositions and functions in the three-dimensional world. Natural selection at the level of the three-dimensional products would see to the rest: harmful combinations would be weeded out and favorable or neutral ones would be preserved, either to be used outright in cellular functions or to serve as pieces for future compositions.

The narrative of this little excursion into old history pivots merrily on "mays," "ifs," and "woulds." I beg indulgence of the reader for that, though I am consoled by the thought that every scientist would have been in the same fix. A writer of history may take an occasional license in his reconstructions of ancient events, but a scientist must warn his listener when he is on hazardous ground. Subjunctives are the customary signals here, and a string of them is the equivalent of a red flag. As for the hazards, I make no excuses. Taking risks goes with the turf of old history, and the history of Circus connection is too important to be left out of a book on biological information.

The Intercellular Functions of the DNA Multiplex

Let us now change lenses to take in the multiplex apparatus as a whole. What does this apparatus mean for the cell and what does it mean for the connected system of cells? First and foremost, it is an instrument for gene control. This is its intracellular function, a function that dates back to single-cell times. Second, it is an instrument for cell subordination. That function, an intercellular one, developed later when the remote-control switch made its debut on the multiplex circuit board, offering certain cells the wherewithal to gain mastery over the genes of others. Both functions are gene-specific, because each gene or set of genes has its particular type of multiplex, a type made of a unique combination of proteins.

Now, as we mount the phylogenetic ladder, the number of genes gets larger and larger, and quite a few of them may have the same type of multiplex. Thus, in cells which don't manufacture the same multiplex proteins—as is often the case with cells of different type in higher organisms—different sets of genes can be brought under the control of a distant cell. This adds combinatorial possibilities for remote gene control (on top of the ones intrinsic in the mosaic of the multiplex) in higher organisms; in our own, for example, cortisone, a hormone produced by the cells of the adrenal cortex, causes the production of one set of proteins in a liver cell, another set in a skin cell, and yet another one in a muscle cell.

So, we may assign two basic intercellular functions to the multiplex apparatus: cell subordination and cell differentiation.

Demons with an Esprit de Corps

These functions, including the primordial one of selective gene activation, result from interactions between the protein constituents of the multiplex, cooperative interactions. Molecular cooperativity is a familiar phenomenon to biochemists; the subunits of many cellular household enzymes cooperate with one another, and such synergistic interactions are easily spotted by their higher-order kinetics. But it's one thing to deal with an interaction between two or three subunits and another to do so with a whole host of them. When a fivesome or, as often as not in the multiplex, a dozen full-fledged protein demons get together, they can do the most amazing things.

Indeed, the demon teams in the multiplexes have their sights set on nothing less than the making of higher-order cellular constructs: organs and tissues. This, in the first place, means making the cells in an organism different from one another, that is, having them produce different molecules. And that is not an easy thing to pull off with a system bound by a covenant of conservation. It took Evolution a good while to fashion the first organismic cell differentiations of some stability. She eventually succeeded by using a strategy where parts of the cellular information reservoirs were shut off to the three-dimensional world and the outflow was brought under intercellular control. That fateful strategy got launched at the beginning of the multicellular organismic age. It toed the line of the old covenant; her creatures continued to accumulate information in their DNAs, just as the ones did in the bygone age, but the flow from the DNA, except for the information concerned with the day-by-day cellular household, now was controlled by intercellular signals.

The successes of that strategy are in evidence everywhere in the living world today. Those who rose in that world have their cellular information reservoirs so reapportioned that parts are rather permanently off limits to the transcription machinery. Those forbidden parts can be quite large. Take our own cells, for instance. At most, 10 percent of their translatable genome ever gets translated in the adult organism. Our organism starts with one cell, a fertilized egg which contains all the genes. That cell divides and divides and divides, giving rise to a variety of embryonic and, eventually, adult types—cells of skin, hair, tooth, liver, muscle, nerve, heart—which produce different molecules and serve different functions. Thousands of genes are held incommunicado in the course of this development, while only a relatively small set gets transcribed—the set that is specific for each cell type.

It is in specific gene activations of this sort, activations engendering cell differentiations, where the remote-controlled multiplex apparatus plays its pivotal role. There probably exist other intercellular mechanisms for cell differentiation too—the intercellular network is vast and we presently grasp but a fraction of it—but this fraction, small as it is, gives us an in. It allows us to follow the information flux throughout a whole hormone communication channel—from information emitter to user (from the machinery sending out the steroid message to the machinery implementing gene transcription)—giving us the first insights into that old mystery of biologists, cell differentiation.

A lucky circumstance here helped to let the daylight in: a natural magnification of gene transcription that occurs in some insect embryos, the larvae of the fruit fly (*Drosophila*) and the midge (*Chironomus*). In these larvae nature proffers a gene showcase, a rare display where one can see the transcription machinery springing into action. The salivary-gland cells in these embryonic organisms are giants—some get to be a tenth of a millimeter across. And their chromosomes are giants—each is made of about a thousand identical strands of DNA (chromatin) neatly lined up side by side, magnifying the transcriptional events at a gene a thousandfold. These chromosomes get visibly puffed up at regions where genes are transcribed. So, when the steroid ecdysterone is delivered to a salivary-gland cell, within a few minutes, six chromosome regions can be seen transcribing new RNA and, after a few hours, about another hundred regions follow suit, while now some of the early-response regions shut off.

A progression of gene activities, thus, is here up to view—two waves, as it were: an early wave which is attributable directly to the hormone, and a later wave attributable to the gene products of the first wave, amplifying the hormone message. A simple picture of serial gene activation, then, becomes apparent, *a scheme where the gene products of the first wave of gene activation are molecular integrants of the multiplexes controlling the second wave*. So we get, with minimal information outlay and largely by dint of a specific multiplex DNA topography, a cascade of gene activities that are ordered in space and time: first, the telecontrolled multiplexes switch their respective gene loci on, and then another multiplex set, in turn, switch theirs on as the gene products from the first wave reach the latter multiplex set by diffusion and are incorporated there (Fig. *10.5*).

The linchpins in this serial gene hook-up (and ensuing cell differentiation) are the multiplexes. They specify the spatial order, the connectivity between gene loci; and the whole kit and caboodle comes at a bargain basement price. Evolution saved information in the development of the multiplex protein components by the usual shuffling and recombination of old one-dimensional information pieces. But here she gives an encore in the somatic realm: the three-dimensional transforms of the recombined information pieces, the protein products, are

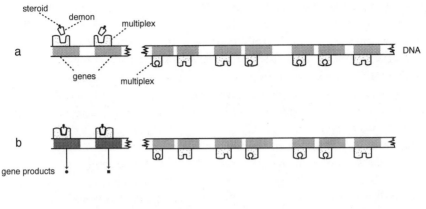

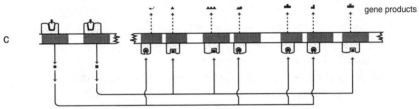

Figure 10.5 The multiplex principle of differential gene amplification. A schematic representation of a stretch of DNA undergoing serial steroid-induced gene activation. (*a*) The steroid-carrying demons are on their way to the DNA—the multiplex system is in stand-by. (*b*) The demons have arrived and plugged into two multiplexes—transcription starts at the corresponding genes. (*c*) The protein products of these two genes (●, ■), after diffusing from their cytoplasmic assembly lines to the nucleus, have plugged into fitting multiplexes and their genes now get transcribed too. Activated genes are darkly shaded; the lines and arrows indicate the information flow.

recombined, in turn, into various pastiches and used again and again at many gene loci—a three-dimensional epiphany of the economy principle she is pledged to in the one-dimensional realm.

A Smoke Signal

This rounds out the telecontrolled multiplex model and concludes our account of the steroid channel. But before we drop the curtain on hormones that breach the cell membrane directly, we will call notice to a variant where the intercellular messenger is a simple inorganic gas: *nitric oxide (NO)*. This came to light only a few years ago and took everyone by surprise. Not only is the molecule so tiny and ordinary, but it is so reactive. The molecule was better known as a villain—a pollutant of the atmosphere. Nitric oxide is produced in car engines from oxygen and nitrogen; and once it gets out into the air, it reacts with oxygen to produce a brownish gas, nitrogen oxide, aka smog.

The direct union of the two atoms of nitric oxide requires the high temperatures of a combustion chamber, but cells manage to do this in the cold with the aid of an enzyme, the nitric-oxide synthase. Quite a few cell types in our body can turn this trick—cells in the blood vessels, brain, liver, and pancreas. The enzyme joins oxygen with the nitrogen from arginine, an amino acid which abounds in our cells, and the resulting oxide is given off as a messenger. In blood vessels the messenger issues from the cells that form the inner lining of the vessels, the *endothelial cells,* to the overlaying muscle cells, telling them to relax. This is why our blood vessels open up when it is hot or when we sleep or meditate; in the absence of the messenger, the sleeve of contracted muscle cells keeps the vessels constricted. This little messenger is also responsible for the benefits of that old and trusted friend of people with heart disease: nitroglycerine. This chemical causes nitric oxide to be released in the blood vessels throughout the body, including those supplying the heart; so, a nitroglycerine tablet put under the tongue relaxes the heart blood vessels, quickly relieving the chest pains produced by their constriction.

The explorations of the physiology of this messenger have only recently begun, but important functions are already mounting. Apart from the basic actions in blood vessels mentioned above, nitric oxide has been found to play a critical role in penis and clitoris erection (by allowing their blood vessels to fill), stomach and intestinal sphincter relaxation, and in immune responses. Moreover, as an intercellular messenger in the brain (hippocampus and cerebellum), it seems to have a hand in memory and learning. Healthwise, there is so much at stake for us regarding this little messenger that one is easily tempted to go off on a tangent. However, we will not wander off, but stick to the information transfer here. This transfer has an interesting lesson to offer regarding the informational benefits of molecular magnetic fields.

First, let us ask a general question: Why did Evolution choose nitric oxide as a messenger? Of the millions of molecules on Earth, why did she pick this one, one of the ten smallest? A perusal of the physical properties of the molecule may help us here. The molecule is made of just one atom of nitrogen and one atom of oxygen, and it is electrically neutral. The small size and neutrality are good qualities for a molecule that has to penetrate the cell membrane. Indeed, nitric oxide readily worms its way in between the phospholipid molecules of the bilayer, going through the membrane even more rapidly than the steroids do.

But apart from that, the molecule has a more unusual quality: it has an electron with an unpaired spin. There are altogether 15 electrons in its orbits—the nitrogen atom brings 7 electrons to the union and the oxygen atom, 8—an odd number that leaves 1 electron unpaired. Because of that, the molecule tends to be drawn into magnetic fields. This property—*paramagnetism*—is rare enough and offers us a hint about the targets of the nitric-oxide message: they may be paramagnetic molecules themselves which the nitric oxide, by virtue of the

Heme

Nitric Oxide

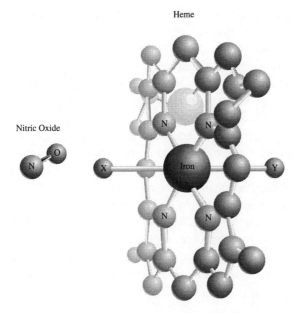

Figure 10.6 The smallest intercellular messenger and its target. The nitric oxide and heme are shown side by side. The heme skeleton is seen in lateral view, displaying its planar 24-atom porphyrin ring with iron at the center. Four of the six ligand positions of the iron atom are occupied by the nitrogens (N) from the porphyrin ring and the other two, by atoms (*X* and *Y*) from the protein guanylate cyclase, wrapped around the heme (the protein is not shown; other members of this protein family—like the cytochrome C, porphyrin, hemoglobin, and the photon-trapping chlorophyl—use different *X*s and *Y*s). The eleven double-bond electrons in the porphyrin ring are delocalized over the entire ring and shared with the iron and, depending on the energy splitting, one or two electrons are unpaired. Nitric oxide (*left*) has one unpaired electron.

unpaired electron spin, can react with. There are not too many molecules in cells that fit that bill. All are proteins that bear a metal atom—some of the best known ones have an iron atom in a porphyrin ring, a flat 24-atom ring called *heme* (Fig. *10.6*).

Indeed, one nitric-oxide target of that sort recently has been identified: *guanylate cyclase*, an enzyme that carries a heme. This protein occupies a strategic position in the blood vessels and brain; it catalyzes the synthesis of *cyclic GMP* (guanosine monophosphate), the signal that ultimately makes smooth-muscle cells relax and brain cells synthesize their glutamate signals. Nitric oxide turns the guanylate enzyme on, which suggests at once how the nitric-oxide molecule might transfer its message: by an interplay with the iron-containing porphyrin ring—a paramagnetic interaction flipping the allosteric structure of the enzyme to the "on" position.

Such an elementary molecular relay would save information. Indeed, every-thing about this molecule reeks of economy. Just consider how it forms. It only takes two gases, two of the most abundant ones in the atmosphere, plus some heat—which could hardly have been a problem in the stygian environment of geysers and foamy oceans on the primeval earth. Thus, the molecule may have been a messenger long before the enzyme was on hand—a primal messenger, a relic from the very distant past.

The economical virtues of nitric oxide also show up on the destruction side of its message. The nitric-oxide information, one way or the other, must be destroyed or else the blood vessels would stay relaxed forever or the signal pro-duction in the pertinent brain target cells would go on and on. With more stable hormones, like the steroids, and even more so the other polypeptide hormones we shall come to later on, the annihilation of the information is implemented by special enzymes. This costs a pretty penny, but the cells using nitric oxide save all that. They let oxygen, a molecule as ordinary as the message molecule, do the killing job, and this one does it for a price Evolution could not refuse.[2]

Cyclic AMP

We turn now to a communication modality where the intercellular messenger doesn't enter the cell, but only knocks at the door, as it were. This is the second modality on our list of evolutionary membrane-breaching strategies (Fig. *9.4 b*). The messenger, in this case, goes no farther than the outside of the target-cell membrane; there the message is received and carried across the lipid bilayer, to be subsequently reissued in larger editions for distribution in the cell interior. The first three functions—reception, portage, and retransmission—are the purview of three protein demons inhabiting the eukaryotic cell membrane. The fourth function, the intracellular distribution of reissued information, is implemented by a couple of lightweights: *cyclic AMP* and *inositol phospholipid*. I will be dwelling mainly on the former. That was the first hormonal retransmitter signal found— Earl Sutherland discovered the cyclic AMP in the 1950s—and three and a half decades of chemical and genetic dissection of such hormone communication channels have given us a reasonably clear picture of how the information flows.

Cyclic AMP—which stands for adenosine monophosphate —is found in vir-tually all organisms. It stems from the equally ubiquitous ATP, the adenosine triphosphate. It is the product of an enzymatic reaction where ATP loses two of its phosphates. The remaining phosphate forms a ring with the carbon atoms to which it is linked (hence, the name "cyclic AMP"). It weighs barely 300 daltons,

---

2.   The very property (paramagnetism) that allows this gas to transmit its information also allows that information to be rapidly annihilated; the molecule reacts with molecular oxygen (O_2), a paramagnetic substance which has two unpaired electrons, and forms nitrite (NO_2) or nitrate (NO_3)—quieter ions than the nitric oxide (NO).

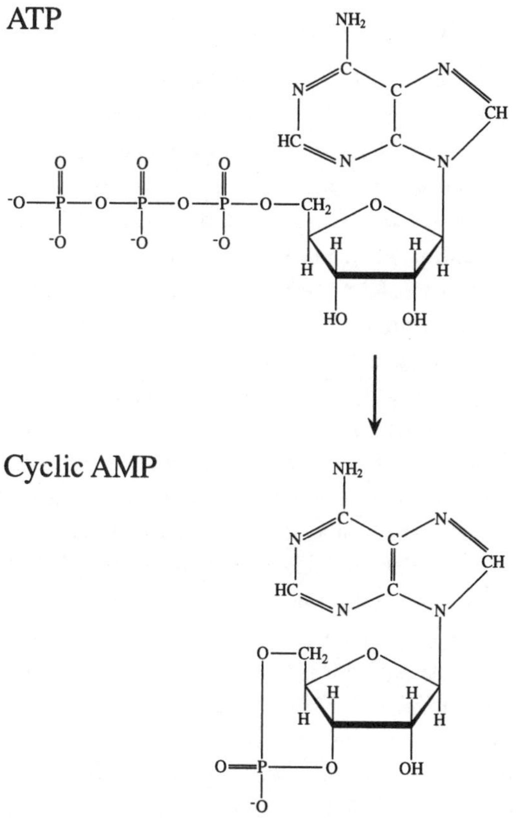

ATP

Cyclic AMP

Figure 10.7 Cyclic AMP and its progenitor ATP. Cyclic AMP forms from ATP by the catalytic action of adenylate cyclase, an enzyme in the cell membrane. The ATP loses two phosphates (pyrophosphate); the remaining phosphate links up with a carbon atom in a cyclic configuration. The idiosyncratic functional groups of cyclic AMP's sugar base are instrumental in second-theorem encoding.

but the functional groups on its sugar base (Fig. *10.7*) make it stand out from the ATP and the rest of the nucleotide crowd, and that's enough to be recognized by a competent protein demon and to satisfy the Second Theorem.

Now, add to that a few more virtues, and you see why Evolution would entrust this lightweight with such weighty business: it is rather stable in the cell water (even heat-stable) and its diffusion in this medium is fast—it literally zips through the cytoplasm, covering physiologically meaningful distances inside cells in seconds or fractions of seconds. And if that wasn't warrant enough, here is something that would make any scrooge's mouth water: the molecule can be bought for a song—its basic structure falls out of ATP, with only the expense for ring formation (Fig. *10.7*).

From Hunger Signal to Inter-Circus Signal

Evolution took this cyclic nucleotide under her wings actually long before start-ing her multicellular enterprise. It served as a hunger signal in single-cell times. It still can be found in this role in many single-celled beings today. Bacteria, for example, use it to report the exhaustion of glucose to their genes; when this food source gets depleted in their medium, the cyclic AMP concentration inside the cells goes up, causing the genes for other sugar-cleaving enzymes (those for lactose, maltose, or galactose) to be switched on. This permits bacteria to adapt themselves to their environment and live on alternate food sources — an enviable ability of fast genetic adaptation. This is what multicellular beings had to give up when they acquired social graces and climbed the ladder of success.

Eventually, a social function was tacked on to that elementary one of survival. Here, too, we need not go far for living examples. The mildews and molds that dwell in our closets and the amoebas that thrive in damp and dark places present archetypes where cyclic AMP still plays this early social role. The slime mold *Dic-tyostelium*, for instance, uses the cyclic nucleotide as a bell to toll for cell congre-gation. This creature has two modus vivendi: in times of food abundance it slithers about as an individual amoeba, and in times of want it gives up its inde-pendence and lives in communities. And cyclic AMP is the communal signal, the signal for want; the hungry cells release the cyclic nucleotide to the medium as an attractant, causing thousands of them to flock together and form an organism.

This foreshadows the inter-Circus role of cyclic AMP at the higher phyloge-netic rungs where it brings cellular genomes under intercellular control. There, the communication job is cut in two: cyclic AMP is relegated to the intracellular stretch of the communication channel, while the intercellular stretch is allotted to a hormone. Thus, in higher hormonal communications we distinguish three channel sectors: (a) a hormonal sector, the segment between the cells, (b) an intracellular sector, the segment served by cyclic AMP, and (c) a membrane sec-tor that bridges the two.

The Information Relays in the Cell Membrane

We take aim now on the membrane sector. This boasts a remarkable trio of demons who, by dint of their cognitive and metamorphic talents, relay the hor-mone message to the cell interior. The trio consists of a receptor of the hormone message, a retransmitter of the message, and a runner between the two (Fig. *10.8*).

The receptor stretches all the way across the cell membrane. Its polypeptide chain makes seven passes through the membrane, forming a pocket for the hor-mone on the outside — the demon's cell-external cognitive domain. The runner, called *G protein* (because a *g*uanine nucleotide is an integral part of it), is imbed-ded in the inner (cytoplasmic) leaflet of the lipid bilayer of the membrane and

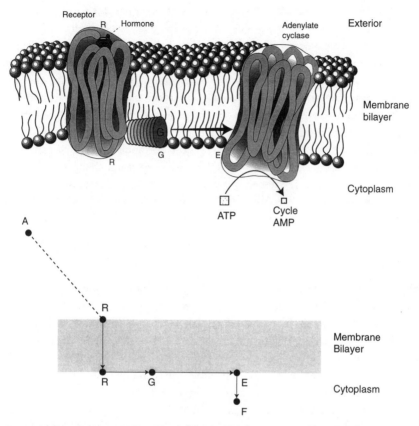

Figure 10.8 The G-protein information relay. *Top:* The receptor (*R*, with its seven transmembrane segments) and the adenylate cyclase are cartooned during transfer of the hormonal message; the hormone has occupied the receptor pocket and the G protein (*G*) is shuttling to the cyclase (*E*), which makes cyclic AMP from ATP. *Bottom*: Diagram of the information flow (*A*, the information source, the endocrine gland secreting the hormone).

moves sideways in it. It shuttles between the receptor and the retransmitter, a traveling distance of the order of 10^{-6} meters or less. The retransmitter here is the enzyme *adenylate cyclase*, which makes cyclic AMP from ATP. This ancient, membrane protein has two cognitive domains: one for the G protein and the other—the catalytic domain—for the ATP.

With this thumbnail sketch of molecular hardware, we can go on to the relay action. But lest we leap before we look, consider that these hardware pieces are internally mobile and that, as in other allosteric instances we have seen, this mobility is triggered by the electromagnetic interactions at the cognitive domains. So, we may envision a relay picture where the trapping of a hormone in the receptor pocket sets in motion the following sequence of events: (1) the receptor undergoes a conformational change that propagates to its G-protein domain on the inner membrane side, causing it to lose its affinity for the G protein; (2) that protein separates from the receptor and changes conformation; (3) now free to move inside the membrane, the G protein travels to the adenylate cyclase and (4) attaches to it; finally (5) the adenylate cyclase, in turn, changes conformation, and (6) its catalytic domain is switched on, generating cyclic AMP.

The overall result of this multistage relay action is a flux of information from the outside to the inside of the membrane, the cytoplasmic side (Fig. *10.8, bottom*). What gets across the membrane here is the hormone message, not the hormone molecule itself, and the relayed information quantum changes as often as the carriers do. And this, stripped of the many fascinating mechanistic details, is the fundamental difference between these hormonal communication channels and the ones operating with nitric oxide or steroid messages.

The Alpha Demon

The foregoing relay scheme applies to many types of hormones—adrenaline, noradrenaline, antidiuretic hormone, luteinizing hormone, parathyroid hormone, and so on. In all these cases, the hormone receptors exhibit similarities of design; they all have seven transmembrane segments and display some amino-acid sequence homology. Also the G proteins show similarities. These demons we will take now under the loupe; they hold the secret of how the various hormonal messages get to the right addresses in the membrane.

But before we launch into this network, let's take a bird's eye view and see how things hang together in such a communication channel. Take, for instance, the channel between our adrenal gland and our liver or our skin fat—a busy line whenever we exercise or are under some stress. Follow the information trail from the source to the user (Fig. *10.8, bottom*): the information leaves the adrenal cell (*A*) in the form of adrenaline, reaches the liver or fat cell via the blood and extracellular medium, enters the cell membrane through the adrenaline-receptor protein (*R-R*), continues laterally along the membrane in the form of G protein (*G*), exits the membrane on the cytoplasmic side through the adenylate cyclase (*E*), and is finally distributed in the form of cyclic AMP (*F*) through the cell's interior; there, the cyclic AMP signal causes glucose and fatty acids to be set free, supplying our body with needed fuels.

The wheels in this information shuttle are the G proteins. Each G protein is made of three subunits—α, β, and γ. The first, the largest subunit, pays watching closely, for it is the one that holds the key to the second-theorem encoding in the membrane sector. That sector is rather crowded. The whole membrane space is only of the order of 10^{-20} cubic meters in a typical human cell and is crisscrossed by various G-protein types and other information carriers. There is, thus, quite a bit of potential informational interference here. But the G proteins are well-prepped for that. Their largest subunit is a true Alpha among demons: he boasts three cognitive sites and then some, and an automatic timer, to boot!

The Alpha Merry-Go-Round

Alpha belongs to an old family that goes back to the Cell Age. The members of that family are everywhere today. They are ensconced in the membranes on the cell surfaces and in the membranes inside the cells. On the cell surfaces, they relay hormonal information, as we already said. But some of them still serve older functions, such as the most basic biological functions of all: the capture of photons. These demons are able to draw from the photons in the environment the information necessary to engender the all-important ATP. One of their back numbers—bacteriorhodopsin—can still be found in the membrane of bacteria; the capturing of photons by these demons drives the weels of the machinery that makes the ATP. A younger cousin—rhodopsin—resides in the membranes of the sensory cells of our eyes, and it is his Maxwellian power that makes us see. Other members of the family populate the internal membrane of cells, where they relay messages to the basic machineries of protein synthesis and translocation.

I could go on, but these examples are enough to illustrate the important position this demon family holds in the living world. It owes its prominence to an information quantum in the DNA, a quantum which runs through the entire phylogenetic tree. This information translates to a valuable thing in the three-dimensional world: a protein cavity that can catalyze the splitting of GTP (guanine triphosphate). This triphosphate is an energy source, like ATP, and when one of its phosphate tetrahedra is split off, it sets free a sizeable package of energy from which a demon can draw information.

The cavity structure, as resolved by X-ray diffraction in two members of the family, displays two conformations—one fits a molecule of GTP and the other, a GDP (guanine diphosphate, the product of the splitting of GTP). Each of these conformations pertains to a distinct overall Alpha configuration: one dovetails with the receptor, and the other with the cyclase. Thus, the demon, in alternate interactions, can play with both. We may envision him to go through three states as he goes through a catalytic cycle and shuttles between the two playmates: (*1*) GDP-bound, (*2*) empty, and (*3*) GTP-bound (Fig. *10.9*). In state *1*, he takes on the form that binds the hormone receptor. In state *2*, he shifts allosteric gears and his cavity is vacant (this state lasts only for the wink of an eye because GTP is so abundant in cells and will instantly occupy the cavity). In state *3*, he shifts gears again to assume the form—the hormone message-carrying form—that binds the adenylate cyclase. And so over and over it goes—a self-contained merry-go-round which, for a good part of a turn, has the hormone message as a rider.

But why the union with GTP? Questions like these willy-nilly come up, though they are not easily stomached so bare-boned because of their teleological

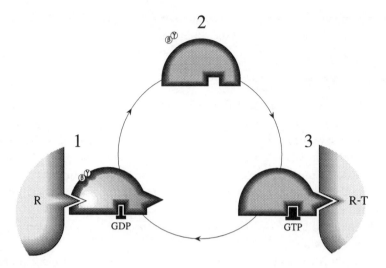

Figure 10.9   The α cycle. Spurred through interactions with the hormone receptor, GTP, and GDP, the α demon cycles through three conformational states. The cartoons show the demon at the start of each state: *1*, binding GDP and interacting with the hormone receptor (R); *2*, empty, after that interaction, *3*, binding GTP and interacting with the retransmitter, the adenylate cyclase (R-T). The direction of the cycle is determined by the irreversibility of the GTP → GDP breakdown. State *2* lasts only a fraction of a second; and state *3*, about 10 seconds in a typical adrenalin communication channel.

flavor. But now and then one has a run of luck and can reformulate them in information terms. And with Maxwell demons we know what the right question is: What resets their memory—or, specifically here, what recocks the Alpha cycle?

So, let's cut to the chase and keep our eyes on the entropies. The essence then shows through all the legerdemain: the demon rescues a little of the energy that gets released into his cavity, energy which, without him, would have been a flash in the pan; the demon draws, in effect, negative entropy from the GTP, as he breaks the high-energy bond of the GTP's third phosphate. And this is why he joins up with this molecule—the why and wherefore and, in a sense, the how of the union. One might at first think that the GTP's energy is the reason. That phosphate, it is true, carries a good punch. However, it's not energy that the demon hankers after, but negative entropy—or, written in capitals: INFORMATION.[3]

3.   This is just as true for the demons for ATP, CTP, UTP, or other nucleotides that serve as phosphoryl donors. The name "high-energy" or "energy-rich" phosphate, commonly used in biology, may be somewhat misleading. There is no *special* kind of bond that releases energy here; as with any chemical bond, energy actually must be expended to break it. What distinguishes these phosphate compounds is their propensity to transfer phosphoryl groups to a substrate or, to use the term coined by Fritz Lipmann, their high "group-transfer potential." Lipmann was the first to see the general role of these compounds in cellular metabolism, and

But we are not quite through yet with this demon. There is something else to his act, concerning the coveted information: he controls the timing of the information delivery. His catalytic cavity has a built-in timer for the release of the phosphate energy; that timer starts ticking the moment the GTP gets into the cavity (which effectively coincides with the demon departing from the receptor) and goes off after about 10 seconds—in good time for the (random-walking) demon to get to and interact with his other playmate, and for some foreplay.

It must have taken aeons to perfect this cabal. The honing of the cavity to get the time just right for the splitting of the high-energy bond, and the concomitant release of information, probably took a good part of the Age of the Cell, not to speak of the fitting of the GTP and GDP. But the end result could hardly be topped: a molecular demon who achieves control over his own moving parts and sentient activity.

The G-Show Libretto

But even an Alpha demon has his limits. Though uncannily self-sufficient, he can't go the whole hog alone. No one can under the law. Apart from the GTP's negative entropy, the extrinsic information he provisions himself with in-house, so to speak, he needs to interact with the receptor. This playmate gives him the electromagnetic push that sets him in motion—the very field effect that implements the transmission of the hormonal message.

I have said nothing yet about β and γ. Alpha, no doubt, is the kingpin here, but the other members of the trio are not precisely chickenfeed. We know much less about them, though this much is clear: they bind to α in state *1*; they seem to stabilize his binding to the (hormone-carrying) receptor, ensuring an efficient information transfer.

Now, putting this together with α's cycle, one may envision the following steps in the transmission of the hormonal message across and along the cell membrane (Fig. *10.10*):

the hormone molecule docks at the receptor // the α demon (in state *1*) attaches to the receptor, assisted by β and γ // the receptor transfers the hormone message to α // α switches to state *2* where he releases β, γ, and GDP, and separates from the receptor // α now diffuses in the plane of the membrane // GTP occupies the α cavity // α attaches to adenylate cyclase // α transfers the hormone message to the cyclase (which makes cyclic AMP) //

the group-transfer potential (a thermodynamic parameter) refers to the relative position on the phosphoryl-acceptor scale of metabolites, where water (a forbidden acceptor) is ground zero. Their high position in this scale makes the triphosphate compounds *usable* sources of information; it is at the gradings above zero of the scale, where the competent demon can rescue some of the energy which, if water were the phosphoryl acceptor, would merely dissipate as heat.

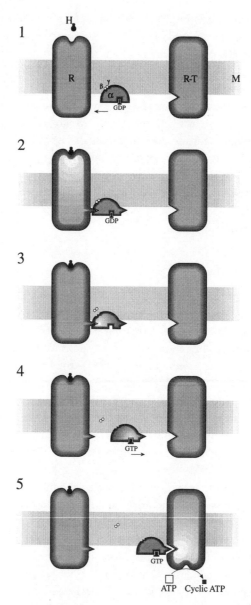

Figure 10.10 The moving parts of the G-relay system in a cycle of transmission of hormone information. *(1)* α, in GDP-bound form, moves, in association with β and γ, toward the receptor *(R)* in the membrane *(M)*. *(2)* As the hormone *(H)* binds to the receptor pocket, the receptor changes conformation; α docks at the receptor. *(3)* As it interacts with the receptor, α switches form; GDP, β, and γ leave α. *(4)* GTP occupies α's cavity. *(5)* α, in GTP-bound form, docks at the retransmitter, the adenylate cyclase *(R-T)*; the enzyme changes to its cyclic AMP-producing form. The cycle starts anew, as GTP is broken down to GDP and α leaves the adenylate cyclase in GDP-bound form.

GTP is broken down to GDP // α switches back to state *1*, separates from the cyclase, and reassociates with β and γ.

And this dance goes on and on, for as long as there is a hormone on the floor.

Crescendos

The foregoing molecular interactions account for the informational continuity in the membrane sector of this hormonal communication channel. But what of the amplification, the obligatory leverage to get the message through a noisy channel? Though the participants are many and the changing of partners in the dance is frequent and complex, there is no reason to expect anything but the usual boosting by carbon copying of the message, the age-old replica method—Evolution is not one who'd give up tradition. So, let's see how things stack up overall in this channel segment; all we need is a nose count:

1 hormone, 1 receptor, 10 Alphas, 100 cyclic AMPs

One receptor molecule passes on the message from a hormone molecule to about 10 Alphas, and one adenylate cyclase cranks out more than 100 cyclic AMP copies during the message-carrying time of an Alpha—overall a step-up on the order of 1,000, enough of a crescendo to be heard on the other side of that membrane wall.

But there is not much waste either—there never is on Evolution's stamping ground. Not only would this be unseemly in her circles, but an overamplification can be deadly, as we shall see. What keeps things in kilter is mainly Alpha's timer, and it does so with little operating expense. This timepiece sets a limit to the transfer of information in the last membrane stage and, hence, to signal amplification. So, it is the demon who puts the soft pedal on the crescendos—and, when all is said and done, that's what his timer is for.

Perhaps nothing shows the importance of this device more clearly than the disease cholera. The cholera bacterium produces a toxin that tampers with Alpha's timer. The bacterium lives in polluted waters, like the river Ganges, where the disease is endemic, and gets into the human intestine with the drinking water. Its toxin attacks the Alphas in our intestinal cells and puts their timers out of commission; it effectively clips the demons in state *3*.

Now, with the Alphas stuck in that state, all hell breaks loose: the adenylate cyclases run amok, the cells are flooded with cyclic AMP signals and pump large amounts of sodium and chloride ions into the lumen of the intestine; water follows suit—a flood tide running at about 1 liter per hour into the lumen—causing a nonstop diarrhea where, in a matter of hours, several liters of body fluid may be lost.

An experience like that will convince anyone of the value of good timing in intercellular information transfer. But to the doubting Thomases who can't take their libations from the Ganges, biochemists offer an equally persuasive treat: a synthetic analog of GTP with a modified phosphate bond (a P-NH-P bond) which Alpha can't cleave. The demon swallows the bait, gets trapped in state *3*, and, as in cholera, causes the adenylate cyclase to run nonstop and the affected cell to burn the candle at both ends.

The Cast of Characters

At this point some broadening of our perspective is in order, lest we lose sight of the membrane network as a whole. The adenylate cyclase is but one of several retransmitters in the G-network—there are many branches carrying hormonal information into cells. However, the same relay principle (Fig. *10.8, bottom*) is used all over the G-map—no need to nose, therefore, into every branch. But let's take a role call of the demons, to get an idea of the network's expanse: *Hormone receptors*—over a dozen types (each specific for a hormone) and over a

Table 2   Receptors, Retransmitters, and Signals of the G-Network

Receptors for

Adrenaline
Noradrenaline
Glucagon
Antidiuretic hormone
Luteinizing hormone
Folicle-stimulating hormone
Thyroid-stimulating hormone
Parathyroid hormone
Acetylcholine
Endorphins
Angiotensin
Prostaglandins
Histamins
Photons
Odorants

Retransmitters and Signals

Adenylate cyclase	:	Cyclic AMP*
GMP phosphodiesterase	:	Cyclic GMP (photoreceptors)*
Phospholipase C	:	Inositol triphosphate**
		Diacylglycerol**
Phospholipase A$_2$	:	Arachidonates**
Calcium channels	:	Calcium ions/Electric signals
Potassium channels	:	Electric signals
Sodium channels	:	Electric signals

* Nucleotides; ** Membrane-lipid derivatives

hundred subtypes; *G proteins*—some twenty types; *retransmitters*—seven types. The Who's Who is in Table 2, though there are new entrants every year in this fast-moving field.

A word of comment on the retransmitters and their signals: The *phospholipase C* and *phospholipase A_2* are as ubiquitous in higher organisms as the adenylate cyclase is, and so are the *inositol triphosphate, diacylglycerol,* and *arachidonates,* their catalytic products (derived from fatty membrane moieties). The *GMP phosphodiesterase,* a retransmitter of lower profile, occurs in visual cells; *cyclic GMP* (derived from GTP) is its catalytic product.

Ion Channels as Maxwell Demons

The rest of the cast (the three last retransmitters in Table 2) have a different modus operandi. They are not enzymes—at least, not in the usual sense. These proteins retransmit hormone messages by moving small electrically charged inorganic molecules—calcium, potassium, or sodium ions—across the membrane; they make use of the strong electrical fields existing at the cell surface.

The key to the transmitter operation of this class of proteins lies in their tubular configuration. They are made of subunits symmetrically deployed about a central axis—four, five or six subunits—forming an aqueous passageway (Fig. *10.11*). That form is ostensibly designed to funnel ions through the cell membrane, but if you peer a little deeper, it acquires a Maxwellian stripe. Indeed, these proteins are capable of selecting their ion passengers, and those with four or five subunits are rather good at that. Their central passageway is just wide enough to let certain inorganic ions through—it has tightly associating α-helical (or β-pleated) stretches. But that passageway is not always open; in fact, most of the time it is not. And here is where the G-system comes in: some of these tubular protein types interact with G proteins, and this interaction switches them to

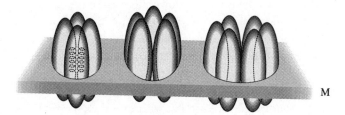

M

Figure 10.11   Ion channels in the cell membrane are made of four, five, or six protein subunits. The narrowest, most ion-selective channels are made of four; the diagram displays the α-helical stretches from neighboring subunits, which are assumed to form the ion-sieving narrow strait (in some channel types the lining of the strait may be β-pleated rather than α-helical). *(M),* the lipid bilayer of the membrane. (After N. Unwin, *Neuron* 3:665–676, 1989, by permission from Cell Press).

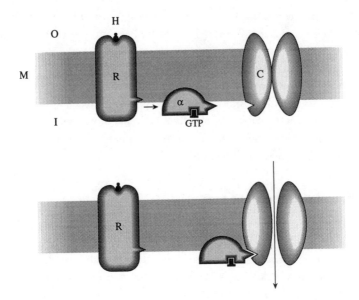

Figure 10.12   An ion channel of the G-network. The diagram illustrates two stages in the retransmission of a hormone message by an ion channel: *top:* a hormone (*H*) has docked at a receptor (*R*), dispatching α who is shown on the way to the channel protein (*C*); *bottom:* α engages that protein, causing it to switch to the open confor-mation (the arrow indicates the direction of the ion flux for the case of a sodium- or a calcium channel. (*M*), cell membrane; (*I*), cell interior; (*O*), cell exterior. (For sim-plicity, the channel protein is shown here with an α-domain. However, the interac-tion with α may not always be so direct, but involve an intermediary step.)

an open conformation (Fig. *10.12*). The G-system, thus, controls the open state of these ion-selective membrane pores and decides what ions will flow through a cell membrane—it opens or closes the gates, so to speak.

The resulting ion flow is mostly downhill because there are steep electro-chemical gradients across the membrane. Calcium ions, for instance, are about one hundred thousand times more abundant outside than inside most cells; moreover, the inside is negatively charged and calcium ions carry positive charge. Thus, when the appropriate tubular proteins open up, massive amounts of cal-cium ions rush in, providing a strong signal for the cell.

This much about the physiology. But what of the cognition? Where, in these tubular entities, is the demon's act and obligatory entropy quid pro quo? We presently don't stand much of a chance to spy on such a demon during the inter-play with the ions—it is hard enough to catch any demon in the act, let alone one who has such pint-sized partners. But a glance at the pertinent electromagnetic fields will clue us in on the general strategy those demons may use for cognition.

To start, consider the passengers—their structure holds no mystery at the molecular level. That structure suggests at once a plausible ion-protein interaction. A long-lived covalent interaction is out; it would impede the transmembrane movement of the ions. This narrows down the possibilities in the case of the sodium or potassium ion. Their puny figure offers little in the way of specific interaction. Calcium has at least a miniature skeleton, but there isn't a ghost of a chance that the other two could be fingered by a protein by means of weak electromagnetic interactions. However, the electromagnetic force can move information from one molecule to another also by repulsive interactions. Such repulsions, we have seen, predominate at distances of less than about 3 angstroms between molecules (Fig. 5.5). And this molecular field effect is precisely what some of these tubular demons exploit to keep certain ions out; they use the charge, the polar amino-acid lining of their passageway, as a screen.

This brings things down to straightforward physicochemical terms: to an electrostatic shielding against the ions—or an electrostatic sieving, if you wish. And so the focus shifts to the dimensions of the passageway and the charged amino-acid residues that guard it. This is something one can lay one's hands on, and it should not be long before a detailed description of the play of forces becomes available for some of these demons.

But no biological story would be complete without an evolutionary perspective. And here the gestalt of these demons gives us some food for thought. That gestalt, and especially its internal polar studding, betokens a design for forcing the ions into contact with the polar amino acids (and I mean here the ions without their water coat). This must have been a long, long evolutionary road to travel. It was not only a matter of arriving at the right concentric subunit configuration and appropriate disposition of amino acids, but, in the context of the whole cell membrane, there loomed the dilemma of how to get both a high-enough ion flux and a high-enough ion selectivity. On the one hand, to get a strong signal, the flux had to be large—and this called for a wide passageway. On the other hand, a wide passageway makes for poor selection—and without selectivity there is no signal (the flux would merely shunt the membrane gradients).

Such damned-if-you-do-and-damned-if-you-don't situations are the current challenges in mathematical biology—they are the very stuff of game theory, as we have seen. A past master of that theory rose to the occasion 2 billion years ago. Precisely how *she* got to the saddle points, we may never know. Her failures—and there must have been many—have been wiped off the record. But her successes are here and offer us ample opportunity to apply the yardstick. They all bear witness to her savvy. I let the measurements of three common types of her four-subunit tubular-protein design speak for themselves: fluxes, of the order of 1,000,000 ions per second; selectivities, in three common subtypes, of the order of 1,000:1 for calcium over potassium or sodium ions, 100:1 for potassium over sodium ions, or 10:1 for sodium over potassium and calcium ions.

And, in this sense, one speaks of *calcium channels, potassium channel,* and so on—and collectively of *membrane channels.*

How to Make Signals from Informational Chicken Feed

Now, what kind of signals do these channel proteins produce? One would hardly expect the potassium and sodium ions to carry information from protein to protein. I know of no protein that could satisfy the second-theorem demands with such tiny customers. Nevertheless, Evolution got round this pickle and, making a virtue of necessity, she turned the channel environment to good account. That environment is largely fatty, and fats make good electric insulators. But this insulator is rather special. The fatty matter here is spread out very thinly on the surface of the cell, just in two molecular layers, and the neatly arrayed molecules constitute a planar crystal which has a respectable electrical capacitance, $1\mu F/cm^2$. Thus, when all the electrochemical forces are in balance at the cell surface, that capacitor gets charged to several hundredths of a volt, and in some cell types, even to about one-tenth of a volt.

This is quite a voltage for such a thin film. Angstrom for angstrom thickness, it amounts to 10,000 to 100,000 volts per centimeter, strong enough to affect the three-dimensional structure of a protein. And this is the force Evolution turned to for signal transmission. Consider, for example, what happens at a G-controlled sodium channel when an appropriate hormone knocks at the membrane door. As the channel opens, millions of ions file through—a massive movement of positive electric charge that will depolarize the crystal in that locale. Such a change of membrane voltage constitutes a loud signal, specially when a bunch of channels snap open in unison. That voltage signal will fade out with distance from the signal source (the open channels), as it would in any electrical transmission line; but many cell types are small enough so that even with some fading, the signal is heard all over the cell.[4] Thus, all that is needed is a voltage sensor. Indeed, there

---

4.   In fact, it often is heard over several cell lengths. Many animal cell types are interconnected by junctions of low electrical resistance, as we will see in Chapter 12; the electrical signals, then, spread far into the tissues. In epithelia and heart, for example, the cell ensembles behave like scaled-down versions of electrical cables with relatively low core resistance and with capacitative and transverse-resistive components distributed over the ensemble membrane. In epithelial cells that spread is a double-exponential (hyperbolic cosine) function of the distance from the signal source. Length constants are typically 10^{-2}–10^{-1} cm. This may not be up to the usual engineering standards for electrical power lines or submarine cables, but it is perfectly adequate for cells; these coefficients exceed many times the length of a cell. In cell sheets, as one would expect, there is more attenuation, but even then the space constants usually are larger than the cell (the voltage decay is fitted by a Bessel function). Nerve and skeletal muscles are the odd men out. Though of microscopic diameter, their fibers can get very long—meter-long in the case of

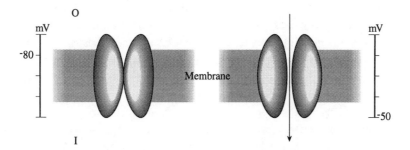

Figure 10.13 A voltage-sensitive ion channel. The diagrams represent conformations of a channel at different transmembrane potentials; the scales give the corresponding voltages (the sign indicates the polarity of the cell interior). The 80 millivolt potential is within the typical range of basal values for animal cells. (*I*), cell interior; (*O*), cell exterior.

are various proteins that fit that bill—some cell membranes, notably those of nerve, muscle, and heart tissues, literally teem with them.

These voltage-sensitive proteins belong to the tubular class, though they are cast in a distinctive mold: their polypeptide chains have (among other peculiarities) a special transmembrane stretch (the *S-4* segment), which contains regularly spaced positively charged amino-acid residues. These polypeptide stretches are imbedded in the membrane-lipid crystal, and so are directly exposed to its electrical field. They are not different in this regard from the rest of their kind. What makes them different is the way they react to that voltage. Their subunit configuration is sensitive to relatively small voltage changes, and this puts the whole tubular conformation under the crystal's heel. Their central passageway generally is closed when the membrane voltage is high and opens up when the voltage falls (Fig. *10.13*). The *S-4* segment presumably acts as the voltage sensor here and sparks off the configurational change.

One can get a rough idea of the channel opening and closing from measurements of electrical conductance across the membrane. Such measurements can be taken on small patches of membrane isolated from a cell, and thanks to this approach pioneered by the biophysicists Erwin Neher and Bert Sackmann, a kinetic picture is emerging, where the channel conformations are ruled by the principles of statistical mechanics—the openings and closings are probabilistic events. Take, for instance, a voltage-sensitive sodium channel on a nerve or heart cell. When the voltage across the membrane—the *membrane potential*—is high,

some of our own nerves — and, given the organic materials, the length constants couldn't possibly measure up to such macroscopic lengths. In those cylindrical structures, voltage attenuates with distance, as a single exponential function, and signal propagation relies on retransmission by voltage-sensitive channels. Such channels are distributed all along the fiber membrane, each channel reproducing the signal voltage to its full extent.

the probability for the channel being closed is greater than for its being open; and, typically, this is what happens when such a cell is in basal conditions and is not engaged in signaling. But at lower membrane potentials the probability shifts, favoring the open conformation. So, in conditions when the membrane gets sufficiently depolarized during the normal activities of a cell, the channel will be open more often (Fig. *10.13, right*).

We now are ready to integrate these fickle, though statistically quite reliable, membrane pores into the communication network. Suppose we have one of them next to a G-sensitive calcium channel, and a hormone interacts with a receptor protein in that locale (Fig. *10.14 A*). The G-sensitive channel then will open and calcium ions will flow in, owing to the α-relay action we have seen. But the effect of the hormone doesn't stop there. Because the flux of calcium ions through the G-sensitive channel brings positive charge into the cell, the membrane potential falls in this membrane locale—and markedly so if we have several such G-sensitive channels operating there. This depolarization now will bring the *voltage*-sensitive neighbor close to the threshold for its open conformation—it will now need only a little push, a small pulse of further depolarization, to open up. Thus, if we have a voltage source generating pulses of subthreshold depolarization in the locale, the voltage-sensitive channel is likely to open up, giving an ion flux of its own. The G-sensitive channel, in effect, then controls the voltage-sensitive one—or, in the language of neurophysiologists, it modulates its activity.[5]

Now, imagine a series of such channels stationed along the membrane, each capable of sensing a voltage change. These then will open in turn, and if they are adequately spaced, there will be a chain reaction—the channels will open one by one, as each gets to feel the voltage change produced by its predecessor in the chain (Fig. *10.14 A,2'*). So, we not only get a strong signal but one that propagates at full strength down the line.

Now, consider a membrane locale with another channel combination: a G-sensitive *potassium*-channel next to a voltage-sensitive sodium channel (Fig. *10.14 B*). Here, upon hormonal activation, the membrane voltage will rise—the cell interior becomes more negative than in basal conditions, as the positive potassium ions leave the cell through the open G-activated channel (in basal conditions potassium is more abundant inside than outside the cell and the prevailing chemical and electrical gradients will drive the potassium ions outward through the membrane pore). The behavior of the voltage-sensing neighbor is now different, but statistically no less predictable: he clams up. Thus, an open

---

5.   The depolarization of a G-sensitive channel is probably too slow most of the time to spark off the voltage-sensitive channels by itself—the probability for a voltage-sensitive channel switching to the open stage depends on the rate of depolarization—but it may be sufficient for modulation.

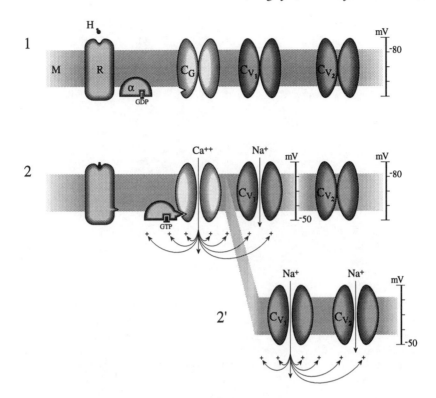

Figure 10.14 A   How the G-network ties into the electric-signal system of the membrane. *A*, coupling between G-sensitive (C_G) and voltage-sensitive (C_V) sodium channels. Diagram *1* represents the basal situation, the condition before the arrival of the hormone (*H*), and *2*, the condition after its arrival and the α-mediated opening of a G-sensitive channel. That opening sparks off a flux of calcium ions into the cell, causing the local membrane potential to drop, which, in turn, brings the neighboring voltage-sensitive channel (C_{V1}) closer to its opening threshold (-50 mV). *2* represents its opening when a subthreshold pulse of depolarization from an intrinsic generator (not shown) summates with the calcium-induced depolarization to the critical -50mV level; *2'*, the situation a moment later when, by dint of the flux of ions now through C_{V1}, the membrane potential drops in the C_{V2} locale and that channel opens up in turn.

potassium channel puts the damper on the electric signaling here; and if we have several G-sensitive potassium channels around, they will do so over a wide locale.

Before we go any further, let me say that these are no mere drills of the mind's eye. Although we have gone together through quite a few thought experiments in this book, this is not one of them. The conditions considered above reflect the actual situations in the heart, many kinds of gland and muscle tissues, and parts of the nervous system (hippocampus and dorsal root ganglia)—to name only those types of tissues that have been thoroughly examined

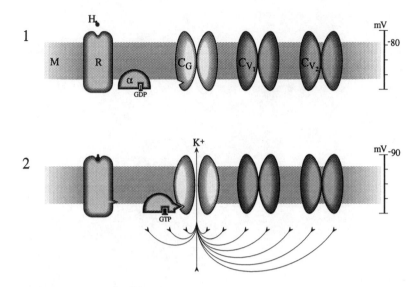

Figure 10.14 B   *B*, coupling between a G-sensitive potassium channel (C_G) and a voltage-sensitive sodium channel (C_V). *1*, the basal condition, and *2*, the situation after the arrival of the hormone. The opening of this G-sensitive channel leads to an outflux of potassium ions, leaving the interior more negatively charged; the membrane potential rises here, bringing the probability for the opening of the voltage-sensitive neighbor even further down than in the basal condition, now silencing the electric-signal activity completely in that locale. The voltage scales give the corresponding membrane potentials.

so far. There, one finds the sort of channel juxtapositions considered above plus the intrinsic electric generators, and the hormonal messages have the predicted modulatory effect and bipolar signal outcome. The schemes of Figure *10.14* give only the barest outline of the plethora of interactions between the G-network and the electric signal network. The resulting modulations serve to fine-tune many of our organismic functions—they adjust the activity of our heart to the needs of our body, steer our nervous system, make us go from sleep to wakefulness, from inattention to attention, and are probably behind our swings of mood—and I could go on and on.

It is the positioning of the two kinds of ion channels together that does the trick of getting the hormonal message from one network to the other in many of those cases. Or perhaps, instead of positioning, we should say conjugating, to point out that what is essential is not the contiguity of the channels, but their informational coupling, a yoking by information derived from the electrochemical energy field at the membrane. This at once brings the membrane potential to the front. Those 70–80 millivolts across the membrane (which is about the average in animal cells) mean 70,000–80,000 volts per centimeter of membrane, an energy field with quite an information potential. But to tap that potential requires cognitive talents which only the channels from the electric network have—those channels from the G-network are deaf to the electrical language, as it were. And this skewness has an important consequence for the operation of the allied networks: the information flow between the two is asymmetric—it goes from the G-system to the electric-signal system, but not in reverse.

This one-way communication scheme probably mirrors the evolution of the networks—the electric-signal system may have developed as an extension of the G-network. And if we follow this train of thought, we begin to see the voltage-sensitive channel as an adaptation for transmitting the hormone message farther afield. And what a splendid adaptation it is to get such a tubular protein looped in with the electrical membrane field! At the heart of that adaptation was a rather short stretch of DNA encoding the S-4 segment of channel protein (or its likes), the transmembrane segment with regularly spaced positive amino-acid residues. Indeed, this voltage-sensing polypeptide stretch is the key element in the loop that propels the chain reaction of channel openings we have seen. It closes a local electric circuit between the depolarized membrane spot with open channels and the polarized membrane regions with closed channels—a loop which is reestablished again and again at each channel down the line, giving rise to a self-regenerating wave that moves laterally through the membrane. And riding this never-attenuating wave is the hormonal message . . . which gets far, indeed, this nice and cushy way.

The first electric extensions of this sort presumably appeared in the Cell Age, and they may have been reserved for elongated cells, where the unregenerated electric signal of G-sensitive channels would fade along the membrane. But the advantages in speed of this mode of information transmission were not over-looked by Evolution when she launched her intercellular enterprise, and we shall see further on how she managed to extend the electric-signal reach. But just as important were the economical advantages. Indeed, the information in the energy field at the cell surface was there free for the asking, and tapping it was a scrooge's dream. So, under the combined evolutionary pressures of information economy and communication speed, the electric "extension" would get longer and longer with the enlarging organismic cellular mass. In the higher organisms it eventually would acquire vast proportions and, in their brains, almost a life of its own. But at bottom, there always is that age-old G-network which, even in the long-necked giraffe or the brainy humans, never lets go entirely of its hold.

The nature of this hold is no mystery. In fact, we can pinpoint the control spot with precision—it hasn't changed in a billion years: the membrane zone of conjugation of the G-sensitive and voltage-sensitive channels, the junction between the two networks. It is here that the decision is made whether the activity in the electric-signal network is to go down or up. That decision is the pre-rogative of the G-network and depends on the ion selectivities of the channels it employs (Fig. *10.14*). Thus, not only information transmission along the membrane, but ultimately also its regulatory effects, are owed to the Maxwellian stripe of these channel proteins.

Now that we see the cloven hoof, let's zoom in on these demons. Two things then leap at once to the eye: the disproportion of the number of demons in the two networks and their hierarchic disparity. Whereas, typically, there are but a handful of demons operating on the G-side, there are many more on the other side—and the disproportion gets huge in nerve cells with inchlong or even yard-long dendrites or axons. And just as striking is their hierarchy; the demons in one network are functionally subordinated to those of the other one. Indeed, there is an unabashed class system here: hundreds or, in some cases, millions of voltage-sensitive demons dance here to the tune of a select few—the charmed circle of G.

The whole class of tubular demons is somewhat thrown in the shade by the other membrane demons, the giants we have met before. However, their cognitive power leaves nothing to be desired; they are good at picking out their ion passengers (the four-subunit demon types are especially good). The tiny inorganic passengers are nickels and dimes by biological information standards. But the demons here are old *noblesse*, and saving information is *oblige*. In fact, they pull off a fabulous stunt with these information baubles: they make signals out of them. And if we look close enough, we see their sleight of hand: they use the membrane lipids to store up that small currency until it reaches signal value— the 1μF-per-square centimeter capacity of the lipid is more than enough for that. So, if you thought for a moment that the tubular demons, as retransmitters, were inferior to the other ones on the membrane map, you underrated Evolution. In her game plan these demons are not really the free-standing units we make them out to be in our analytical ploddings; rather, they form a functional whole with the membrane lipid. And this channel protein–membrane lipid entity integrates the bitsy ion information, generating signals for cellular communication with great information economy.

A Small Ion with Cachet: The Calcium Signal

Having seen how the sodium and potassium channels of the G-system tie into the electric-signal network of the cells, we now turn to the third member of this coterie, the calcium channel. This class of demon ties into the electrical network too—and by way of the same lipid crystal—but they have also another option: they can use the calcium ion itself to carry the hormone message into the cell.

The reason doesn't lie in any special quality of the demon here, but in the ion. Informationwise, calcium is in a different league from the sodium and potassiom ions; though still quite small, it holds enough information to carry a hormone message. Crystallographers have known for some time that calcium is a cut above the rest of the small inorganic ion crowd: it has a flexible crystal field—bond distances and angles are adjustable, with coordination numbers that can vary from six to ten—and it has higher ionization energies. The adaptable coordination sphere permits a wide variety of cooperative packings, giving the ion an advantage in the cross-linking of crystal structures, both inorganic and organic ones. Proteins—domains with, say, six carboxylate residues—can accept the ion rapidly and sequentially and twist around it, and all this specifically and reversibly.

Thus, we may rightly say, this ion is within the proteins' grasp. Indeed, a number of proteins are found inside cells that make a profession of fingering the ion. These proteins have six to eight oxygen atoms (each carrying a negative charge) so positioned as to form a cavity that fits the calcium crystal radius (0.99 Å); or more precisely, they bring these atoms into position for this fit (they are allosteric proteins of the inducible sort). And this is every bit as discriminating a fit as those in the organic world we have seen; no other inorganic ion comes close to this radius (the nearest, magnesium, is 0.65 Å). All this makes these proteins capable Maxwellian entities.

But all those fine qualities of the calcium ion and its allosteric partners would be beating the air if the ion's background level weren't low inside the cells. In fact, it is very low: at rest and with everything in balance, the cytoplasmic calcium-ion level is generally under 10^{-7} M. That level goes up (transiently) some ten- to hundredfold when a bunch of calcium channels open up in a cell, constituting a loud and clear signal. It is this large differential that makes calcium such a good signal; the high signal-to-noise ratio enables the system to use the individual calcium molecule straight as it comes. None of the other common inorganic ions in cells could hold a candle to calcium in this regard.[6]

How did this low calcium-ion level come about? That level is an age-old cellular feature which may have been linked to the primordial development of cellular energy metabolism, the development that made ATP and other phosphates the energy sources of cellular life. The abundant calcium in the terrestrial medium must have been a big hurdle in the way: phosphate and calcium ions tend to bond, forming insoluble crystals; so with calcium-ion concentrations like those in sea water (10^{-3} M), not many phosphate ions stay free. Thus, to get a useful phosphate level inside cells, the calcium-ion level had to be brought down. From the concentration of free phosphate prevailing in cells today (10^{-2} M) and from the binding constants, one can estimate by how much: at least two orders of magnitude.

In fact, Evolution went two orders of magnitude better. Anybody who has prepared a calcium solution and tried to bring it down that much for any reasonable length of time will appreciate the difficulties involved; not many buffers are up to the mark. Evolution knows every buffering recipe in the chemists' books,

---

6.   The cytoplasmic background levels for sodium or chloride ions, for example, are 10^{-2} and 10^{-3} M, respectively—only one order of magnitude lower than the concentrations outside the cells.

but no amount of buffering alone would do with such steep gradients. To keep up with all the ions flowing in through opening channels and those trickling in through membrane leaks, something with more zing was needed. And Evolution came up with tubular proteins of a special sort that could select the calcium ions *and* transport them against the gradients. These are membrane proteins distinct from the ion channels; they do mechanical work against the electrochemical gradients. Driven by ATP—which here supplies both the energy for the mechanical work and at least part of the negative entropy for the selection—these protein entities literally pump the calcium ions out of the cell. The pumping rates are of the order of 100 ions per second—not a match locally for the sudden influx when channels open in a cell (1,000,000 ions per second), but enough to keep the cytoplasmic calcium concentration under 10^{-7} M in steady state. Thus, by investing in the right type of protein, Evolution got two birds with one stone: a low calcium-ion background for energy metabolism and for signaling.

Signal Mopping: An Economical Cure for Cross Talk

But how did she get privacy of communication with such a ubiquitous signal? By virtue of its idiosyncratic physicochemical properties, the calcium ion, as we have seen, can specifically interact with proteins, and so makes the grade for second-theorem encoding. But even the best encoding won't do if the same signal is used by more than one information channel. And this is precisely the dilemma here. Calcium is one of the most frequently used signals in cellular communication—so frequent that it is not uncommon for it to carry different hormone messages to different addresses in one and the same cell. So, how do such messages get to the right addresses? How are they prevented from getting mixed?

This is the old question of how to keep information under wraps. There are various practical solutions to this problem. A familiar one is the straightforward method used by telephone engineers to channel information through solid-state gear. When they furnish an office building with private telephone communications, engineers use insulated wires, often one wire for each telephone on a desk. Each wire here is an extension of a cable in the trunk line from the telephone central. Thus, calls from the outside can come to desk 10 independently of desk 21, and without cross talk. But Evolution doesn't go in for such bone-dry things. In fashioning her communication systems, she had to make do with the material on hand, including the most abundant one—water—and she likes to surround her paraphernalia with plenty of it. So, when the time came to make a signal out of calcium—and this happened in truly primordial times— she had to find a way for separating the signals in the water of the cells.

Easier said than done. Short of solid-state partitioning, there really aren't many ways to keep such swift little molecules apart. With larger and more ponderous ones, one might rely perhaps on the square law of dilution; but this is hardly practical with so readily diffusible molecules—calcium ions cover significant cell distances in fractions of a second. There is, however, another possibility: active dampening of the signals, namely, cutting the number of calcium ions down *after* their entry into the cytoplasm. This, in principle, is the method cells often use for the larger molecular carriers of information; those carriers are then destroyed by an enzyme at some point in the communication cycle. But calcium is too sturdy for that. There is, however, another possibility for active dampening: to compartmentalize the information. And this, in fact, was Evolution's tack.

I want to be quite clear about this dampening method, for you won't find it in engineering handbooks. It is a method for a wet environment, and your typical communication engineer has no stomach for that. The ion signals in the cell water are essentially mopped up; they are sucked into cytoplasmic organelles—little membrane-bound compartments within the cells. The membranes of these organelles are endowed with calcium-pumping proteins that suck the calcium ions up. The pumps are driven by ATP and are not unlike those in cell-surface membranes, but they are oriented inside-out, so to speak—they draw calcium ions from the cytoplasm into the organelle. Each organelle is, in effect, a calcium sink.

You can actually watch those sinks in action if you put a substance into a cell, which lights up with calcium. There is a substance tailor-made for that: aequorin, a protein made by luminescent jellyfish; this protein glows with calcium concentrations in the physiological range—the signaling range. The jellyfish gene can be transplanted into a cell of a higher organism or the protein itself can be injected into such a cell; the protein will then report the changes in cytoplasmic calcium concentrations occurring during G-activated signaling inside a cell. Thus, the entry of calcium ions into the cell through a cluster of hormone-responsive channels can be seen in the microscope as a luminescent dot, a sphere of about 1 micrometer in diameter or less, brightly lit up at the center and rather abruptly fading toward the periphery (Fig. *10.15, left*).

It is the abruptness of the fading that tells us that the calcium gets mopped up. If the calcium ions were allowed to spread freely through the cytoplasmic water, they would fade more gradually. This is readily demonstrated by blocking the energy supply of the calcium pumps with cyanide. Then, the luminescent dot expands and, eventually, the entire cell lights up—the calcium ions now have the run of the cell (Fig. *10.15, right*).

Normally, this wouldn't happen, of course; there are scores of energized calcium sinks inside a cell, and every time a calcium ion enters, it must run this gauntlet. Not many ions get far. In short, these sinks see to it that the general calcium-ion background stays low in the cell—the necessary condition for intracellular signaling (and the reason why cyanide is so lethal).

After this little experimental preamble, we are ready to consider what happens at a G-activated calcium channel. As the channel opens, the large downhill flux of ions will cause the calcium-ion concentration to rise in that cytoplasmic locale.

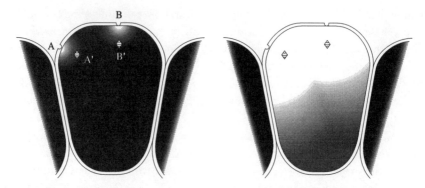

Figure 10.15 Private calcium-signal lines. *Left:* The diagram represents the distribution of calcium ions entering through two separate clusters of channels into a cell, as revealed by the glow of aequorin against a dark background (microscope darkfield). The intensity of the glow indicates the calcium ion concentration on a scale where white is 10^{-3}M (the concentration prevailing in the cell exterior) and black, 10^{-7}M or lower. *A* and *B* are the calcium influx sites and *A'* and *B'*, calcium-sensing proteins, the receivers of the two calcium messages in the cytoplasm. Because of calcium mopping, the calcium-message spheres don't overlap—messages from *A* get to *A'* and messages from *B* get to *B'* without cross talk. *Right:* After blocking the (ATP-dependent) calcium mopping with cyanide, the message spheres overlap; the diagram represents the distribution of calcium ions a few seconds after the channels at *A* and *B* had opened. (After Rose and Loewenstein, Science 190:1204–1206, 1975.)

With a bunch of channels opening simultaneously there—and they needn't be many—the calcium-ion concentration will (transiently) reach signal value. Thus, we now have a signal source. If this were a water compartment pure and simple, the signal would decay with the square of the distance from the source. But because of the presence of energized calcium sinks in the cytoplasm, the signal falls off more steeply; in fact, so steeply that barely a micrometer away from the signal source—a distance on the order of one-twentieth of the diameter of an average mammalian cell—the calcium-ion concentration will be down to that of the cytoplasmic background.

Now, this dampening gives a leg up to the communication enterprise, as it allows, without further ado, multiple calcium messages to be carried inside a cell; it provides a means for getting information privacy and, no less important evolutionarily, an affordable means. To see how this works, let's put a second signal source a few micrometers away from the first one and place a calcium sensor close to each source. Either source now will produce a calcium signal, but, thanks to the mopping, each signal reaches only the nearest sensor—the one lying within the signal sphere of a source (Fig. *10.15, left*).

Thus, the messages from two distinct hormones can be carried by calcium ions to distinct cytoplasmic addresses, without message interference. To engineers, this may seem a strange cure for cross talk, though the new generation striving to be more in harmony with nature could learn from that. Indeed, this dampening method is in a class all by itself, because of its information economy. The cell here gets two or three communication lines (a larger cell, perhaps even more) with a minimum investment in communication hardware and signals—with one species of signals, and a cheap one at that, it gets what otherwise would demand two or three.

That signal segregation, of course, doesn't operate in a void. It relies on the low calcium-ion background inside the cells, which has its own information cost in pumps and other membrane equipment. But this information capital, we might say, has already been amortized, as it served the cellular genesis of ATP, a genesis that over the long run has paid back its own good return.

Thus, this modality of cellular communication betokens frugality from end to end—it shouts it from the rooftops. However, such an extreme should come as no surprise. It is precisely what the principle of information economy predicts for something, like the calcium signal that goes back to that far side of Nowhere, the dawn of life!

The Shoo-in Called *Calcium*

Now let's give the once-over to that whole signal operation and ask why this ion was the Chosen one. If you think I am laying it on, just look at yourself. There is hardly a function in your body that doesn't involve calcium: heartbeat, muscle contraction, gland secretion, signal transmission in the nervous system, signal transmission between nerve and muscle, cell division, cell motility, biolumines-cence, cell adhesion, and so on and on. So it behooves us to ask why this ion occupies so many physiological niches.

If we used physiology as our sole basis, the answer would be tenuous and hedged around with qualifying clauses. But we can lean on information theory, and this may work to our advantage when a large physiological and pathological inventory is available, as in the case of functions involving calcium. Physicians have long known how important calcium is for our health, and physiologists have investigated the reasons now for over a hundred years. Their efforts began to bear fruit during the first half of this century (when many of the calcium functions mentioned above came to light). Indeed, there was a time when hardly a year went by without a discovery of a new calcium role. Everybody seemed to burn with calcium curiosity—whole journals were, and still are, devoted to it— and who could remain impassive at such omnipresence? I personally will never forget my talks with one of the masters of the early epoch, the physiologist Otto Loewi—the discoverer of chemical signal transmission in the nervous system— who spent the summers of his later years at Woods Hole. Our conversations there—this was in the 1950s—invariably would end by him saying: "Ja, Cal-cium ist alles."

Today, I could tell him why. Through the information lens we can see the ion's unique virtues for specific and reversible interaction with protein. And we can envision the ion on the primordial scene: a molecule with a modicum of information—a firstcomer who had everything going for himself and who would take up the available niches. So, we take it in stride that the ion has so many roles in the developmental plot . . . or with a little tongue-in-cheek we might say, calcium was a shoo-in in the evolutionary contest.

A Streamlined G-Map

Let us now return to things of here-and-now. Having checked out some of the individual G-pathways, we will try to compose a map. One may think at first, perhaps, that there can't be much of a map, as everything that matters here takes place in a membrane of barely molecular thickness. But that doesn't reckon with that film's other two dimensions—an area on the order of 10^{-6} cm^2 in vertebrate cells—a considerable surface in molecular terms. Indeed, the membrane is speckled all over with members of the G-troupe; there is, after all, as much information transmission along the membrane as there is across, and what with the retransmitters and G-go-betweens, that's not a minor crowd. But there is more: some of the retransmitters (the ion channels) tie into the electric-signal line of cells. Thus, the G-system casts a net far and wide.

Only a few members of the G-confrerie, it is true, are represented in any given cell; but those who are, come in many copies. Thus, there are intercom lines (actual and potential) crisscrossing all over the membrane map—higher animal cells number them by the hundreds. One could draw a detailed blueprint; enough is known nowadays to do this for at least some cells. But we'll pass this up—no need to swirl our senses—the Second Theorem proffers us a way to cut through the jumble; we can assort the communication lines according to their three-dimensional encoding marks—their complementary molecular interaction domains. We actually only need to heed two pivotal ones here: the receptor-to-Alpha and the Alpha-to-retransmitter domains; and even where the three-dimensional structures are not yet known, the chemical affinities tell us enough about the complementary fits to guide the assortment.

By this expedient, the essentials of the G-network stand out: the hordes of receptors, retransmitters, and middlemen reduce to a handful of subsets, and the intricacies of connectivity come down to a few forks and confluences (Fig. *10.16*).

Analytical expedients are not without their pitfalls of distortion. But there is no undue twisting here (nor in the bare-bones scheme of Fig. *10.8, bottom*). We really are doing no more than what the principal players in this information game do naturally when *they* sort themselves out. Thus, the vicissitudes of the information flow are predictable from the players' encoding marks: *when an Alpha fits more than one receptor type, there will be convergence of the message lines*

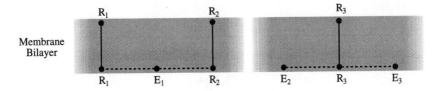

Figure 10.16   Schematic of the G-network. The network in a cell with three hor-mone-receptor types *(R)* and three retransmitter types *(E)* has been reduced to its essentials by lumping the many receptor and retransmitter copies into subsets, according to type. The information transmission down the transmembrane receptor structure, the 100-angstrom stretch where the information flows *across* the mem-brane bilayer, is represented by solid lines; the transmission via α, the longer stretch where information flows *along* the membrane, is represented by dashed lines. *Left:* convergent information lines; two receptor types use the same α (α is not shown). *Right:* divergent lines; two α types fit the same receptor. Terminology and abbrevia-tions as in the diagram of Figure *10.8,* but the G-intermediates are omitted.

(Fig. *10.16, left), and when he fits more than one retransmitter type, there will be divergence* (Fig. *10.16, right) — the man in the middle has the whip hand.*

T Junctions and Organismic Harmony

Information forks and confluxes of the above sort are hallmarks of the multicel-lular enterprise. They are responsible in large measure for the good behavior of the cellular masses. Take our liver, for example, where the cells—huge masses of them—store the glycogen fuel for our body and, in due time, break it down to glucose, the currency for our energy transactions. This sugar production is per-fectly in tune with our energy needs—a well-balanced automated operation which entails a wiring of convergent lines of precisely the aforedescribed sort, namely, G-lines coming from adrenaline- and glucagon receptors, which share the same Alpha type. The lines intersect at the retransmitter *adenylate cyclase,* where the two hormonal messages are integrated, giving rise to the emission of cyclic AMP signals—the final signals controlling the glycogen breakdown. Or, to consider an instance of divergent wiring, take our salivary glands or pancreas, where the masses of cells secrete a starch-cleaving enzyme. Here the retransmit-ters are potassium and calcium channels, and the hormone message in a cell is forked to both retransmitter types, producing a balanced secretory activity.

The books of physiology and medicine are replete with examples of self-regulating operations like these. In their three-dimensional detail, such organismic mechanisms are as kaleidoscopic as are the organismic forms, but under the look-ing glass of information they all come down to this: a flux that ties the Circuses of cells together into a whole. Such information fluxes and confluxes are responsible for much of the cellular orchestration in our body—a veritable cornucopia of har-mony. And so rich is that cornucopia, so harmonious the orchestration, and so

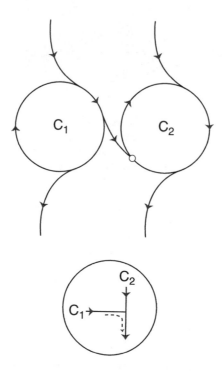

Figure 10.17   A T Junction. *Top:* Two interconnected Circuses (C_1, C_2) and their junction, the encoded T junction (*T*). *Bottom:* An enlarged view of the T junction implementing the information conflux. (In the two kinds of Circus connections via the G-signal system considered in the text, the junction is constituted by calmodulin or by protein kinase A.)

unfailingly appropriate the resulting cell behaviors to the environmental circumstances, that, when the whole is considered out of its evolutionary context, one may well get the impression that it is owed to a design with an end in view. This may explain why even nowadays, 150 years after Darwin, we come upon educated people holding such an idea—a relapse, in one disguise or other, to Aristotle's old tenet of a design of nature with a telos (τελος), a goal. Under the information lens, those will-o'-the-wisps dissolve in thin air and all that remains are the Circuses, whose information rolls smoothly from one to the other (Fig. *9.1*).

Such Circus mergers are the result of numberless trials and tests in the laboratory of Evolution . . . and of the unsuccessful ones, the terminated experiments, we have no record. We saw the four tactics that brought these mergers to a happy issue (Fig. *9.4*), and to put strategy *b* into practice, she deployed the whole phalanx of demons now before us. We tracked that phalanx all through the cell membrane and now, moving our lens over the cell interior, we will bring its last member into focus, the one charged with the Circus liaison.

This liaison demon acts as T junction of the intercellular information flow (Fig. *10.17*). In schematic renderings, such junctions look deceptively like the Ts of water pipes or electric lines. But theirs is a sap that doesn't bend so easily around corners. Like the other members of the molecular chain, this final link both receives and passes on the hormone message. But it does more: it melds that foreign message with a well-encrypted information world, the Circus. And this is something only a demon could do; it calls for communication deftness, an understanding of the language of two Circuses.

Snapshots of Two T Demons

Those, then, are the demons who pick out the signals—the calcium ions, the cyclic AMP, inositol triphosphate, diacylycerol, and the rest of the crowd (Table 2)—which are released on the underside of the cell membrane. We now will have a go at two T specimens, the demon for the calcium signal and the demon for cyclic AMP. For the former, the available crystallographic data afford us a glimpse into the cognitive act, and for the latter, the chemical data let us in on the melding of the signals with the Circus stream.

The demon for calcium is called *calmodulin*. Identified chemically in 1967 by Wai Yiu Cheung, it is rather small by demon standards—a single chain of 148 amino acids, which folds up into four domains with negatively charged carboxylate groups, each making a perfect nest for a calcium ion (Fig. *10.18 A*). This polypeptide chain lacks the usual props (disulfide bridges or hydroxyproline braces) that proteins are generally shored up with, and so is quite nimble. It can fold around the calcium ion, housing it like a ball in a socket. And when the cytoplasmic calcium ion concentration is at signal level (above 10^{-7} M), four such sockets form, each one with a calcium ball. In reverse, when the concentration returns to background level, the socket opens as the polypeptide chain unfolds and the ion disengages.

That is the sum and substance of the demon's cognitive act. The performance is nicely balanced. The ion, with its unique crystal field, provides a template of sorts; the demon's polypeptide chain folds around it, link by link—old demon and old ion making a good match. And not to forget the overall picture, the folding is always in (thermodynamic) harmony with the cell interior—the act has been on center stage for some 2 billion years! As for the number of demons, there are some 10 million calmodulin molecules in a typical animal cell, more than enough to take care of the calcium ions when they come on in signal strength and to statistically ensure the continuity of the information flow at this T.

Let's pry a little more into this ancient cabal. We are fortunate that there are X-ray pictures at hand—snapshots of the protein and its ion partner—to anatomize the interaction further. To begin with, the snapshots tell us something about how the calcium ion gets fingered and seized: the demon does this by means of a loop of twelve amino-acid residues which are flanked by two helices positioned nearly perpendicular to one another (*E* and *F*; Fig. *10.18 A*). This helix-loop-helix motif, or *E-F hand*, as its discoverer Robert Kretsinger calls it, is

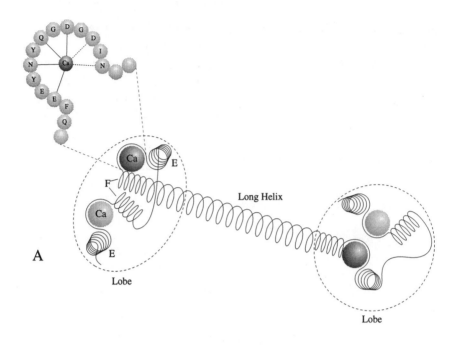

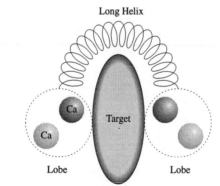

Figure 10.18 The structure of a calcium demon. *(A)* Calmodulin with four calciums in position (a schematic diagram drawn after X-ray crystallography performed by W. J. Cook and colleagues). Each calcium nests in a loop flanked by two short α helices *(E and F)* positioned at right angles to one another. The concluding helices of each pair are joined without interruption by a long exposed helix. (*D*, aspartate; *E*, glutamate; *F*, phenylalanine; *G*, glutamate; *I*, isoleucine; *N*, asparagine; *Υ*, valine; *Q*, glutamine.) *(B)* Interaction with the target (a free rendering of a model proposed by Persechini & Kretsinger). As the long helix flexes, calmodulin, like a pair of tweezers, grasps the target between its lobes, completing the T junction.

found in a large family of calcium-sensing proteins that function as T junctions. More than twenty members of that family are presently known. Calmodulin carries four such motifs—a pair of them, back-to-back, on either end.

Perhaps I am harping a bit long upon this gestalt, but it's not every day that one gets to look into a demon's innards. So bear with me for one more spin: as the calcium ion binds sequentially with the six amino acids forming a socket and the demon contorts himself around the ion, he seems to undergo yet another important configurational change: the E and F helices in his two lobes move with respect to each other, exposing large sticky (hydrophobic) patches on the lobes. In this conformation the lobes probably attach to the target, completing the melding of the hormone information with the Circus. And here is where that lissome helix between the two calcium-holding lobes comes in. This long exposed α helix—an unusual thing in proteins—is, by itself, inherently unstable (most α-helices are stabilized by lateral contact with other parts of the protein). But this very instability is what gives it flexibility; it can bend and, to judge by X-ray crystallographic and nuclear magnetic resonance data, this pliancy seems to be instrumental in the interaction with the target. A current model has the demon clasping the target between his lobes (Fig. *10.18 B*).

Thus, we may envision the following picture: the binding of the calcium ions triggers an allosteric movement propagating from the E and F regions to the long helix of the demon and culminates in a pincer movement; the information, thus, would flow from the four calcium ions to the demon's lobes, and from there to the target, insensibly continuing with the Circus stream.

Now to the other exemplar. This one has more body: four good-size protein subunits (Fig. *10.19*). The demon belongs to a family which biochemists call *protein kinases*—enzymes that catalyze the transfer of a phosphoryl group (like the terminal phosphate of ATP) to certain proteins. In the present case, the phosphoryl group is transferred to serine or threonine amino-acid residues of the target protein, and this is what melds the hormone message with the Circus here. These serine and threonine phosphorylating enzymes, which were discovered by Edwin Krebs and Donald Walsh, go under the names of *protein kinases A* or *cyclic AMP-protein kinases*.

These demons consist of two pairs of subunits: one pair receives the cyclic AMP signal and the other, the smaller pair, does the catalytic job of transferring the phosphate information. The act begins when the cyclic AMP level in the cytoplasm reaches signal strength (10^{-8} M or higher). Each receiver subunit then binds two cyclic AMP molecules—this is how the mechanical statistics pan out. The binding sets off a conformational change, causing the catalytic subunits to come apart (Fig. *10.19:2*), which, in that free form, now can engage their targets and transfer the phosphate information.

The information here flows from cyclic AMP to receiver subunit to catalytic subunit to target. The first and last transfer steps are done the jog-trot electro-

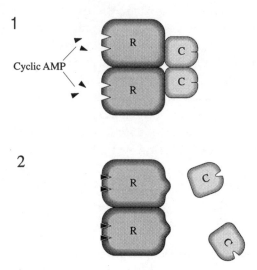

Figure 10.19   The demon for cyclic AMP, *protein kinase A*. (*1*) Stand-by conformation: the receiver subunits (*R*) and the catalytic subunits (*C*) are held together by
weak forces. (*2*) Active conformation: the receiver subunits have bound four cyclic
AMP molecules and switched conformation, causing the catalytic subunits to dissociate; these subunits are now free to diffuse to their cytoplasmic targets, completing
the Circus T junction. (The receptor subunits are 49,000 daltons, and the catalytic
subunits are 38,000 daltons.)

magnetic way. But there is something extra here, a trick the likes of which we
have not come across before: this demon splits itself up. Why? It surely couldn't
be for the sake of signal amplification; there is no signal gain in this schizy trick.
The reason must go deeper, and if we dog the footsteps of this demon, we see
that he is really after something else: a wider audience.

Now this is a yen we can empathize with, and to get to the bottom of how
the demon manages to indulge himself seems worth a replay. So let's reel off the
scene again where the two catalytic subunits come apart. Up to this moment, the
demon was in a fixed position somewhere inside the cell. Like many other cytoplasmic macromolecules, he is probably tethered to the cytoskeleton, the lattice-
work of filaments crisscrossing the cytoplasm. His information reach thus is
barely four angstroms—not much of a chance that his message would have a
receiver. But this demon is of modular build. His two modules are held together
only by weak forces—a trigger-happy situation where it takes but a slight push
to shift the balance, and the structure comes apart. And this is precisely what
happens when cyclic AMP binds to the demon. The demon then jettisons his
two catalytic subunits into the cytoplasmic water—his subunits fly the coop, so
to speak, vastly stretching the demon's formerly miserable horizon.

The G-Pathway in Synopsis

At this point we pause for a retrospect of the main episodes in the transmission of the hormone message through the G-network:

Picked up by the membrane receptors, the messages are distributed by Alphas to retransmitters in the cell membrane, and from there are sent out in hydrophilic form to the cell interior, where they are received by demons that tie in with the target Circus. So, in six transfer steps,

$$\text{hormone} \rightarrow \substack{\text{hormone} \\ \text{receptor}} \rightarrow \alpha \rightarrow \text{retransmitter} \rightarrow \rightarrow \substack{\text{retransmitter} \\ \text{signal}} \quad \text{T demon} \rightarrow \text{Circus}$$

the inter-Circus linkage is completed.

eleven

Intercellular Communication Cheek-By-Jowl

We now shift our line of inquiry from hormones to the other two forms of intercellular communication. Willy nilly, we make a fresh start here because the membrane G-system is not bound up with these forms. The cells here take advantage of the cheek-by-jowl situations prevailing in tissues and make their Circus connections right at the points of contact. There, certain proteins in the apposing membranes come within reach of one another and interact directly across the intercellular gap. These interactions—they all are of the short-range electromagnetic sort—tend to hold the membranes tightly together. The membranes, thus, form a sandwich; and these sandwiches, tiny (and far between) as they are, are Lucullan feasts of information. In fact, so much information is concentrated at these contact spots that entire organs and tissues are under their sway.

Or, one might almost say, under their spell, for these are seats where demons dwell. The membrane sandwiches house a whole host of demon types, each one with a special talent. But two are a cut even above that crowd: one type makes an information bridge across the intercellular gap—a direct Circus-to-Circus linkage; and the other type forms a tunnel across the membrane sandwich through which intracellular molecules can flow directly from cell to cell.

Those two are the dramatis personae on the communication stage we are now about to visit. They are formidable Maxwellian entities, as we shall see. But let me say at the outset that we don't know yet much about their cabals; indeed, on the whole, our picture of the pertinent communication channels is not as complete as in the case of hormonal communication. The studies of hormones have had a hundred-year head start and are by and large technically simpler—it is a lot easier to follow the information trail in the blood or the liquid outside the cells than in the tight squeeze between membranes.

Well, we will have to bear that cross. But there is no inherent limit to analysis here—no more than in any solid-state problem. It's a question of sharpening present tools and of a little gumshoeing in the sandwich. And that can be rewarding, especially if one has the law on one's side—truly a high one in this case, the Conservation Law, which guarantees that nothing in this universe vanishes without a trace. So, there is bound to be some residue of energy or matter for the sleuthhound—some telltale sign of the information wake in the membrane sandwich.

Membrane Hot Spots: The Syncretic Communication Unit

We start with a membrane speck where the demons directly link contiguous Circuses. The communication unit here consists of a pair of proteins of the transmembrane sort which interact by their complementary protruding ends across the intermembrane gap. One of the proteins acts as the donor of information and the other, as the receptor (Fig. *11.1, top*).

Such binary units tend to cluster. The clusters are self-generating when the donor and receptor molecules can diffuse laterally in the membranes and enough copies of them are on hand. Then, the formation of a few stable units may maintain a local cell contact long enough for more individual donor and receptor molecules to flock to the contact region, forming what we might call a "hot spot" (Fig. *11.1, bottom*).

There is something mesmerizing about such spots: the way they grow, the sprightly foregathering of the units, the clockwork regularity of the zippering of the membranes—all this keeps one captive. But once you get over that, you spot the true lodestone: the information economy. From a one-time investment of information in a pair of ligand molecules, the cells reap a full-fledged membrane organization, an organization of macromolecular proportions! It is this austere elegance of self-organization, the getting of so much information for so little, that seizes one with wonder. But as with Evolution's other ploys we have come across, the trick turns out to be not so arcane. This one can be spelled out in four words: choose the right ligands. The accent is on "right" here; what you must choose is a protein pair with complementary domains that stick out of the lipid bilayer the right amount and lower the free energy in the membrane system by the right amount. To come up with just the right ligand combination may take some time . . . give or take a billion years . . . but then, you have it made: things will roll downhill when cells get close enough together—constrained by the membrane bending moment, communication unit will beget communication unit and, before you count to ten, a full-blown macromolecular communication structure materializes before your eyes (Fig. *11.2*).

The momentousness of it all sinks in as soon as we brood over the things that might spring from the information flowing through that structure—a finger, a

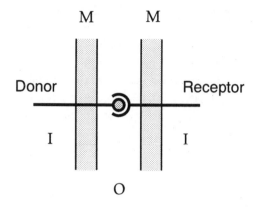

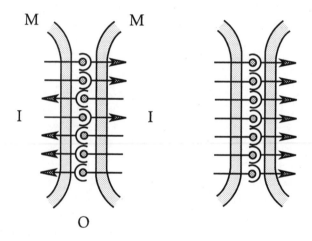

Figure 11.1   *Top:* The syncretic communication unit: a pair of ligand membrane molecules, an information *donor* and an information *receptor*, bridge the gap between a pair of adjacent cells, directly interconnecting the cytoplasms. *Bottom:* Clusters of syncretic units or "hot spots," at regions of cell contact. On the *left*, a bidirectional hot spot (either cell expresses donor and receptor molecules); on the *right*, a one-directional hot spot (donor and receptor expression is asymmetric). *(M)* Cell membrane; *(I)* cell interior; *(O)* outside medium.

hand, an eye . . . ? But before we can hope to view such lofty heights, we must see through the lower levels. So, first the question: In what direction will the information flow? This is something we should be able to handle without too many "ifs" and "woulds," for the direction obviously must depend on the unit makeup of a hot spot. We may even squeeze out an answer in terms of cell type, as the pertinent genetic information for the unit components in the two cells must determine the symmetry of the units: *with cells of the same type, where the two sides of a hot spot are roughly symmetric, the information traffic will be two-*

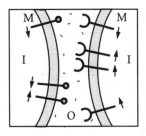

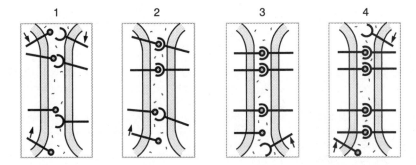

Figure 11.2 The formation of a hot spot. Moving randomly (laterally) in the lipid matrix of the membranes *(top)*, opposite donor and receptor molecules come within range of electromagnetic interaction of each other at a region of cell contact *(bottom)*. The diagrams *(1–4)* illustrate successive stages of syncretic-unit self-assemblage in a minimal-information model: as donor and receptor molecules wander into the contact region, they get trapped by their mutually attractive forces and the bending moment of the membranes; so, with time, syncretic units accumulate in the contact region. (See Chapter 12 and Figure 12.2 for a more detailed account of this trapping, a model originally advanced for another type of intercellular molecular connection, the cell-to-cell channel.) *(M)* cell membrane; *(I)* cell interior; *(O)* outside medium.

way; but with cells of a different type with sufficiently asymmetric unit-component expression, the traffic will be one-way (Fig. *11.1, bottom*). The first condition is likely to be common inside a tissue, and the second is the sort of situation one expects at boundaries between tissues.

A vocabulary matter must be wedged in here. I held this off as long as I could, but obviously our stock of words is becoming too low to discourse upon a communication form that hasn't yet a proper name. So, keeping the union of membrane ligands in the center of our attention, we will coin the term *syncretic communication*, borrowing from the Greek *synkretismos*, "union."

The Case for Syncretic Communication

This field of communication is relatively new, and we don't yet know too many cases, but in all cases of syncretic communication so far known, the communica-

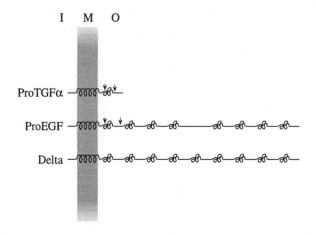

Figure 11.3 The EGF insignia. Three representatives of the EGF family displaying, in various repeats, the clan motif—a stretch of polypeptide chain puckered by disulfide bonding between three pairs of cysteines (the points of contact between the loops represent the three disulfide bonds). The arrows indicate where the membrane-anchored molecule can be enzymatically cleaved off, an option exercised when the insignia information is used for hormonal, instead of syncretic, communication (the cleavage points in Delta are unknown). *(M)* Membrane; *(O)* outside medium; *(I)* cell interior. ProTGFα, proEGF, and Delta are all membrane glycoproteins; the first two are instrumental in controlling cellular growth in vertebrates and the last, in guiding nervous-system development in some invertebrates.

tion units are made of glycoproteins—transmembrane proteins whose polypeptide chains have sugar chains attached to their extracellular domains. The sugar chains are hitched on after the polypeptide comes off the ribosome assembly line. They are made of glucose, galactose, mannose, glucosamine, or other carbohydrate units. Unlike the polypeptide chains themselves, the sugar chains have branches, and this adds to the information bulk of the communication units.

The best studied syncretic units belong to a protein family which biochemists and cell biologists have been conversant with for some time: the *growth factor (GF)* family, though they were then unaware of their syncretic communication function. Its members have a characteristically looped polypeptide chain—a stretch 30 to 40 amino acids long with three pairs of disulfide-bonded cysteines—a sort of family insignia that is displayed in various repeats (Fig. *11.3*). For instance, the precursor of the epidermal growth factor, *proEGF*, has nine repeats and that of *proTGFα* has one. EGF and TGFα, the growth factors themselves, are cleavage products of these membrane-anchored precursor molecules and serve as messengers in hormonal communication; after their cleavage (Fig. *11.3, arrows*), they are secreted into the extracellular medium and, following the hormonal paradigm we have seen, they eventually attach to specific receptors on target cells to which they pass on their growth-stimulating message. This is what was known to biochemists and so was the growth-stimulation effect on cells of the diffusible cleavage products. These products reach the responding cells via the intercellular medium or the blood; so they were appropriately classified as hormones (hence, the name *growth factors*).

But it turned out that the precursor molecules will bind with high affinity to the EGF receptors too, and this binding occurs while both the precursor and the receptor molecules are anchored to their respective cell membranes. So, what we have here is a pair of molecules binding to each other across the intercellular gap, transmitting information from one Circus to another—in short, a syncretic communication unit. Over the past four years, Joan Massagué and his colleagues made the case for this through a tidy bit of genetic engineering. They introduced a proTGFα gene into cells that ordinarily don't express it, and paired them with a cell type that expressed the EGF receptor. Two questions, thus, could be asked: Will the genetically doctored cells adhere to the cells containing the EGF receptor (ordinarily they don't), and will this stimulate cellular growth? The answers were positive to both. In fact, the cells gave the full gamut of known physiological reactions to EGF receptor activation—a raised level of cytoplasmic calcium ions, tyrosine phosphorylation, DNA replication, cell proliferation—and all this in the absence of diffusible EGF (or TGFα) messenger.

These genetic experiments provide a solid stick to beat down the argument that the interaction across the intercellular gap may be merely mechanical. We face a special problem here because of the local protein inhomogeneities; apart from the syncretic communication unit, we have to reckon with another class of proteins. It is as respectable a class as the syncretic one—they stick to each other quite selectively (and act in pairs). But some may not do more than that. Thus, one must go to some length to make sure that one is dealing with an act of intercellular communication and not merely with something that goes with the sandwich territory, like cell adhesion. Short of tracking the Circus-to-Circus information flow, the proof of the pudding lies in the reacting of the cells, their pertinent physiological responses. And in the present case, these responses tell us that the proTGFα/EGF-ligand interaction is indeed part of an inter-Circus pathway, and not just a dead-end mechanical alley.

How Syncretic Information Flows

The question now before us is, where does the pathway lead to? Or, more immediately, where does the syncretic unit tie in? The initial stage of the information trajectory, the stretch in the extracellular portion of the unit, is reasonably clear: the flow goes from the outside portion of the TGFα-polypeptide containing the family insignia—the information-donor domain of the unit—to the receiver

I   M   O   M   I

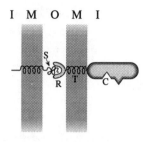

Figure 11.4   The proTGFα/EGF syncretic pathway. The diagram shows, on the *left*, the TGFα information donor molecule with its family insignia (*S*); and on the *right*, the EGF receptor molecule with its receiver domain (*R*), catalytic domain (*C*), and α-helical transmembrane segment (*T*).

domain on the outside portion of the EGF-receptor protein (Fig. *11.4*).[1] This much we can make out without too much trouble; it's the intracellular trajectory, which is the question.

So, let's put our loupe over the intracellular portion of the receptor protein. There, jutting out into the cytoplasm, is a familiar gestalt: it has contours fitting an ATP molecule and the accoutrements for meting out ATP's phosphate to protein—a gestalt of the sort we have seen at hormonal T junctions. However, on closer scrutiny, it turns out to be but a look-alike: the contours fitting protein are somewhat different; they fit a tyrosine residue, instead of a serine or threonine. In short, we have here a catalytic domain that phosphorylates proteins on tyrosine. But what further catches our eye is that this contour rapidly changes depending on what happens outside the cell, at the receiver domain of the molecule. And this provides us with our first clue, for the catalytic domain does not become active until the receiver domain is occupied by the donor half of the syncretic unit (or the EGF diffusible ligand in the case of hormonal communication).

Thus, we may envision the following sequence of events during a syncretic act: (1) the hemi-units bind to each other across the intermembrane gap; (2) the extracellular portion of the receptor protein, the receiver domain, changes conformation; and (3) the cytoplasmic portion of the protein, the catalytic domain, changes conformation.

But how does the information get from one domain to the other? Mind you, this is not your run-of-the-mill allosteric demon who can't give a nod without folding over from head to tail. This one has a still section, an α-helix crossing the cell membrane in one pass. Such short helices (20 amino-acid residues) are

---

1.   A dozen carbohydrate chains are attached to the extracellular aspect of the EGF-receptor polypeptide chain. This amounts to a substantial portion of the molecule (about one-fourth in weight). These carbohydrate chains and their branches supply information for the folding of the 1,186-amino-acid-long polypeptide and for its shuttling to the cell surface, but there is no reason to think that they are involved in the flow of syncretic information.

rather inflexible—more like rods than springs—which makes it unlikely that the conformational changes of the two domains are mechanically coupled. Granted, the "rod" might slide to and fro if it were by itself. But it isn't by itself. It is surrounded on all sides by phospholipid polar headgroups that impose severe constraints on perpendicular movement.

Thus, we'll put our bets on another transmission mode where information propagates sideways in the membrane, between receptor molecules. This will require some detective work, but I can promise that it will be worthwhile: we will get onto one of Evolution's clever tricks, her information crossover gambit. But first we need to do some legwork; we must track the information spoor some ways into the receiver cell. Not that we need to saunter far; there is every reason to expect the Circus T junction to be within earshot of the cell membrane. The logical place to start, then, is at the receptor's cytoplasmic domain. Well, the search is on in various laboratories and, even as I write these words, several syncretic modalities are under investigation; it may take awhile until a generalized scheme can be drawn. But at least two or three modalities are approachable right now. In particular, the proTGFα/EGF-receptor syncretism is amenable to analysis because the recipient of the receptor's catalytic information has been brought out of the closet. This recipient is *phospholipase C-γ 1*.

This protein belongs to a family of demons who produce lipid signals. One of his cousins we already have met; he sat at the end of a hormonal G-path, sending out diacylglycerol and inositol-triphosphate signals (Table 2). Now, the demon sitting at the end of the syncretic path emits exactly the same signals. And here we are provided with our second clue, for that emission is driven by the information from the EGF receptor's cytoplasmic domain.

So, a picture of the information flow in this syncretic modality may be roughed out:

1	2	3
Donor hemi-unit (proTGFα)	Receptor hemi-unit (EGF receptor)	phospholipase C-γ1 $-<$ diacylglycerol / inositol triphosphate

where stage 2, namely, the phosphorylating interaction between the receptor's catalytic domain and phospholipase C-γ1, constitutes the T junction, and the diacylglycerol and inositol-triphosphate signals carry the syncretic message into the Circus stream of the receiver cell.

The Information-Crossover Gambit

We are now ready to take on the question of how the information flows *within* the receptor hemi-unit. As mentioned before, there seems to be no suitable mechanical link to propagate the conformational change from the receiver to the catalytic domain of the hemi-unit. How, then, does the information get across that membrane-wide transmission gap? Here we will take our cues from polymer chemistry. From test-tube experiments with EGF receptor molecules, it is known that the molecules tend to stick to one another side by side, forming dimers. Such a pairing is common among long electrically charged molecules. But what is uncommon is that such molecules, which don't do any obvious mechanical work, mete out ATP phosphate to each other. And that's the tip-off—somebody here is doing an entropy balancing act and transferring information.

However, one cannot extrapolate just like that from test tube to membrane. This would press the argument too far; there is no guarantee that the receptor molecules will also phosphorylate one another in the lipid bilayer—they may not get close enough. We need to know how such molecules behave when they are in place in the membrane. The answers are in, thanks to experimental work with EGF receptors and their mutants. This work brought to light that the receptor molecules do phosphorylate one another, indeed, in their natural habitat—they do so at their carboxyl ends in the presence of EGF-diffusible ligand—whereas carboxyl-end mutants fail to do so; moreover, phospholipase C-γ1 does not get phosphorylated with these mutants nor do lipid signals then issue forth.

These experiments provide us with our third clue—indeed, jointly with the test-tube work, they let the cat out of the bag. They tell us (a) that there is transfer of information between receptor molecules in situ—a lateral transfer between dimers; and (b) that this transfer is a requisite for the events down the line.

So, extrapolating from membrane to membrane-sandwich, our syncretic scheme expands to:

	1		1a
Donor hemi-unit →	Receptor hemi-unit extracellular domain	→	Receptor hemi-unit cytoplasmic auto-phosphor domain

1b		2		3
→	Receptor hemi-unit cytoplasmic catalytic domain	→	phospholipase C-γ1	− <	diacylglycerol inositol triphosphate

where the syncretic message bypasses the α-helical transmembrane stretch of the hemi-unit receptor and goes laterally from one hemi-unit tail to another inside the cell.

But we are not yet out of the woods! There remains the question of the mechanism of the bypass—how the information gets from head to tail. There is no question that the tête-à-tête outside the cell is the pivotal act in this communication. Just block stage 1—a mutation of the donor or receiver domain may be enough—and everything down the line grinds to a halt. So how does this act at the head of the receptor molecule elicit the phosphorylation at the tail if there is no allosteric link between the two portions? Here we will borrow from a model that the molecular biologists Joseph Schlessinger and Axel Ullrich have advanced for receptor activation by the binding of its hormone. Backed by experiments with genetically engineered receptors, Schlessinger and Ullrich proposed that the receptor molecules in their unbound form cannot get close enough to one another for autophosphorylation; their head conformations hinder them from doing so. But upon hormone binding, these heads switch to a conformation allowing them to draw closer. Though originally formulated for hormonal communication, this idea may well also be applicable to syncretic communication; the diffusible and syncretic ligands carry the same insignias (Fig. *11.3*).

Thus, a conformational change of the outside region of the receptor molecule would mediate stage 1a, and this change would lift the steric (or electrostatic) constraints for their mutual lateral approach. In this way, the syncretic communication scheme becomes coherent: *it consists of cooperating pairs of units where information flows from donor to receiver head and reemerges inside the cell, but crossed over* (Fig. *11.5*).

Viewed from the top, the syncretic information flow now cuts a classy silhouette: χ. This is quite a production for such a little receptor demon, especially if you consider that this one cannot twist or crane. But there are two demons in this act, and what *one* can't do alone, a pair can, through cooperation.

This doesn't, of course, still our sense of wonder and, if experience is a guide, there should be a cool sleight of hand in store for us. Well, if you are still game, here is one more script for the demons' playbook:

(once activated by their syncretic ligands and freed of their head constraints), the demon pair glide the thermals on the membrane track (they slither ever so slightly to and fro on that well-oiled track) to make one.

The entire act will take only a fraction of a second, but it will get the demons exactly what they want: information continuity (Fig. *11.5, bottom*).

A crossover may seem a strange way for getting continuity, let alone communication; but not if you are on a shoestring budget and are short of Lebensraum. Indeed, decussating the information lines was one of Evolution's clever gambits, which permitted her—pardon the double entendre—to make ends meet. Consider, for instance, the operating expense of the demons' sleight of hand. The pair merely dance to the tune of the thermals here, while the membrane lipid crystal guides them to their link-up. The dance is for free, and the crystal—that jack-of-all-trades—has already been paid for several times over. Also, the demons themselves are sensibly priced; their helical necks are as short as they could be and still reach across the membrane. Contrast this with the receptor demons of

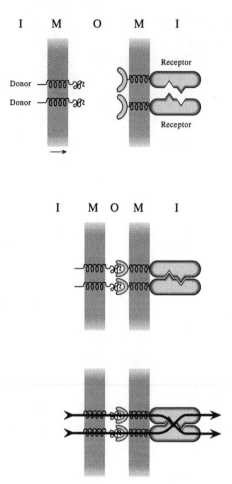

Figure 11.5   A crossover of syncretic information. Two stages in a model of syncretic receptor dimerization: *Top,* two receptor molecules are standing by. *Middle,* their extracellular portions have undergone a change in conformation upon interacting with the donor molecules and, after a final stage of lateral diffusion in the membrane, the receptor molecules have formed a stable lateral union that allows them to phosphorylate each other's cytoplasmic portions (phosphate acceptor domains are represented as notches). The syncretic information flow is sketched on the *bottom* diagram; the phosphate acceptor domains mark the points of decussation. *(M)* membrane; *(O)* outside medium; *(I)* cytoplasm.

the hormonal system, whose helices snake seven times across (Fig. *10.8*). Such long polypeptide chains have more allosteric snap, but they burden the system with spiraling inflation.

As for lebensraum, a small neck-size has its obvious advantages in the politics of a crowded membrane district. But the whole decussation gambit wouldn't be worth a damn if the χ were asymmetric. Indeed, there is not much leeway here;

even the slightest asymmetry would bring on crosstalk between lines. That interference could be remedied by signal processing—such cures are implicit in
Shannon's Theorems—but Evolution avoided those information-costly things
by the simple expedient of stamping the cooperating demons from the same
mold. All she did was use the tried and true copy method—a frugal habit from
the Cell Age she would never wean herself of. As for the result, that could hardly
be improved upon: not only did she get symmetry, but identity of decussating
information lines.

Metamorphosis

Now that we have its measure, where else should we look for the gambit? We'll
have to rummage a little through the nooks and crannies of the membrane sandwich. Those are inviting places for all sorts of demons to rest their heads. Among
the ones carrying the EGF insignia, there are: proAR, proVGF, notch, delta,
crumbs, lin-12, and gp-1, apart from the standard-bearer himself. The list gets
longer if we include the characters going incognito under prolix names, like
"colony-stimulating factor," "tumor necrosis factor," and so forth.

This is as far as I dare to go now with this form of intercellular signaling. But
before we tackle the next form, let's pause a moment to seek out the common
denominator between the last two communication modes we dealt with. So far,
we have put the differences in signal transmission in the foreground, but regarding their molecular structure, the syncretic and the protein-hormonal forms have
much in common. We only have to hold up the two together in the EGF paradigm to see that they share a modular piece (the family insignia) on the donor-
cell side and the entire receptor molecule on the receiver-cell side. All this bears
the unmistakable marks of Evolution's shuffling hand, but here she seems to
have gone a step beyond genetics and reached into the realm of protein. Indeed,
she capped her endeavor of combining one-dimensional information pieces, as
we shall see, by breaking up three-dimensional ones.

The strategic element in this splintering is a bond in proTGFα, a covalent
bond between an alanine and a valine, located close to the outer face of the cell
membrane. Split it, and you break a good chunk off proTGFα (Fig. *11.3, first
arrow*). And this is precisely the bond Evolution went after. The alanine-valine
pair is part of a motif of small polar amino-acid residues, which fits into the cavity of one of her enzymes, where the bond is stressed to the breaking point. So
in that cavity, the head of this molecule gets chopped off, losing its anchorage in
the membrane.

This enzymatic act, which takes place in many organs and tissues, is fraught
with more drama than my phrasing could convey. The reasons are all in that
head. They are spelled out in two words: *EGF information*. Though not much
by itself, this information adds up to a mountain when there is a Circus with

EGF receptors around. That Circus then provides the lion's share of the information, the context (see p. 198). Thus, severed from the membrane, the head becomes a means for long-distance communication—the syncretic communication apparatus metamorphoses into a hormonal one!

Well, wonders never cease in Informationland. Here a syncretic molecule metamorphoses into a hormone . . . there a caterpillar, into a butterfly . . . over there a girl, into a demoiselle—they are all but chips off the old block: the immense hunk of molecular information coherently deployed in two realms. But it is at the base of that block—the lower communication levels—where those wonders start. And down there, things are not yet too churned up to see the whirligig at the bottom. This is the same kaleidoscope—half wheel of information, half wheel of chance—we have seen bringing forth one creation after the other in our voyage along the intracellular stream. Here at one of the primary intercellular branches of the stream, we just saw it functioning in the dissociative mode, and (because of the coherency à outrance of what it whirls) its creativity continues undiminished. The syncretic molecules, like the hormone molecules (and the molecules that make up the cell-to-cell channels which we will take up next), are the information links that tie the cells of an organism together. But if we can tear ourselves off that physiological three-dimensional stereotype and think in four dimensions, all these molecules turn into links of another sort, phylogenetic links that tie the multicellular rungs together.

Intercellular Communication Via Channels in the Cell Membrane

The Cell-to-Cell Channels

Now to the other membrane demons with a penchant for cramped spaces, the ones who implemented Evolution's fourth strategy for Circus union (Fig. *9.4 d*). These demons are of the six-subunit tubular-protein sort (Fig. *10.11*) which boasts a wide central passageway (16–18 angstroms) that admits not only the inorganic molecules, but also quite a few of the organic ones inside cells. Not that this passageway is devoid of polar amino-acid residues—it is lined with them just as the passageways of the ion channels we saw before are, or else it wouldn't let the water through. But the electrostatic field, the shield associated with such a wide aperture, makes itself felt only to ion passengers with some body to them, charged molecules of a mass of 300 daltons or higher—the small inorganic ones just zip through.

So, it is in a relatively high molecular-size range, a range concerning only organic matter, that these tubular entities show some Maxwellian bent. But this is only part of their savvy. The other part, the lion's share, is for their own macromolecular kind. That savvy is built into their ends facing the intercellular space. It enables them to put that space under surveillance and to spot their opposite numbers across the membrane gap—a mutual act of cognition that culminates in two partner demons interlocking end to end, forming a 12-piece tube connecting two cell interiors, the *cell-to-cell channel* (Fig. *12.1*).

This channel is thus a composite to which each cell in a junction contributes one-half the genetic information. That information quantum-codes for a protein called *connexin*. It is a quantum that has been well conserved through evolution, and so we find this intercellular connector all over the zoological map, from coelenterates to humans.

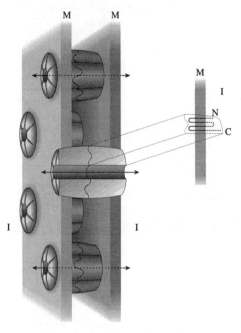

Figure 12.1   Intercellular membrane channels. The diagram on the *left* represents a cluster of channels at a junction between two cells. Each channel consists of two interlocking halves, one half from each cell membrane *(M)*. Each half is made of six subunits. The cutout on the *right* represents the polypeptide chain of a subunit; the relatively conserved portions are shown as solid lines—*C*, carboxyl end; *N*, amino end. The arrow indicates the aqueous passageway between the two cell interiors *(I)*. (After W.R. Loewenstein, *Physiol. Rev.* 61:829–913, 1981, by permission of the American Physiological Society.)

How They Form

The cell-cell channel is well-adapted to the vagaries of the wafer-thin world it inhabits. Like the other membrane proteins we have seen, it is in thermodynamic harmony with the membrane lipid and in tune with its constant flux. But, in addition, it is in tune with another flux: the shifting contact relation between the membranes. Anatomical diagrams notwithstanding, these relations are rarely static in a living tissue. Cells slide alongside one another, breaking contacts and making new ones, they swell and shrivel, and they die and get replaced by others— an endless sea change which calls for a highly adaptable connection device. But what could possibly maintain an intercellular connection in the face of such utter contact instability? If you asked an engineer to make a connector that would hold up in such a quagmire, he would tell you to take your dreams to the psychiatrist's couch. There is certainly nothing in his considerable bag of tricks that could even remotely fill that bill. Yet precisely this sort of thing, one-and-a-half-billion years ago, trundled out of Evolution's protein workshop. Well, where else would a thing like that come from, something with a quality so truly Protean that our dictionaries have no word for it?

The thing is a Maxwell demon, of course, and this one is as rich in information as it is nimble: it recognizes its counterpart across the intercellular gap and, in the twinkle of an eye, dovetails with it. Virtually all cells that form cohesive tissues are endowed with demons of this sort; their junctional membrane regions are literally studded with them. And this is why they take the contact whims in stride, for their demons see to it that there is a degree of intercellular communication at all times.

To see how they accomplish that feat, we will take a brief look at the experiments that originally brought the demons out of the closet. A major clue came in 1969 when Shizuo Ito and I measured the electrical conductance across the forming junctions between embryo cells. We used early newt and frog embryos because their cells are large and easily isolated in the test tube. We put microelectrodes inside the cells and manipulated them into contact, while monitoring the electrical conductance across the forming cell junctions. Our results can be summarized in few words: a cell-to-cell conductance developed after contact, rising gradually, over a few minutes, to a plateau.

These findings bore a strong message. Just three years before, probings of various adult tissues with fluorescent molecules had led me to postulate the existence of the cell-to-cell channel. Thus, in this light, the gradual increase of conductance implied that such channels accreted at the region of cell contact during the rising phase of conductance. But we were not able to detect the individual channels. That had to wait another nine years, until we could boost (by means of phase-lock amplifiers) the resolution of the measurements to the order of magnitude of the single-channel conductance, 10^{-10} siemens.

With that resolution, Sidney Socolar, Yoshinobu Kanno, and I were now in a position to eavesdrop on the forming channels. Indeed, the nascent channels showed themselves individually as quantum steps of conductance. And as we followed the process from the ground up, the first quantal jumps of nascent channels appeared within seconds of cell contact, and over the next minute or two, the conductance kept rising, step by quantal step, as more and more channels formed.

The next question was whether the regions where these channels formed were fated. This point was tested in a variant of the experiments, where we broke the cell contact and remade it elsewhere. The cells here were pushed together at randomly chosen spots—spots chosen at the whim of the experimenter, not of the cells. The answer was clear: channels formed wherever the cell contact happened to be. They did so even when their protein synthesis machinery was blocked, as Judson Sheridan and colleagues showed—which suggested that there was a pool of channel-precursor material ready and waiting in the unjoined membranes.

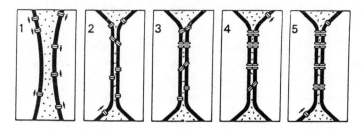

Figure 12.2   A model for intercellular channel formation. (*1*) The protochannels—the closed precursor channel halves—move randomly in the membrane lipid matrix. (*2*) At a membrane junction, they get within (head-to-head) attraction range of each other, interlock, and open up, forming the complete cell-to-cell channels. (*3–5*) With time, channels concentrate at the junction, as more and more protochannels get trapped by their mutual attraction energies and the membrane bending moment. (From W.R. Loewenstein, *Physiol. Rev.* 61:829–913, 1981, by permission of the American Physiological Society.)

All together, the experiments now offered us a reasonable number of constraints to construct a model where the (six-piece) channel halves are the precursors. The model is simple: *the channel halves diffuse sidewise in their cell membranes until they come within end-to-end attraction range with their vis-à-vis at the regions of cell contact; then, they interlink across the intermembrane space, trapping themselves* (Fig. *12.2*).

Underlying the model is the assumption that the channel halves are not open in the unjoined membranes (they don't make the cells leaky). The channel halves are virtual channels or *protochannels*—allosteric elements that open up only upon end-to-end interaction. This cognitive interaction makes the system self-organizing. The channels self-assemble from their halves and cluster by the tens or hundreds at the regions of cell contact. Images of such clusters can be seen in the electron microscope. The channels are large hunks of protein and their clusters stand out from the membrane background. Such images go under the name of *gap junctions.*

The heart of the model is the protochannel trap. The protochannels get trapped in the space between the connecting membranes (a 30–40-angstrom-wide cleft between the two lipid bilayers)—they stick out into that space just the right amount to electromagnetically interact and interlock. And once a pair interlocks, it is forever caught in that space. The thermal energy, the force that got it in, can't get it out; it is far too weak against the energies that pin it down, the mutual electrostatic attraction energies of the protochannel pairs and the elastic energies (the bending moments) of the membranes.

This macromolecular trap, indeed, the whole channel-formation mechanism, is glaringly economical. No special membrane markers here, no bells tolling for channel congregation—just a pair of humdrum membranes, and . . . abracadabra, in a matter of minutes, a full-fledged macromolecular organization materializes in the space between the membranes! All that is needed is some phosphate currency to pay the entropy tithe for the demons' rounds of cognition, as prescribed by equation 3. The rest is plain footwork. The trick, of course, is to constrain that random walk, so that the probability for vis-à-vis encounter rises sufficiently at the spot of membrane contact. And that's the intrinsic beauty of this trap; it makes that probability soar—from zero, where membranes are not in contact, to virtual certainty, where they are. So, once two cells are brought in apposition, the formation of the channels is largely automatic and their accumulation only a matter of time.[1]

The trick is Evolution's, of course—and it has her economy marks all over. Indeed, if we look at the trap mechanism from her point of view, she got a good bargain here. With a rather small information investment for a cognitive protein site—a site for mutual cognition—she obtained a splendid macromolecular organization. That cognitive site seems to be her only new investment. The membrane part of the mechanism (and this includes the double sliding track) was there from an earlier age and so was probably much of the channel protein itself. The mutual cognition site represented a quantum leap over the channel-protein information of yore, but it's something we may imagine to have been affordable by the time of the Grand Climacteric.

The experiments I recounted above perhaps tell something also in this regard. The cells used in these experiments were from an early embryonic stage, and so their membranes may have mirrored the frugal beginnings of the multicellular organismic age. We recall: upon contact, these cells formed cell-cell channels seemingly anywhere on their membranes. However, one should keep in mind that the protochannels may not always have the run of the whole cell membrane. In highly differentiated adult cells things may be more restrictive. We know that they are for other tubular membrane proteins. For example, the proteins that pump sodium ions out of epithelial cells are often located only on one side of these cells, the side they are shuttled to (by the endoplasmic reticulum). The proteins here seem to form microdomains stabilized by electrostatic interactions with charged membrane lipids. Thus, it would not be surprising if in differentiated cells the membrane territory of the wandering protochannels would be somewhat more restricted—a billion years of conservative management makes an information capital grow enough to pay for a little finery.

---

1.   Computer simulations of the model, where protochannels are initially, randomly, and sparsely scattered in the unjoined membranes, yield times for junctional channel saturation (hexagonal packing) nicely matching those for junctional conductance saturation observed in the experiments with embryo cells. And the computed steady-state channel distributions, including details of pattern, like the haloes around channel clusters when the protochannels get depleted, are in satisfying agreement with the electron-microscopic data, too.

How Cells Are Excommunicated and When

Let's now put these channels in the organismic context. They are universal fixtures in organs and tissues—as universal as anything ever gets to be in this cellular polychrome. By virtue of their internal continuity with the cytoplasmic water phase, these channels tie the cells into communities which are continuous from within. And this is their primary function—they are communal devices. Such interconnected cell communities get to be quite large at the higher phylogenetic rungs. In our own organs and tissues—skin, glands, gut, liver, kidneys, lungs, heart—thousands or millions of cells form communities. But there is a price to be paid for such a sweeping connectivity: the community becomes vulnerable as a whole; the slightest puncture can make that community leaky—the injury of even one cell may spell doom for all!

Evolution evidently had to find a safeguard here—a safeguard that must have been in place early on in that canalicular intercellular enterprise of hers. One could hardly imagine that that enterprise could have prospered, or even come off the ground, without some means to hold a damaged cell incommunicado. How else could our skin endure a bruise—even the tiniest one—our liver bear up to a sore from a drinking bout, or our heart survive the death of even one cell?

Evolution didn't have to start from scratch to build a safeguard. Like all her products, this one is more concoction than creation—an amalgam of old information. Indeed, there was a whole informational assortment here on hand in the old repository from single-cell times. The ion channels of one-celled organisms (and their nonjunctional successors in multicelled organisms) all can close down—in fact, they spend on the whole a lot more time closed than open—this is what enables them to maintain the vital electro-chemical gradients across their membranes. Some of these closure mechanisms are controlled by the electric field at the cell membrane, others are controlled by specific molecular signals (ligands)—we have already come across specimens of either sort, voltage-sensitive and ligand-sensitive ones.

Thus, information for channel shutter mechanisms was not wanting at the Grand Climacteric, when the bells tolled for cell congregation. And based on that old genomic information, various shutter designs that were suited to intercellular communication arose in due evolutionary course. We will concern ourselves mainly with one design here, a mechanism of the ligand-sensitive sort where that old signal crackerjack, the calcium ion, plays a key role. This design is found in a class of cell-cell channels present throughout the zoological scale, and there are both structural and electrophysiological data on hand to help us figure out how such complex molecular pores get closed. These channels, Birgit Rose and I showed, respond to 10^{-5} M calcium concentrations, which is well above the usual background level of the ion in the cytoplasm. They react rapidly to the ion. Indeed, they literally snap into the closed conformation, as we shall see.

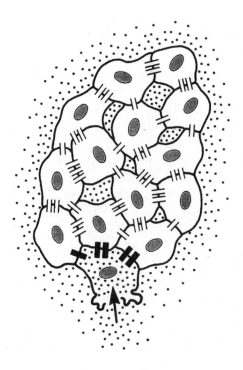

Figure 12.3 Channel closure by calcium: the cell community seals its borders to an unsound member. The cartoon illustrates the situation in which a cell in a tissue has sprung a leak. The consequent influx of calcium ions (dots) from the external medium causes the intercellular channels to close up, holding the cell incommunicado. Open channels, *white*; closed channels, *black*.

This calcium-sensitive design follows an old trend. It makes use of the low level of calcium ions in the cytoplasm. That level is normally very low, under 10^{-7} M— and is held there, as we have seen, by ATP-driven pumps in the cell membrane that expel the calcium ions against a ten-thousand-times-higher concentration on the outside. Thus, when a cell springs a leak or its calcium pumps give out, the cytoplasmic calcium level will rise at once and the canalicular junctions seal off (Fig. *12.3*). The pumps or their fuel supply are among the first to go when cells get sick or reach the end of their life span—calcium makes a good alarm signal.[2]

But what of the ordinary calcium signaling that goes on all the time in cells under the influence of hormones? Will those pips set off the alarm and seal the junctions? Here the calcium mopping machinery comes into play in its ancient role as purveyor of communication privacy. Just as, by dint of this mopping, dif-

2.   Another alarm signal may be electric. The collapse of the membrane potential—a change of almost a tenth of a volt—that is associated with cell injury may be such a signal for the voltage-sensitive cell-to-cell channels. Those channels close in response to depolarization of the cell membrane or to transjunctional potentials.

ferent hormone messages are kept separate from one another and prevented from reaching the wrong intracellular targets, so are those messages kept separate from the alarm signal and its particular target (Fig. *10.15*). The message sources—the calcium channels of the G-network—are for the most part in non-junctional membrane regions. Thus, given the efficiency of the calcium mopping and the relatively high calcium concentrations needed for channel closure (10^{-5} M), the hormone messages, in effect, are not within earshot of the cell-cell channels in many types of cells.

So, the alarm won't be set off by every little calcium blip; it takes a good gush, like the one in the wake of an injury. And it is for contingencies like that, circumstances which put the whole cell community at risk, that the community exercises its last resort: excommunication.

How to Achieve Control in One's Birthday Suit

Thus, the channel shutter mechanism provides the cell community with a safeguard against local injury. The community insulates a hazardous member by sealing off its borders. This control is well in tune with the organismic environment, a milieu where nothing is forever cradled in the lap of stable luxury. Cells get unwell, die, or come loose, ion pumps wear out and membranes frazzle, but the cell community endures because it never loses control over its borders.

One is caught between wonder and dismay by the simple elegance of this mechanism. But perhaps by now we ought to be inured to the wonderwork of demons. And this particular demon sports no less than two cognate sites: one looking toward the cell exterior, with which he senses his partner across the gap, and another one accessible from the cell interior, with which he senses the cytoplasmic calcium-ion concentration. That second cognitive act leads to the closing of the channel and gives the cell community the power of controlling its compartmental borders.

This control—not precisely your run-of-the-mill cybernetic engineering loop—needs some clarification. A glance at the channel situation from inside a cell with ligand-sensitive channels will tell us much of what we need to know. The channels by and large normally stay open, and this gives the junctional regions their peculiar stamp: they look more like sieves than membranes—they are regions punched full of holes. But that situation changes radically when the cell is injured or takes ill. Then, these regions turn into walls—as if on command, every one of its channels closes up. In information terms this means that the channels have a signal line that keeps them informed about the health of the cell pair they connect, and automatically go into the closed state in a community crisis signaled by calcium. And this prodigious cybernetic manifold is achieved with no more than calcium and water.

This is cybernetics utterly stripped to the buff. But it is the sort of thing we have come to expect by now from the products of the evolutionary workshop—a workshop that unswervingly abides by the principle of information economy.

The Channel Shutter

This much about communication control. What of the shutter mechanism itself—*how* does the channel close? How does a tubular protein, whose core is lined with hydrophilic amino acids, suddenly become impervious to the cell water and its smallest solutes? Here we move into largely uncharted territory. However, some progress has been made by X-ray and electron diffraction of channel molecules crystallized in membranes as planar or helical arrays. This is still a somewhat hazardous art, especially as applied to membrane junctions. The inherent structural disorder in the membrane sandwich has been keeping the resolution down to about 18 angstroms so far, which precludes seeing atomic details of the protein configuration. But the broad outlines of the molecular architecture can be gleaned by three-dimensional electron image analysis.

But before we take up this point, let's ponder what to expect of a closing channel structure on purely physicochemical grounds. Or perhaps, to warm up, we should start with what *not* to expect. There is first of all the delicate thermodynamic balance between lipid, protein, and water in cell membrane, which took aeons to be achieved; the closing of the channel must not upset that venerable apple-cart. Therefore, we would not expect the channel protein to undergo changes in conformation on a scale that would seriously perturb the surrounding lipid structure or greatly alter the proportion of hydrophilic and hydrophobic subunit surfaces in contact with lipid and water. This, from the outset, eliminates a number of possibilities—large-scale contortions, distensions, or contractions of the individual channel subunits—and serves as a useful guide for what to look for.

As for positive expectations, topping the list is one concerning the extent of the channel closure: the pore closes all the way, that is, to the point of exclusion of all cytoplasmic molecules—including the small inorganic ions like potassium and chloride—regardless of their electrical charge. This follows directly from the results of experiments in which we probed the permeability of cell junctions with neutral fluorescent-labeled molecules and measured the junctional electrical conductance while the cytoplasmic calcium-ion concentration was elevated. And these experiments also showed the other side of the coin: the channel promptly reopened when the normal calcium-ion concentration was restored.

In short, *the channel here seems to get blocked by actual matter and not just by an electrostatic field, and the blockade is reversible.* All this narrows down the number of possibilities quite a bit and leads one to suspect that the channel closure is due to an internal structural rearrangement where nonpolar amino-acid residues may get deployed into the channel water space.

With these a prioris, we can get down to anatomical brass tacks and take stock of the X-ray diffraction and electron images of channel crystals. Nigel Unwin and his colleagues have obtained suitable images by rapidly freezing the membrane material in a thin film of water medium and observing it at less than

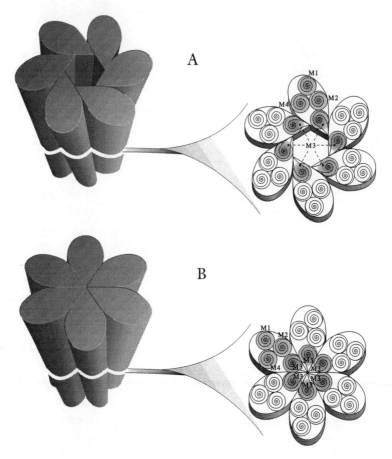

Figure 12.4   Architecture of the cell-to-cell channel. (*A*) A view from the cytoplasmic side of a channel half in the open and closed configuration. The six subunits delineate an aqueous passageway down their center in the open configuration. The cutout represents diagrammatically how the four M-helices (the transmembrane segments shown in Fig. *12.5*) may be packed in a subunit and how the M3 helices may be deployed around the passageway. (*B*) In the closed conformation, the M3 helices have rotated, swinging out hydrophobic amino-acid residues (see Fig. 12.5)into the former aqueous space. [The diagrams are free renderings after models proposed by P.N.T. Unwin and G. Zampighi, *Nature* 283:545–549, (1980); and by Unwin, *Neuron* 3:665-676 (1989) by permission of Macmillan Magazines and of Cell Press.]

–130°C where the ice remains stable and the channel protein can be viewed against the lipid and water background. A schematic representation of the channel half based on Unwin's reconstructions is shown in Figure *12.4*: a structure with six columnar 75-angstrom-long subunits symmetrically deployed and slightly tilted, delineating a central water pore. The whole looks somewhat like the iris diaphragm of a camera, where the *concerted* rotary movement of the unit pieces opens or constricts the aperture.

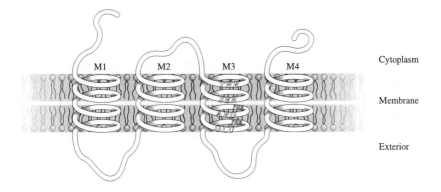

Figure 12.5 The polypeptide chain of a cell-cell channel protein and its deployment in the cell membrane. The four M-helices of one of the columnar subunits in Figure *12.4* are displayed here in a plane. The third (M3) exhibits a ridge of polar residues down the helix (*shaded*; this ridge borders the aqueous passageway in the open conformation of the channel in Fig. *12.4 A*). The ridge is flanked by two large nonpolar residues (PHE). The set of residues illustrated is from the sequence of rat-liver connexin. (THR = threonine; SER = serine; ARG = arginine; GLU = glutamic acid; PHE = phenylalanine)

And here we come to our main point: when calcium is added to the medium, the channel-half changes conformation; the subunits overall seem to straighten up and—all part of the same motion—rotate and slide radially, closing the pore (Fig. *12.4, B*).

This makes for a smooth operation. At each individual subunit, the movement is rather small, causing minimal electrostatic and mechanical disturbance to the surrounding membrane lipid. But the movements of all six subunits are concerted, and so their displacements sum up.

A Channel Yin-Yang

The shutter thus emerges as an element built into the columnar channel architecture—a six-piece rotary shutter wrought into a six-piece channel. The prospecting of the channel innards awaits methods of higher resolution. Meanwhile, we will fill in some detail, based on connexin DNA sequences determined in the laboratories of David Paul, Norton B. Gilula, Dan Goodenough, and others. Some stretches of that DNA have remained unchanged over aeons, providing useful clues about the molecular channel architecture, including the shutter.

Consider first how the connexin polypeptide might be deployed inside a channel subunit. The subunit, as delineated by image analysis, is a column with a cross section of about 500 square angstroms; and the polypeptide chain, as indicated by the hydrophobic and hydrophilic qualities of its amino acids, folds up into four transmembrane segments (Fig. *12.5*). Four such segments in α-helical form, bundled together, would fit nicely into such a column (Fig. *12.4, right*). This would also stand to reason from the mechanical point of view; α-helices are quite stiff (they have a hydrogen bond between every fourth peptide group), and such a bundle of helices would be well-suited to transmit, or even amplify, a ligand-induced conformational change through the membrane.

Consider next which of the four helices might provide the lining of the channel. There isn't really much of a choice. Only one of the transmembrane segments, the third traverse (*M3*), has a significant number of hydrophilic residues; the other three are quite hydrophobic. The *M3* chain is strung out so that there is one hydrophilic amino-acid residue in every fourth amino-acid position. But you really begin to get your ducks in a row if you work this out in three dimensions: the hydrophilic residues are in line, forming a polar ridge down the α-helix (Fig. *12.5*); and six such ridges round the circle (together with their backbone CO and NH groups) would make a good channel lining, indeed.

Now, on the offside of that polar ridge, there is a set of nonpolar amino-acid residues. These are large chunks of hydrophobic matter that would be large enough to blockade the central water space if they were to swivel as part of the rotary-shutter movement. The two phenylalanines in liver connexin, for instance, which in Figure *12.5* are shown to flank the polar ridge, would completely obliterate the water space in a rotary movement of the α-helices (they extend 9 angstroms from the axis), while the small polar-ridge residues would get buried at the sides. Unwin has advanced a hypothesis in these terms, elegantly rounding out the channel model: *the nonpolar residues would swing into the former polar space by the rotation of the subunits during their concerted iris-diaphragm movement.*

This makes a plausible model also for some nonjunctional membrane channels. The M3 motif—a polar ridge with nonpolar wings—is found in a variety of ligand-sensitive channel types, such as the acetylcholine-, GABA-, and glycine-receptor channels of the nervous system, which have very different functions from those of the cell-cell channels. Thus, this rotary-shutter design—a Yin-Yang based on amino acids with opposite affinities for water—may have been used for quite a few cellular portholes.

What Goes Through the Cell-Cell Channels

To consolidate concepts, let us take stock of the main things that have transpired so far about this modality of intercellular communication: The unit of cellular interconnection is a bipartite membrane channel made of twelve subunits. The channel closes in exigencies, thanks to an intrinsic shutter mechanism; but normally it stays open a relatively long time. Thus, in organs and tissues, where such channels are abundant, the cells form systems which are continuous from within.

But continuous with respect to what? What precisely are the signals that go through these channels? The intercellular signals here are not as tangible as they were in some of the other communication modalities we dealt with before—the

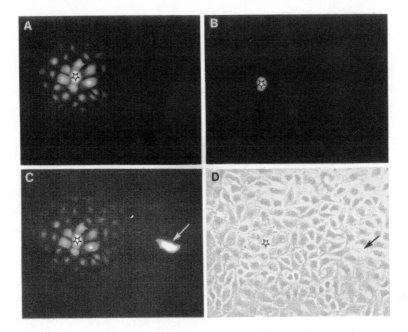

Figure 12.6 Probing the cell-cell channels with fluorescent molecules. The photographs illustrate a set of experiments in which a mixture of green-fluorescent glutamate molecules (500 daltons) and red-fluorescent maltotetrose molecules (1,200 daltons) was injected into one cell of a liver culture. The *top-row* photographs show the intracellular fluorescence 3 minutes after the injection through filters that display either the green fluorescence of the small molecules (*A*) or the red fluorescence of the large molecules (*B*). The small molecules have spread from the injected cell (marked with a star) to many neighbors, while the large molecules do not go beyond the borders of the injected cell. Next, the calcium-ion concentration inside the cells was elevated by a chemical agent (diacylglycerol), and the molecular mixture was injected again into one of the cells to the right of the field (*arrow*). Photograph *C* shows the green fluorescence three minutes after the injection. Now the small molecules also stay within the confines of the injected cell (on the left is the cell field of the previous injection). Photograph *D* shows the cell culture in bright-field microscope view.

cell-cell channel is but the conduit, not the receiver, here. The receiver molecules—and there are bound to be quite a medley in this communication form—are still largely unknown. We are not entirely in the dark, however; we know the size of the channel pore and this gives us some useful clues. Indeed, with a little help from information theory, we get a fix on the general molecular category of the signals, as we shall see.

Let's begin with the pore size. This was relatively easy to measure. We prepared a series of fluorescent-labeled hydrophilic molecules in our laboratory—a graded series of known dimensions—that could be injected into cells, and used these molecules as yardsticks. The spread of the fluorescence to the neighbors of the injected cells told us whether the molecule in question would go through the cell-cell channels or not. Figure *12.6* gives a sample of an experiment in which the channels were probed with a molecule small enough to fit through the channels (500 daltons) and one too large for that (1,200 daltons). This experiment also gives us an idea how far and fast a cytoplasmic molecule of 500 daltons would spread through a population of cells and what happens when the channels are shut down by the calcium ion (Fig. *12.6 C*). Now, with a wide-enough repertoire of such fluorescent probes, we could put the yardstick to the channel aperture with reasonable accuracy, and the upshot was an effective chan-nel diameter of 16–18 angstroms in mammalian cells and a little over 20 angstroms in some invertebrate (insect) cells.

That is quite a hole in the membrane. It makes the ion channels look like pin-pricks—even the largest ligand-sensitive ion channels are only about half that size. It is wide enough not only to let all the cytoplasmic inorganic ions through, but also the nucleotides, amino acids, small peptides, high-energy phosphates, and basic cellular metabolites—all but the kitchen sink . . . though not quite. This aperture still has some cytoplasmic sieving ability, and precisely where it counts: in the macromolecular range. The selectivity starts at the level of about 1,000-dalton molecular weight. So, the channel waves off the DNA, RNAs, and proteins—the carriers of genetic information.

Viewing the Signals Through the Looking Glass of Game Theory

Now that we have the channel's measure, we can take the question of the com-munication signals under active consideration. No need to look among the cyto-plasmic proteins here—given the above pore size, we can skip the whole macromolecular dominion and concentrate on the smaller fry. This cuts short the quest, though not enough to make an actual molecular chase appealing; that small fry is still a fair-sized crowd. But is there perchance something else besides the pore size—a hidden clue—to help us narrow down the number of molecu-lar possibilities? When the plain and plumb facts fall short, as they are bound to do some time or other in any scientific enterprise, one gently avails oneself of any useful clue that comes along, and says thanks for footprints, waves, or wakes, and other telltales left here and there in nature. Those tip-offs—and we have seen some examples in Chapters 2 and 6—are the bonuses from a coherent and conserving universe, which a scientist can lay claim to on occasion. A special feel-ing goes with that, which is hard to put into words. Einstein once put it this way: "The Lord is subtle, but he is not mean."[3]

Well, with such things one can never be quite sure, but I think that there may be such a tip-off concealed in the pore size here. It takes the looking-glass of information, though, to bring it out. So, we'll fall back for a moment on infor-mation theory—*reculer pour mieux sauter*. Then, those 18 or 20 angstroms of

---

3. "Raffiniert ist der Herrgott, aber boshaft ist er nicht."

channel aperture, which allow one to bracket the cytoplasmic macromolecules, acquire a deeper meaning. As we put those orifices in a developmental context and consider what they do let through and what they don't, they begin to show some of the earmarks of the "minimax and maximin" gambits of game theory of Chapter 9, for 20 angstroms is about as far as Evolution possibly could have pushed the cell-cell channel aperture, without compromising the genetic individuality of cells. Just a little more, and the RNAs and proteins would go through—which would mean the kiss of death to cellular informational selfhood. So, an 18- or 20-angstrom hole has every appearance of one of those Solomonic saddle-point compromises: it minimizes the shunting of the macromolecular information between cells while maximizing that of the nonmacromolecular information, the molecular small fry.

The exact formulation of saddle-point strategies for the games in the evolutionary arena is still beyond our reach—not enough genomic knowledge is on hand for bidding time return a billion years. But given the end-state of these channels and their bores, we may sketch out an analogue of such a game, where tubular proteins are the players and cytoplasmic molecules are the balls, under the following rules: (1) the balls must go through the players for the players to score; (2) the throughput scores are kept in information bits; (3) while playing, the players remain passive—the balls are impelled by thermal energy; (4) all the players may do is to adjust their bores from time to time, but they may not bring other information to bear, for the game must be played with minimal information—the ground rule for all evolutionary sports.

Now, a useful boundary condition for designing a winner's strategy is implicit in the limit of macromolecular information throughput here—the taboo of intercellular transfer of genetic information we discussed above. So, we can see our way to a saddle-point prescription (the minimax-maximin for the highest throughput score) which, stripped of mathematics, would read like this: "make your bore as large as the nonmacromolecular balls with the highest information density."

Thus, through the looking-glass of game theory, we get a glimmering of what Evolution may have been after when she went for a tubular protein with an aperture so much wider than that of the ion channels of the Cell Age. And looking to the practical experimental side, we now also have our clue about where to fish for communication signals inside the cells: *among the nonmacromolecules with the highest density of information.* This means first and foremost, small peptides—those are the ones that pack the most information. The experimental data from channel probings with fluorescent-labeled peptides circumscribe this some more: short chains of up to about three small (and not too charged) amino-acid residues. And going down the informational pecking order, next in line would be the nucleotides and their short strings, and so on. So, with a little luck, the signals should soon be out of the woodwork, and then it may not be harder to read the intercellular messages than Hammurabi's Code in Babylonian cuneiform.

How Cell Communities Stay on an Even Keel

At this point we shift our line of discussion from the signals to the physiological roles of this canalicular communication form. Here we are not entirely in the dark. Although the identity of the signals still escapes us, some functions may be inferred from the broad-scale cytoplasmic shunting in tissues and from the way cells behave when that shunting is disturbed. And some other functions, if not strictly inferable, leap to the eye from several teleonomic angles. I use "functions" in plural here on purpose, because it would be surprising if such a wide and old channel between cells had not adapted to more than one function.

We start with the most basic physiological role: the equalization of the non-macromolecular content in the cell community. This function is an immediate consequence of the shunting by the intercellular channels, their capacity of short-circuiting the interiors of the cells; in many types of tissues, especially epithelial ones, there are so many channels in the cell junctions that there is virtually cell-to-cell continuity for the nonmacromolecules. Thus, any gradient of these molecules in the cell community, be that chemical or electrical, will be evened out.

One can get an idea of the speed and extent of this equalization from microinjections of small fluorescent molecules into cells, as exemplified in Figure 12.7. Such injections give rise to palpable local changes in intracellular chemical concentrations in a tissue, imposing substantial local chemical gradients. But the gradients are soon evened out, as the substances spread to the cell neighbors and differences in concentrations are undone. For 200- to 400-dalton molecules, the size range of most metabolites, the equalization is virtually complete within a few minutes in a population of thousands of cells.

This smoothing out of local gradients is just the consequence of interconnecting large masses of cells—a buffer action, pure and simple. But for the physiology of the cell community, indeed, for the entire organism, this dampening is of the highest importance: it keeps the cell community on an even keel, despite the inevitable local variations due to genetic differences or environmental perturbations.

E Pluribus Unum

In 1878, the physiologist Claude Bernard, impressed by the stability of our body temperature and the constancy of the contents in our blood and body fluids, introduced the concept of homeostasis, which became one of the cornerstones of organismic biology. He called it *homeostasis of the interior milieu*, the tendency of organisms to keep their internal environment stable. This concept has set the tone for physiological research for more than a century. Homeostasis, it turned out, is due to a large series of coordinate dampenings ranging from simple chemical mass buffering to cybernetic controls, the latter reaching the highest degree of sophistication in the nervous system.

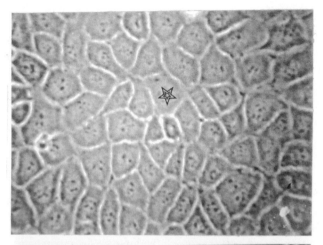

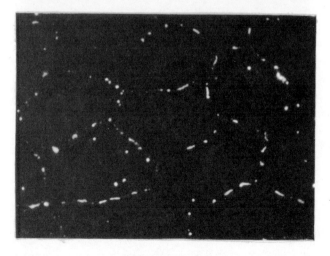

Figure 12.7 Cell-to-cell spread of tracer molecules. Photographs of a sheet of rat liver cells in culture. The cell marked with a star was injected with fluorescent molecules of 300 daltons (which falls in the size range of many cellular metabolites, nucleotides, and other biomolecular building blocks); the *middle* photograph shows the cells 30 seconds after the injection. This fast and far-flung intercellular diffusion is typical of many epithelial tissues; the cells in these tissues make cell-cell channels with all their neighbors. The *bottom* photograph (taken at a higher magnification) shows the distribution of the cell-cell channel protein in the cell junctions of such an epithelium, as revealed by fluorescent antibodies against the cell-cell channel protein, connexin43 (courtesy of Birgit Rose).

But little did Bernard (and the next three generations of physiologists) know that inside that interior milieu there is still another one with the tenure of the most basic homeostasis: the milieu constituted by the communal cytoplasmic space, the communal nonmacromolecular compartment. That compartment, the most interior of interior milieus, keeps itself constant without extrinsic help, and does so with a minimal outlay of information. In fact, insofar as it relies on the mass-buffer action of cells, it doesn't even use cybernetics in the conventional sense. Yet it achieves good stability in the face of transient, or even long-term, local deviations. It matters little how the deviations come about—whether through physiological activity of individual cells or aberrant conditions—any deviations are wiped out so long as the cells are linked by enough channels.

Consider, for instance, what happens when a cell goes on a binge and burns up much of its ATP and metabolites. If it were on its own, that cell might soon scrape bottom. But, in fact, except perhaps for a little dip, its ATP and metabolite levels stay put, thanks to its neighbors; the ATP and virtually all metabolites easily go through the intercellular channels (Fig. *12.8 a*). Or consider a cell whose sodium pumps give out. Its sodium-ion level, instead of rising to deadly levels, will hardly change; the surplus of ions is diluted out by the cell community even as the surplus forms, as these small ions breeze through the channels (Fig. *12.8 b*).

As a further example, consider what happens when a cell is stricken by a genetic mutation that puts a member of the molecular small fry at risk. This is something that happens not infrequently in large organisms. Not only are the mutation rates high, but the mutation of one single enzyme can give rise to deficiencies in whole strings of molecules of the small fry. We will use an instance where such a mutation interferes with the production of a (channel-permeant) building block for DNA—an inheritable human disease called "Lesch-Nyhan syndrome." This is a disorder caused by a mutation of an enzyme which supplies the information necessary for recycling DNA and RNA (hypoxanthine-guanine phosphoribosyl transferase); the enzyme is needed for making, from the purine base hypoxanthine, the ribonucleotide *inositate* that serves as a nucleic-acid building block. The genetic defect that renders the enzyme ineffective is inherited as a recessive, sex-linked trait (males only). The enzyme deficiency manifests itself in a number of tissues, most tragically in the nervous system of children, where it leads to mental retardation. But we'll see the brighter side here: the rescuing of the stricken cells through communication. Such rescuing takes place in the skin of patients, where not all cells are afflicted by the disease. The mutant cells are then bailed out by the normal ones—the mutants, as discovered by John Subak-Sharpe and John Pitts, receive the missing ribonucleotide from the normal cells via their intercellular channels (as in Fig. *12.8 a*). Thus, thanks to the channels, these genetically flawed cells get cured somatically and live happily ever after.

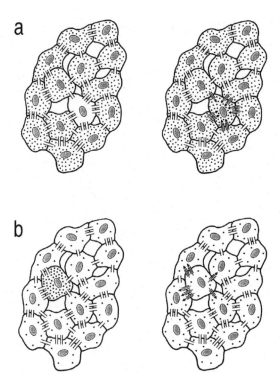

Figure 12.8 The homeostatic function of the cell-cell channels. (*a*) Offsetting a local molecular shortfall. The diagrams represent the situation where a cell in a population has exhausted its store of a molecule that can go through the channels—say, ATP—or lacks a metabolic molecule because of a mutation. Flowing downhill through the channels, the molecules from the neighbors buffer the shortfall. (*b*) Offsetting a local molecular excess. Here a cell in the population has accumulated a surplus, say, in sodium ions, because its pumps have worn out. The downhill outflow through the channels evens out the molecular levels in this cell and its neighbors.

This spontaneous cure is by no means a special instance. A wide variety of human diseases (cancer included) are nipped in the bud this way—though we are rarely aware of it. These little channels see quietly to our health—at times, even to the point of snatching us from the jaws of death—and the girding of aberrant cells by communicating ones in our tissues constitutes one of our primary mechanisms of defense.

This low-keyed strategy of communal defense is of quite a different tenor from what we are used to in our contrarian human societies. The society of communicating cells here keeps *itself* on an even keel, despite all sorts of external and internal perturbations and individual differences. It achieves this by means of a coordinate equalization. This is an equalization of a quite wonderful sort: inequalities are leveled out among the cells without jeopardizing their individuality. Lady Evolution knew exactly where to draw the line; by using a channel with a 16–20-angstrom bore for the leveling job, she, in effect, exempted the stuff of cellular individuality, the cytoplasmic macromolecules, from the equalization. Thus, she brought off that rare thing, the perfect union, a union where many units merge *part* of their information to form a larger one—a union of individuals . . . E Pluribus Unum, as it says on the Great Seal of the United States.

How to Ferret Out the Functions of a Channel

Having dealt with the most basic function of the intercellular membrane channel, we now turn to the question of what higher functions it may subserve. By higher functions we mean genuine communications in the sense of the second theorem of information theory (p. 182), namely, transmissions of information that overcome the noise by virtue of their encoding. Surely, among the multitude of molecules using this pathway between cells, there must be some that were elevated to the haut monde of encryption—the information class biologists call *specific signals*. So long as those signals remain anonymous, we have to rely on macroscopic cell behavior to get at the higher channel functions—we'll just have to do as well as we can with the undissected, polychrome channel throughput. This is a handicap, no doubt, but it doesn't exactly reduce us to twiddling our thumbs, either. There is much to be learned here from plain cell behavior— the whole edifice of physiological science was built that way. Indeed, how cell communities behave or misbehave when their channels are absent or gummed up can be a valuable guide to us. And it isn't too hard to engineer a wholesale channel blockade by inducing mutations in the connexin genes and their viral bêtes noires—there are even some naturally occurring connexin mutants on hand.

However, that genetic road to physiology is not without pitfalls. The information streams inside cells don't just run from A to B—we have already had an eyeful of how well endowed they are with branches. Cell-behavioral anomalies, therefore, are rarely sufficiently unique to allow one to make straightforward inferences of functions from malfunctions. So, it is well to go the road in both directions: first, in search of the channel-blocking mutations and the resulting (communal) cellular malfunction(s) and, then, in pursuit of the cure—gene manipulations that restore the viability of the channels. The proof of the pudding is in the restoring of the function here.

Just as I write these words one such double approach has been successfully completed with connexin mutants, opening a view to a function that is as old and ubiquitous as the cell-to-cell channel itself: the regulation of cellular growth.

The Regulation of Cellular Growth

The regulation of growth of our cells is one of those inborn things we take for granted and don't pay heed to. Nor do we have to, for the steering is fully automated. The control of growth starts right after the egg cell has been fertilized—in the same breath with our life, we could say. The aboriginal cell divides into two, into four, into eight, and so on, to a vast number. And all this cell proliferation is always orderly, disciplined, and harmonious. It actually doesn't take all that many division rounds for cells on such a doubling spree to reach large or even astronomic numbers. A tale from Indian folklore tells us why.

Once upon a time, the story goes, there was a lovely kingdom by the sea, which was in mortal danger. But at the eleventh hour, disaster was averted and the kingdom saved—and all this by the power of the wits of one man. The grateful Maharaja offered to pay the wise man with his weight in gold. But the man humbly said: "I am but a poor mathematician with little use for gold. Perhaps, Great Raja, you could give me some rice instead"—encouraged by the half-gleeful half-incredulous mein of the Maharaja, he pointed to a chess board in the hall— "pray, have the squares on this chess board filled for me, starting with a grain of rice on the first square, two grains on the second, four grains on the third, eight grains on the fourth, and so on." The Maharaja (who was not renowned for his liberality) put his golden seal on the man's wish on the spot. Only the following day, when his Chancellor of the Exchequer brought him the scroll with the numbers worked out (too late to have heads chopped off), he saw what was hidden in that "and so on": by the tenth square, he would have to dish out 512 grains, by the twenty-first square, about a million—all quite modest—but by the fortieth square the reward would have grown to billions, and by the sixty-fourth square, to a truly astronomic number, enough to cover all of India with rice many times over.

What the wise man knew was that when you start with a nonzero number, the doubling rule makes the numbers grow quickly. So, you can easily work out that it takes just 60 rounds of cell division—to come back to our population of dividing cells—to get from 1 to 10^{17} cells, the estimated number in a human body. But what concerns us here most is that, throughout all rounds, the growth, astronomic as it is, is always harmoniously controlled. Indeed, it is so well coordinated and so perfectly balanced that one expects the helmsman to come out of his hidey-hole any moment.

This steering of the cell divisions is most palpable when an organism forms. You can experience this firsthand if you watch an embryo grow. In early summer, our seashores teem with starfish embryos that are beautifully transparent. Inside those fledgling creatures you can actually see the cells dividing. Not much gear is needed: a microscope and a video system to enhance the image contrast.

All we have to do, then, is to put a starfish egg cell in a dish with seawater and let a few sperms loose on it, and we will be regaled with the spectacle of our life: an organism unfolds before our eyes. All within a few minutes, one round of cell division follows another, and a thousand-and-one wonders come to pass. But what keeps one riveted from the start is the synchrony of cell replication: the replicative activities of all cells—the DNA replications, the formation of the chromosome spindles that ushers in each cell division, and the cell divisions themselves—all occur at the same time in all cells of the embryo!

Clearly, the replicative activities here are steered. We may not see the helmsman, but his presence is felt everywhere in the coordination of the cell divisions—the beat of the cellular growth. In due time, when the cells become visibly differentiated and form the primordial tissues, that coordination becomes less global. Then, each incipient tissue, and eventually each cell community in a tissue, has a staccato of its own. But the control never stops; it is only less centralized. From the first breath of an organism to the last, the replications are tightly regulated. Once *that* helmsman takes charge, it is for life!

When the organism finally reaches adulthood, the beat quiets down. In some places, notoriously in the brain, it stops altogether. This augurs badly for brain injuries and the natural wasting of our brain cells as we get older; there isn't a prayer of a comeback once that precious gray is destroyed. However, in most of our tissues the cell matter that is lost is not beyond recall. Cells continue to divide in skin, liver, kidney, stomach, and many other places—nothing like the explosive growth in the embryo, though enough to keep up with the cells that die. And should the need arise in injury or infection, the cells can go back to the embryonic mode: a quick shift of gear, and they are proliferating exponentially, making up larger losses.

This, then, is how we grow from one cell to a trillion, and all our parts get to be the size they are and in proportion with one another. When we talk about growth, we mean an increase in cell number; our body mass is largely the product of the number of cells in it. That number, astronomic as it is, is finite and its various subsets, the numbers of cells in our organs and tissues, are finite too and are strictly held in rein.

That is why our fleshly self stays within bounds. When poets and clergymen complain about the weakness of the flesh, they are barking up the wrong tree—they mean that grey mush in the head. Well, pound for pound, at least, our flesh is highly disciplined. And we are no better or worse in this regard than the other beings on Earth, including the plants, the creatures of utter sangfroid and self-restraint, where the saying goes, "The trees don't grow into the sky."

But why don't they? What is behind that discipline? Who holds the reins in every seed and shoot, every leaf and blossom, every shrub and tree and . . . in every organ and appendage of the human body? This question—the problem of the control of cellular growth—is one of the major challenges in contemporary biology. It is an age-old problem, to be sure, though one that has been especially slow to cede its mysteries. But perhaps the time has come to turn the tide.

However, first of all, the problem must be recognized for what it is: *a question of automated control and, hence, of information loops.* Strange as it may seem, this is not how the problem is generally dealt with in biology. The workers in the field of cellular growth or the related field of cancer have largely focused on growth-stimulating and growth-inhibiting proteins and their genes, as if cellular growth were an insular problem of genetic information, rather than of such information tied into control loops. They have made some deft stabs, indeed, revealing a host of such protein molecules, but we have yet to learn how and where these fit into a cybernetic picture. In the following sections, we bring the problem down to the appropriate reductionist unit, the information loop, and examine it in the light of information theory.

The Cellcosm-in-the-Dish

But before everything, we have to decide what sort of molecular information to train our sights on. There is no dearth of growth-regulating molecules —if anything, we have an *embarras de richesses*. The current journals of cell biology are brimming with reports about growth-stimulating substances or "signals," as they are loosely called, which are produced within the growing tissues themselves or in far-away places, exerting telecontrol.

A prominent example of the tele-sort has been known for nearly a century: the influence of the pituitary over our body growth. That gland is located at the base of our brain; its front end produces the hormone somatotropin, a polypeptide of 191 amino acids, which regulates the growth of the cells in our bones. Broadly speaking, the more of those signals the bone cells receive, the more these cells grow. Pituitary and bone cells are tied by a regulatory information loop that sees to it that the growth of our bones normally stays within bounds (though occasionally the loop gets out of kilter or is geared to the wrong standard, producing a giant or a dwarf).

But for all its physiological importance, this is not the sort of control we are after. What primarily interests us here are the controls that are present in all growing cell populations, regardless of type. And these controls, the most basic ones, are local; they are contained within the growing cell populations themselves.

We can speak with some confidence here, because when cells are isolated from an organism and put into culture, they don't run out of control. It is easy enough nowadays to isolate cells from organs and tissues, including human ones, and transplant them into a glass or plastic dish. The cells will adhere to the bottom of the dish and some will survive and multiply in a simple salt solution plus some nutrients and a few essential molecules. Such cultures may go on for days, weeks, or even years. Typically, as such cells divide, the population first grows exponentially for a while, then the growth tapers off and runs at a small trickle, and finally dries up—a test-tube caricature of the behavior of the cells in an organism (Fig. *12.9*).

The time when the growth stops will depend on how many cells are initially put in culture. But the interesting thing here is not the time, but the number of

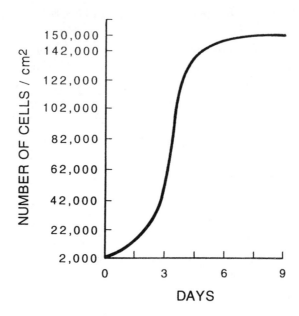

Figure 12.9   The growth of a normal cell population in culture. The cells here—epithelial cells from a rat liver—were originally seeded at 2,000 cells/cm^2 in the culture dish (time zero), and the density of cells in the growing population was plotted against time. The cells stopped growing when the population reached 150,000 cells/cm^2, the characteristic number of this cell type.

cells. What awakens the sleuth in us is that the cells stop dividing when they reach a certain number. Take, for example, the cells from our liver. These cells form a neat, single, cobblestone-like layer in culture, and so the number of cells on the dish can be determined rather precisely. In such populations all cells stop dividing when the population reaches 150,000 cells per square centimeter (Fig. *12.9*). This number is no happenstance. You can do the experiment over and over, and so long as you keep the conditions constant, growth always comes to a halt at this number. With other cell types that number may be different, but each type has its standard number.

So, we have reason to believe that the cells here, isolated from the organism as they are, contain the basic elements of an information loop for growth control. At first sight, such a bunch of cultured cells may not look like much, but once you catch on to their self-reliant swing of growth, you begin to see them in their full information splendor: a self-regulating universe which controls its girth and steers itself to a certain number.

How Do Cells Know Their Number?

That self-steering alone speaks volumes. It implies that the cells somehow can count their number! This is the inescapable conclusion if we view this growth

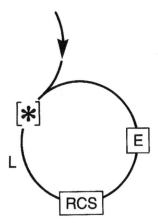

Figure 12.10   The cybernetic loop: (*) *Signals*, (*L*) *signal line*; (*R*) *receiver*; (C) *com-parator*; (S) *standard*; (*E*) *effector*. In the case discussed in the text, *E* is the cellular replication machinery and *L*, the cytoplasmic compartment of the channel-connected cell population (the internal communal space).

behavior in the light of cybernetics, the branch of information theory dealing with automatic control. Thanks to this branch, we now possess, at least to a first approximation, an adequate theory of self-regulation. This theory has turned up quite a few trumps already: the thermostat, the voltage regulator, the automatic speed control for our cars, the self-steering rudder, the gyroscopic pilot, the artificial (feedback) kidney, the cruise missile, the robots.

At the core of all this enginery is the *cybernetic loop*, an information circle not unlike those we dwelled on before but with a special stamp: it steers something to a goal—part of the information flowing in circles is fed back on to the thing to be controlled, the *effector*. Such a loop, consisting of a *signal line, receiver, comparator, standard,* and the mentioned effector, is depicted in Figure *12.10*. The signal line conducts the signals to the receiver which decodes the signal information; the comparator weighs that information against a preset standard and up- or downregulates the effector accordingly. These are the minimal elements information theory sets forth for automated control—any automated control regardless of the nature of the effector, be that a car's fuel injector, a rudder, or a cell's replication apparatus. And when these elements are aligned and in balance, they can hold an effector in rein and steer it to a preset goal— say, a 272° magnetic-compass course, or a 150,000 cells/cm^2 population density.

We will largely gloss over the last three elements of the loop (in Fig. *12.10* they have been lumped into a box, *RCS*) and concentrate on the signal line and the general principle of signal encoding. That will be enough to start up our analysis of the cell-growth prototype. Recall, then, how things stack up in that little cellcosm-in-the-dish: to begin with, the cells divide at a good clip, adding new members to the community, and this continues so for awhile; but at some point, and as if on cue, the cells slow down in their proliferation and, then, skid to a halt when they reach a certain number (Fig. *12.9*).

But how do the cells know that number? How is the information that the cell population has attained 150,000 members per square centimeter—to use the liver-cell population example—conveyed to each of the cells, so that every one of them shuts off its replication machine at the right time? This goes to the heart of the matter. It is actually a twin question: (1) how that feedback information is enciphered and (2) how it gets to the cells. The latter point brings us to the issue that ever since we broached the topic of cell growth has been waiting in the wings: Which form of intercellular communication takes care of the feedback conveyance—the hormonal, the syncretic, or the canalicular form? This is a hard row to hoe without being able to lay our hands on the feedback signals, but because these communication modalities are operationally so distinct, we can make a start by weighing the possibilities of the three forms and see which ones are likely to do the job and which ones are not. In fact, we will get farther this way than we might think offhand; one of the forms turns out to have all the odds on its side.

Indeed, even as we begin to set the three communication modalities against each other, we see that the one that uses the cell-cell channels has all the edge: *its signals operate in a space that is inherently related to the cell number.* Now, this is something quite out of the ordinary. This built-in relationship confers on the communication system the potential to read the cell number—it allows the system to use, without further ado, the signal concentration to impart information about that number.

Let me explain. The space I referred to is the space formed by the connected cell interiors, the continuous intracellular compartment of the cell community— what we might call *the internal communal space.* This is the aqueous space circumscribed by the carapace of the cell membranes and the intercellular membrane channels—a space of a finite volume where all the hydrophilic cytoplasmic signal molecules get diluted. Now, that volume bears a straightforward relationship to the number of cells in the community; it grows in proportion to that number. Thus, any signal produced in that communal space will get diluted in proportion to the cell number—or put in information terms, *the cell number is coded into the signal concentration.*

Parlez Vous Cellulaise?

One could hardly think of a simpler code for a chemical system. A cell in a growing population here—meaning any cell member of the channel-connected community—could keep track of the changing size of the population by merely gauging that chemical parameter, the concentration of the signals. But the full beauty of this mode of transmitting feedback information comes out as we ponder its information cost. Inescapably, there is some cost to information transmission of any sort—the encoding, the conveying, and the decoding, all have their price.

Consider the conveyance first. The signal molecules here amble rather freely through the internal communal space. Such random walking costs no information; it only needs a little heat. Nevertheless, it gets the message speedily to every cell in the community at the biological temperatures. It's the game of molecular seek and find, the oldest game in town.

The encoding is no less thrifty. To encipher the controlling feedback variable (the cell number), the system here merely uses, as we already said, the concentration of the signal molecules in the communal space—the average concentration or some other parameter of the statistical distribution of those molecules. The system, thus, speaks the simplest possible language: the language of chemical concentration. This is but the language we have seen spoken elsewhere inside cells. It is the lingua franca used all along the intracellular information mainstream and its branches—every mechanism in the stream, every enzyme understands it.

I use "understanding" on purpose here in order to underscore the often overlooked second Maxwellian aspect of enzymes. Apart from their well-kenned catalytic aspect, these macromolecules typically are also receivers of information. And this places them smack in the middle of an ancillary information flux that is crucial for biological cybernetics. They receive information from inorganic molecules (often metal ions) or small or medium-sized organic molecules—the "cofactors" and "enzyme activators" in the biologists' parlance—which switch them from an inactive to an *active* (catalytic) conformation. That switching makes them functional kin to the mercury switches or microchips of our room thermostats. But instead of reading a voltage, they read the chemical concentration in their feedback lines and then flip to the on position when that concentration reaches a critical value—a statistical-mechanical minimum called *activation threshold*. And all that cybernetic knowhow is built into the allosteric structure of the enzymes—their own demon selves.

Thus, in a cybernetic circuit, such demons can serve all at once as receivers, comparators, and standards of molecular feedback information. So, we can rightly say that those enzymes understand the language of chemical concentration. And the enzymes involved in cell replication—and they need only be the garden-variety sort—ought to be no exception. That language has been in use since the dawn of life—it runs in the blood, this Cellese.

Where Canalicular Communication Holds the Trump Card

Now, for this language to be of cybernetic use in cell growth, there must be some constant relationship between cell-population size and feedback-signal concentration. Which brings us back to the central question, the role of the cell-cell channels in growth control. These channels are precisely the elements, as we saw, that make that relationship so simple. No other form of intercellular communication could possibly do that. We will check the communication forms out one by one in this regard—this point is too important for just a lick and a promise.

The usefulness of canalicular communication as a conveyance of feedback information is owed to the finite volume of its "signal line," the internal communal space. That space, as we saw, is proportional to the cell number, the controlling feedback variable. Hormonal communication systems, on the other hand, use the space *between* cells. That extracellular space is tortuous and shifty. From whichever angle you look at it, it bears no relation to the controlling feedback variable—it has not even a fixed volume. Thus, the hormonal form of communication hasn't a chance to code the cell number into signal concentration.

Nor has the syncretic form with its penchant for binary interaction. The syncretic signal molecules cannot betake themselves from cell to cell, and the signal concentration—if we may speak at all of a concentration in the case of membrane-bound ligands—is a density in the lipid bilayer. That parameter bears no relation to the number of cells in the population. Nor do the syncretic systems offer much of a prospect for conveying a feedback message to populations of any considerable size; their signal lines are inherently short—they are good for binary communications, interactions between two cells, but not for communications involving a whole cell population.

This much for encoding. As for decoding, the differences are no less glaring. Because the canalicular flow of feedback information uses the cell's mainstream language, it melds naturally with that stream. No special decoding stage is required to couple the two flows—when two speak the same language, they don't need an interpreter.

Thus, the canalicular form of communication holds all the trump cards. It can do both the encoding and decoding of growth-controlling feedback signals at a so-very-much-lower information cost than the other two forms. So, if there ever was an evolutionary contest between the three on that account, it was an open-and-shut case for nolo contendere.

A Model of Growth Control for Small Cell Populations

Let us now put the canalicular mode of communication under a stronger loupe to see how it might do the job of regulating cellular growth. Not knowing the identity of the signals, we will keep our options open and have a go at either control mode, positive and negative.

A control of positive sign comes perhaps most readily to mind, because any canalicular-communication signal will get diluted in proportion to the growing cell number, proffering an intrinsic cybernetic loop for positive (growth-activating) signals. So, let's weigh the potentialities of such a loop.

Suppose that cell replication is governed by an enzyme that requires an ancillary information input, such as considered above; thus, there is replication only in the presence of certain cytoplasmic molecules (signals) and when these molecules reach the enzyme's activation threshold. Suppose further that somewhere in the cell population there is a cell that produces that molecule at a continuous trickle—a signal source emitting more than enough to raise the signal concentration in the internal communal space to threshold, when cell growth is underway. Then, all the elements for an automated growth control are in place: as the internal communal space expands in the growing cell population, the concentration of signals gets smaller and smaller, eventually falling below the activation threshold, and growth stops by itself (Fig. *12.11:1–3*).

We see here a powerful cybernetic principle at work: by dint of the dilution of positive feedback signals, and without further ado, the population regulates its own growth. This principle (hereafter, *dilution principle*) gets its moxie from the pervasiveness of canalicular communication; the feedback information (the steady-state signal concentration in the internal communal space) is carried to every cell in the population by this communication system, and rapidly so. And a single enzyme with built-in receiver–comparator–standard element (RCS box), as discussed above, would do the cybernetic rest, turning the effector (replication machinery) off when the signal concentration falls below the threshold (Fig. *12.11:3*).

But how does one get *one* cell to be a signal source? This presupposes the existence of a cell a cut above the crowd, and such a differentiation runs into money. It is not so much the signals themselves that worries us here—those could be produced economically by the usual molecular tautology, the emission of identical signal copies at a source—but the cascading genetic mechanisms that go with any sort of cellular differentiation. Such mechanisms, even the simple ones we have considered in Chapter 10, amount to a tidy information sum, and it is scarcely to be expected that Evolution had the wherewithal for all that when her whelp was still young.

So, true to principle, we will strike out for the most information-economical model, namely, one where the cells in the growing population are all of a piece and emission takes place randomly in the population. Such an egalitarian population, however, sets dynamic constraints—egalitarian communities are always hard to please. Obviously, there would be no signal dilution if all cells in the population were to emit at the same time. However, in small cell populations, there is room for alternation between signal sources, if the emissions are brief and occur with a low probability; the cell cycle in many mammalian cells is about one day, and diffusion of channel-permeant molecules in the internal communal space is three to four orders of magnitude faster than that. Thus, for example, with signal bursts lasting a few seconds and occurring with a probability of the

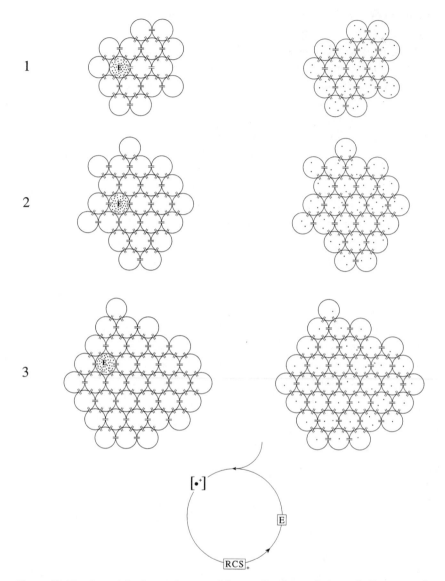

Figure 12.11 A model of growth control for small cell populations. A diagrammatic representation of the signal field in a population where one cell *(E)* is emitting a brief burst of growth-activating feedback signals (•). *(1)* The field at the start of the burst and *(2)* a few minutes after it, when the distribution of signal has reached steady state in the internal communal space of the population. As that space expands in the enlarging population (*1* to *3*), the signals get more and more diluted, eventually falling below the activation threshold of the cell replication machinery. *Bottom:* The information loop embodying this feedback principle. The positive sign denotes the bias; the replication machinery, the *effector E*, is switched off when [•⁺] falls to the level at which *S* is set (symbols as in Fig. *12.10*) [After Loewenstein, Biochim Biophys. Acta, *Cancer Rev,* 560:1–65 (1979), by permission of Elsevier Science].

order of one per day in a population of a few hundred cells, there could be enough emission asynchrony for signal dilution.

Hence, our model—a model for smallish cell populations—comes to this: all cells in the population are capable of generating the signals and all cells can sense the signal concentration (and always do), but the generating probability is such that, at any given time, only one cell is active as a source. This model will self-regulate by the dilution principle (Fig. *12.11*).

A Model for Large Cell Populations

For large cell populations, however, that model won't do. The cell replication cycle sets a limit here to the population size of the order of 10^3 cells. Beyond that, and even if we push the signaling probability to the limit, the dilution principle will run amiss. But there is more than one way to skin the cat, and if you will bear with me for another gedankenexperiment, I can promise another feedback principle rooted in intercellular connectivity, which is capable of continuing the cybernetic game beyond the dynamics limits of the preceding model.

We will start out from the undifferentiated cell society of the preceding model—all cells are potential signal sources and all are sensors—but now several cells may be active concurrently as sources, setting up more than one signal field in the larger internal communal space.[4] Analytically, the situation here comes down to signals diffusing from sources randomly distributed in space and time—discrete regulatory centers (Fig. *12.12 a*). Consider three such simultaneously active centers. The signal concentration falls off concentrically with distance from each source, the fields overlapping here and there (Fig. *12.12 b*). The overlap is a bit of a nuisance, but there is enough regularity in the fields to extract information. To this end, we survey the concentration profiles of the fields for minima—an old gambit in statistical mechanics—which define spatial limits *(L)* at which there is no net exchange of signal. These low points mark off the borders of virtual compartments inside the population.

With such a partitioning of the internal communal space, the situation becomes analytically manageable. The cell population, at a first approximation, as Wolfgang Nonner and I have shown, may be treated as a mosaic of virtual compartments. Each compartment, on the average, has 1 signaling cell as the center and n-1 nonsignaling ones—all the rest—as the periphery. In the ensemble average, signals cross compartment borders equally fast in either direction. The situation in the ensemble is then equivalent to that in a single isolated compartment with closed borders. Hence, such a compartment provides a represen-

4.   All cells have the same probability here to become emitters; but at any given moment, only a few cells are active as sources, and the ratio of signaling to sensing cells is constant in the same or successive cell generations.

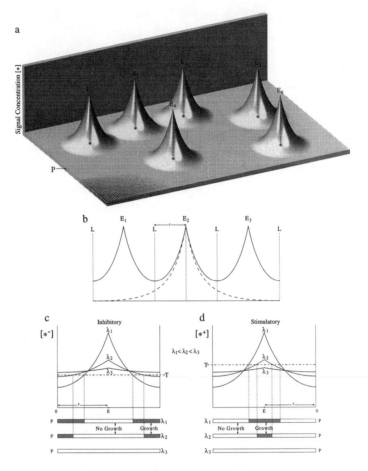

Figure 12.12 A model of growth control for large cell populations. *(a)* The signal field. The diagram represents the distribution of feedback signal (*) in a cell population growing as a single layer *(P)* in a culture dish. Several cells in the population are concurrently signal emitters *(E₁, E₂, E₃ . . .)*; the signal concentration *(ordinates)* falls off concentrically with distance from each signal source. *(b)* Three of such partly over-lapping fields are shown in profile, the *L* lines through the concentration minima demarcating virtual compartments with radius *r* (the dashed profile represents the situation when only one cell is emitting). *(c, d)* Regulation. The diagram illustrates the situations in a compartment when the intercellular connectivity is varied—for negative *(c)* and positive *(d)* signals. As the connectivity rises, the signals spread farther and farther from the source; the signal concentration profiles at steady state are shown for three settings of the characteristic length λ: $\lambda_1 < \lambda_2 < \lambda_3$. *T* are the threshold levels for inhibition or activation of cell replication. The corresponding projections of the signal concentration profiles onto the cell field (represented by the bar diagrams *P*) give the extent of the field where growth is stopped. That field increases with increasing λ with either negative or positive signals. (For operational requisites, see footnote 6.) [*b, c*, and *d* from Nonner and Loewenstein, *J. Cell Biol.* 108:1063 (1989), by permission of the Rockefeller University Press].

tative statistical sample of the whole population—a *pars pro toto* in statistical-mechanics terms (Fig. *12.12 b*).

It pays now to go over that *pars pro toto* with a finer comb to see what makes the mosaic tick. Something of note may be seen at once from the way the field varies with the connectivity. The signal distribution is primarily ruled by the intercellular permeability or connectivity (the sum of the conductances of all the intercellular channels in a cell junction). The higher that connectivity, the farther the signal spreads from the source. Figure *12.12 c* represents this situation graphically by a characteristic length, λ , relative to the compartment radius r. This way, we can see at a glance the lay of the land: *with either positive or negative signals, the signal field increases with increasing* λ (Fig. *12.12 c, d*).

One or two more stabs, and we get to the regulatory heart. Consider first what happens with negative signals, the simpler case. Here we only have to draw in a threshold level for inhibition *(T)*, the minimal signal level at which cell replication is switched off, and the corresponding projections of the concentration profiles onto the cell population *(P)* will give us the boundaries of the field where growth stops. *This field increases with increasing* λ, *and monotonically so* (Fig. *12.12 c*).

With positive signals, things are not as straightforward. Here our graph (with *T* now above the equilibrium concentration of the signals) reels off a relationship that is not necessarily strictly monotonic. But the important point is that *from a minimal value onward, also in this case an increasing* λ *will always increase the field where growth is stopped* (Fig. *12.12 d*).

Thus, whether we use positive or negative signals, our model can regulate in the same direction: a rising λ hushes up more and more cells in a continuous diminuendo.[6] This outcome of our thought experiment is as exciting as it is simple, for it puts us on the track of a second information loop which transcends the one we have seen: a loop that is closed by an inherent rise of λ.

Let me explain. In a growing channel-connected cell population, the intercellular connectivity increases with the cell number over a range; as the population grows and becomes more dense, the average cell gains more and more neighbors (Fig. *12.13*), and so there will be more and more cells in communication with one another. Hence, λ rises. This rise is a feature inherent in the topography of

---

6.   The operational prerequisites of the model are readily specified in terms of infinite λ: in the limit of homogeneous signal distribution ($\lambda \to \infty$), the concentrations of inhibitory signals must be above the threshold for inhibition of replications (Fig. *12.12 c*), whereas the concentrations of stimulatory signals must be below the threshold for stimulation (Fig. *12.12 d*). Formally, the latter condition is like that in the model for small populations, where the steady-state concentration of signal falls to the threshold level by dilution. The model obeys the conservation rule that the total amount of signal in a cluster is constant, due to an invariant balance between signal production and signal degradation/leakage; the concentration profiles in Figure *12.12* are scaled to that effect.

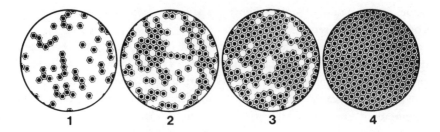

Figure 12.13 The number of first-order cell contacts (hence λ) increases with cell density in a growing cell population. In the hexagonal grid example shown (*1–4*), the average number of first-order cell contacts is 1.5, 3, 4.5, and 6. [From Loewenstein and Rose, *Semin. Cell Biol.* 3:59–79 (1992), by permission of Academic Press.]

the growing population—a gut property of its map, we might say. But there is more to that. On closer inspection, this feature turns out to be loaded with cybernetic potential. All we need here is an *RCS* box (see Fig. *12.10*) with a standard geared to the sensitive λ range, and we have a powerful information circle for the control of growth: λ *increases as the number of cells in the growing population goes up; and as an increase in λ, in turn, reduces the number of replicating cells in the population, growth will stop when the population reaches a certain number.* In each population compartment, then, the field of replicating cells will diminish in a concentrically progressive manner over a range of λ, as charted in Figure *12.12 c* and *d* for the negative and the positive modes of operation of the model.

This feedback principle begins to operate when the population gets to be about 20–30 cells—the lower limit for the topographic rise in connectivity. That is how the statistical topology of the two-dimensional cell clustering in growing (mammalian) populations happens to work out. So, we wind up with a mechanism of automatic growth control good for large cell populations, where the internal communal space is parceled into territories—discrete regulatory centers—and the feedback is by way of intercellular connectivity.

Controls for Two Seasons

Our search for growth-controlling mechanisms, thus, netted us two minimal information models: one for small cell masses and one for large ones. We didn't deliberately set out here to find two mechanisms for the price of one. It just turned out that neither model could do the whole job by itself because minimal cellular differentiation imposes dynamic and statistical constraints. But they can work in tandem. And this raises the interesting possibility that the second mechanism, the one with discrete regulatory centers, might be the evolutionary successor of the first.

Indeed, the models mesh well enough for such a strategy—they are functionally and informationally coherent. The first model, we have seen, can work with populations of up to about a thousand cells—its feedback principle sets that constraint, but it sets no lower limit. The principle of the large-population model, on the other hand, sets a lower limit at about 20–30 cells. So, that mechanism could take over when the population outgrows the dynamics limits of the first mechanism.

On the information side, things seem just as auspicious. The two models have a common information lineage. They use essentially the same encoding for the feedback variable, the same basic language (though their feedback principles are different). This informational coherence—a sine qua non of self-developing systems—would allow a smooth progression from one model to the other.

Thus, the idea of the two mechanisms in tandem can fly, and we may end our musing with the thought that at some point in phylogenetic (and embryonic) development, when the cell population becomes sizable, the second mechanism dons the mantle of the first.

λ Determines Cellular Growth

Models and thought experiments are the life breath of science. But in biology it is no good to stand on them for very long, and better not to try. The cards are stacked against one—that's in the nature of the complex beast—and making models is pretty much like taking chestnuts from the fire. Thus, we had better check our fingers to see whether they are burned.

There are several ways to put our model to the test at this stage. One tack is to vary λ, the intercellular connectivity, and see whether growth stops at the predicted cell number. Our λ graphs (Fig. *12.12 c, d*) tell us at a glance which way that number should go when the connectivity is on the way up: *down*—the field of growing cells gets smaller with rising λ, at least over a range.

Now, for varying the connectivity experimentally, our cell-culture paradigm offers us two degrees of freedom: the number of the cell-cell channels and their open state. Parmender Mehta, John Bertram, and I made use of both. In one approach, the transcription rate of one of the channel-protein genes (connexin 43) was varied. Meddling in the private affairs of cells? Yes, but a gentle meddling, if there ever was one. The agent here was retinoic acid, a native cellular molecule (which cells normally make from vitamin A). This is one of those crafty little lipid signals—it barely weighs 300 daltons—that can zip through the cell membrane, just like the steroids do. It bears the message to the connexin-gene circuitry for accelerating transcription. More channel-protein subunits are thus produced by the individual cells and, hence, more channels are made among them. That increase in the number of cell-cell channels depends on the dose of retinoic acid administered to the cells, offering us a convenient handle for changing λ gradually. The result was the predicted one by the model: as λ rose, the cells stopped growing at a lower and lower cell number.

As for λ's second degree of freedom, David Shalloway, Roobik Azarnia, and I resorted to a bit of genetic engineering. We constructed DNA stretches coding for an enzyme (the src protein) that puts phosphate on tyrosine residues of the cell-cell channel protein, a phosphorylation that flips the channel to the closed configuration. The various DNA constructs were inserted into the cells' genomes together with suitable promoter sequences, yielding tyrosine-phosphorylating enzymes with various degrees of activity; and the higher that degree, the more cell-cell channels got closed. This way, the λ of the cell populations could be lowered several notches below its ordinary level. The results here matched the model as nicely as the results in the upper reaches of λ did: as λ decreased, the cells grew to higher and higher numbers.

Matches are made in heaven they say . . . but, once in a blue moon, also in the lab.

What Is Cancer?

The foregoing results bolstered our confidence in the model enough to try it out on cancer cells. Such cells are notorious for their lack of growth control—they show a relentless drive to multiply. Is their λ decreased?

To explore this point, we resort once more to the little cellcosm-in-the-dish. Though separated from tumor and organism, the cells here still show their aberrant behavior: they double and double, and do so ever and anon, reaching huge numbers (Fig. *12.14*). The contrast with normal cells growing in the same culture condition could not be more striking; after their initial exponential spree, the normal cells bring their growth to a halt.

So, in our little universe-in-the-dish we see at once the match in the powder barrel. The nonstop exponential growth, in the light of control theory, can mean only that there is somewhere a flaw in the feedback loop of cell replication. And this, even with the foggy "somewhere," tells us more about cancer than a thousand autopsies. We will try to get a better fix on this flaw further on, and we will see that several cancer-causing genes—the *oncogenes*—tamper with the cell-cell channels. But, first a word about these genes.

Oncogenes are old information pieces of the cells' genome, which have been well-conserved through evolution. They are tied in with the circuitry of cellular growth, though the specifics still escape us. Better known are the fiendish doppelganger of these genes, who dwell in the virus world. These worm themselves into the genome of cells and foul up the controls. The immune system keeps them at bay to some extent, but there are times when those defenses are weakened or overrun. It is then when the virus presses its attack and, more often than we ourselves would wish, comes out on top.

Once it sneaks its oncogene into the cellular genome, the tiny virus all but ties its colossal quarry to its apron strings. Take the Rous sarcoma virus, for example, a virus which preys on vertebrate cells. This virus bears the *src* onco-

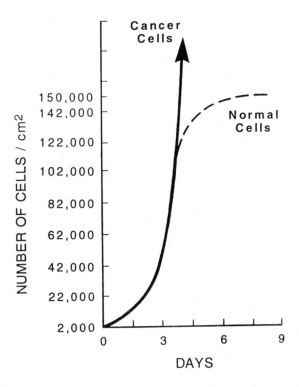

Figure 12.14   The growth of cancer cells in culture. Just as the normal cell in Figure *12.9*, these cancer cells were originally seeded in the culture dishes at 2,000 cells/cm² (time zero). Their population grows nonstop (solid curve), overshooting the characteristic number of their normal cell counterparts, which is reproduced here from Figure *12.9* (stippled curve).

gene, a double of the cellular *src* gene we touched on before. It is a close double (in RNA form), with just a little shorter tail and a few different codons. But this tiny informational difference translates to a large functional one in protein; whereas the cellular src protein has a weak, barely noticeable catalytic (tyrosine-phosphorylating) activity, the viral src protein has a strong one—and that has dire consequences for the cell.

The trouble begins as soon as the virus takes charge of the transcription machinery of the cell. Its RNA oncogene then is transcribed to DNA (by a reverse transcriptase), the transcript is amalgamated with the cell's DNA and, in due time, this information glides down the cell's mainstream, giving birth to an alien protein—a cuckoo in the starling nest.

And what a cuckoo! When this one hatches, the cell changes beyond recognition. Gone are its social grace and cooperative interaction—it gives itself up to procreation. And the misbehavior doesn't stop there, but is continued by the progeny; the viral information is passed from cell generation to generation.

Thus, once that information makes its dreaded cellular debut, it tends to mush-room, occupying more and more of the body that whelped it.

That, in a nutshell, is what is behind the ghastly growth. But where does the villainous information come from? How did the minuscule virus, a relative new-comer on the evolutionary scene, get hold of it? Only in mythology does Athena spring up at once full-grown, breastplates and all. In real life such things take aeons and a membrane—neither luxury did the virus enjoy. The solution to the riddle was provided by Michael Bishop and Harold Varmus, the discoverers of the cellular *src* gene: the virus lifted the information from the cellular genome. This may have seemed a strange thought at first but, once the word was out, it didn't take long before a whole band of such doppelganger were spotted in cells—about three dozen, by the latest count. So, such copycatting must have been going on all through evolution.

Judging by the *src*'s RNA sequence, the virus must have copied the cellular gene from A to Z, changing only a bit here and there. These were very slight alterations of the urtext, but consequential ones, for they brought forth some-thing rather novel: a demon with a high cognition power for tyrosine and phos-phate. And this is the culprit who makes the cells go round the bend. The demon bears down on several cellular proteins, foisting phosphates onto tyrosines which, in the normal scheme of cellular things, are not phosphorylated. Such phosphorylations lead to a multiplicity of alterations inside cells, for which biologists have a buzzword, *transformation*.

λ Is Reduced in Cancer

Among that multiplicity, there is one alteration that is of immediate interest regarding our model of cellular growth control: the cell-cell channels, which normally are open, now are mostly shut. The protein subunit of these channels is one of the targets of tyrosine phosphorylation, and that sort of covalent modifi-cation of channel structure, as we have seen, sends it into the closed configura-tion. Indeed, such closure is one of the earliest detectable cellular effects of the viral src demon. No sooner is this demon launched, then the first channels are phased out and, all within a few minutes, most of them follow suit.

Thus, the chicanery of the virus could not be more effective: λ falls to near zero, rendering the cell incommunicado—and not only the invaded cell, but all its progeny, for the heinous alien information gets reproduced with the cellular genome. Like an Old Testament scourge, the punishment is meted out to the cell's kins and kins of kins.

The Rous sarcoma virus is not the only one who has it in for the channel. At least, four other viral oncogenes (polyomavirus, SV-40, viral *ras,* and viral *mos*) meddle with its function—its strategic position in the cybernetic information flow of growth makes it an inviting target. The information flow from those oncogenes remains largely untracked, and we don't know as yet how it hits the channel. Some of these oncogenes take much more time to impair the cell communication than *src* does—they may foil the channel protein synthesis or channel assembly, rather than the channel open state. But the end-effect we know: λ falls to near zero.

Thus, with such a high leverage on λ, all five oncogenes would be expected to effectively cut off the cells from the feedback flow in our model. But what of oncogenes with less purchase, like the (normal) cellular *src* gene or the adenovirus (E1A) gene. Neither oncogene here, even when overexpressed, causes cancer on its own, though they do in conjunction. Will their λ reductions add up? The answer came from experiments in which the two oncogenes (ligated to suitable promoters) were inserted into the genome of mammalian cells and were transcribed in abundance. When either oncogene was present alone in the cellular genome and overexpressed, λ fell somewhat, though not anywhere near as low as with the viral *src*; but with the two together, λ fell all the way—bearing out yet another prediction of the model.

A Cancer Cure in the Test Tube

The correlation between communication and cancer raises the question whether the curtailing of λ by the oncogenes is the cause of the cells' aberrant growth. Indeed, this point would follow directly from our model, but it is not an easy proposition to prove. The multiplicity of alterations that are typically associated with cancer, the "pleiotropy," makes it hard to distinguish cause from side effect. And this is one of the reasons why, in general, progress in cancer etiology has been excruciatingly slow. The discovery of elements, like the oncogenes, which are sufficient to cause cancer, was a big step forward, to be sure. But it didn't make the problem of pleiotropy go away. That problem—a general problem of genetic physiology—is inherent in the diverging genetic information lines. Take *src*, for example. The information flow emanating from this oncogene breaks up (over stages of tyrosine phosphorylation) into several branches and only one of them ties into the growth-controlling feedback loop. The oncogene itself does not nurse that loop—it does not encode the feedback signals. It sends out a branch that hooks up with the loop, impeding the feedback signal flow, while the other branches tie into the cellular network elsewhere, giving rise to other cellular alterations (which are all part of the cancer syndrome).

Thus, the problem of pleiotropy remains and if we may hope to prove the proposition of causality of λ in cancerous growth, we must find a way to cut across the oncogene information branchwork—short-circuiting it, as it were. There are a number of spots on the feedback loop that one might tap into, but until we know the identity of the feedback signals and learn more about the RCS box, the only place—and the logical one in terms of the model—is the cell-cell channel. And this was the tack that Parmender Mehta, Birgit Rose, David Shalloway, and I took. We used a mouse cell which had a flawed gene for the channel protein (connexin43). The flaw had been produced by a chemical agent (methylcholanthrene) that kinked the transcription process of that gene, without perturbing its translation process or any of the other stages down the line to channel assembly. Thus, the crucial questions could be asked: Will the defective gene cause cancerous growth? And will the correction of the gene defect correct the growth aberrancy?

The answer to the first question came at once: the mutant cells had very low λs and grew nonstop, reaching numbers in the culture dish comparable to those of cells infected with viral oncogenes (and they produced tumors when they were injected into mice).

The posing of the second question entailed a bit of genetic tinkering, namely, the grafting of a connexin 43 gene from a normal cell (rat heart) to the aberrant one. The transplants took, and the result was as electrifying as the first: the mutant cells and their progeny now exhibited normal λs and normal growth—they stopped growing at the same cell number as normal cells do.

The normalization of growth was evident also in mixed company. To this end, two types of cells—the mutants and their (normal) parental cells—were put together in the culture dish; say, the mutants were seeded at various densities on top of a confluent layer of the normal cells. The mutants then would rapidly outgrow their normal counterparts and before long would dominate the culture picture, even when initially there were only a few of them around. For instance, when the mutant cells were seeded sparsely on top of the normal cell layer, they would keep on dividing; you could see them piling up as clones and then, eventually, in typical cancer fashion, tread everything underfoot. But that aggressive behavior melted away with the gene transplants; then not a single clone would grow on top of the normal cell layer (Fig. *12.15, left*).

This restoration of normal social behavior, too, went hand in hand with restoration of normal λ. The recipients of the gene transplants made functional channels among each other, as well as with the normal parental cells underneath—they became one with the normal community.

Our First Line of Defense

It is not often that one has the opportunity to put a genie back into the bottle, let alone such a malignant one. So, we may as well take advantage of this good turn and lean a little more on these channel-deficient clones and their rehabilitated brood, to learn as much as we can about mending the cancerous condition.

One question comes up at once in the wake of our success in the test tube: To what extent do normal cells themselves, by virtue of their communication, influence the aberrant growth? Can they ward off the malignant growth? The communication field of a normal cell extends well beyond the cell borders. Our model gives the scope of this field of influence, the signal field, precisely: it equals λ— meaning some 10 cell diameters in most tissues of our body. Thus, the question is whether this field will deter the development of an incipient tumor. This goes to the very core of the cancer problem, for the malignant growth starts in the midst of normal cells—one stricken cell among normal thousands.

To come to grips with this issue, we will briefly return to the scheme of Figure *12.10* and put the problem in cybernetic terms: What stage in the feedback loop gets stricken? That loop is what normally holds the cellular growth at bay— a genetic strike cutting the feedback anywhere along the line will give the cells free rein. Now, there are in principle three genetically vulnerable points along that line, namely, the three information-transfer stages on the way to the growth effector. So, depending on the stage where the mutation hits, we have three elementary pigeonholes of cancer (apart from the one concerning the effector): signal-production⁻, signal-conduction⁻, and signal-reception⁻ mutants (the last category includes mutations of the cybernetic C and S elements).

This categorization bypasses the usual gestalt taxonomy of our doctors and lets us see straightaway the taproots of the disease. But there is yet another bonus here: we see what cancer category could be deterred by the normal-cell field. Slim chance that the signal-conduction⁻ or reception⁻ mutants would—no amount of extrinsic signal will do any good when a genetic deficiency of this sort holds a cell incommunicado. But the odds are better with the signal production⁻ mutants. Though unable to make the correct signals by themselves, those cancer-cell types will get them from their neighbors via cell-cell channels. So, provided the other two information-transfer stages are OK, their abnormal growth behavior will be suppressed—they will have a hard time to take off in normal cell company.

Situations of such happy portent may not be uncommon. Signal-production cancers constitute a broad category—a multistage mechanism, like signal production, provides ample opportunity for mutation. And given the rate of DNA mutation and the number of cells in our organism, the chances are that there may be one such black sheep right now in our body. Yet, fortunately for us, it is likely that it will be stopped in its tracks by the true fold—cells with long λ are good company!

However, things get to be too close for comfort if the rogue somehow manages to escape and grows a lump. Then, as the lump gets larger than about 20 cell diameters—the 2λ-boundary condition for arrestation *in limine* spelled out by our model—even the best company won't help.

Thus, there was all along a message for our health hidden in the model: *the cell-cell channels are our first line of defense against cancers of the signal⁻ class.* Just how powerful this line of defense can be was shown experimentally with (channel-competent) cells that were on their way to cancerous growth. Originally sent off that way in the test tube by a withering dose of a chemical carcinogen, a high percentage of these cells became malignant and formed cancer clones in culture. But not when they were put on top of normal cells. Then only very few cancer clones would grow, and even those hovered for a while, trembling in the balance.

A Ray of Hope

Our story of intercellular communication is now taking on unmistakable medical overtones—dogging the information trail will get one to all sorts of places! Indeed, the success of the experimental connexin-gene transplants open a window to a treatment of cancer, namely, cancers with cell-cell channel deficiencies—the signal-conduction˙ mutant class. These cancer types we can readily recognize by electrical or fluorescent-tracer probings. They also are among the nastiest ones. Impervious, as they are, to the signal field of the normal cells in their surround, there is nothing to stop them in their infancy, except perhaps for a vigilant immune cell. And this is probably the reason why this cancer class is so common in organisms. Alas, this includes us humans—the tumors of the breast, liver, lung, thyroid, stomach, uterus, and prostate afflicting our race bear that nefarious channel-deficient signature all too often.

And this brings us to our last question: Will our test-tube cure hold up inside an organism? Whether one will or not, the spadework here will have to be done in animals—there is no humane shortcut to the bedside. Well, the preliminaries have already started: the mutant mouse cells, which had received our connexin-gene transplants, were injected into (immunosuppressed) mice. The results inside the organism were as dramatic as those in culture: the original mutant cells (and those which had received mock gene transplants) produced large tumors, whereas the recipients of the transplants produced none over several months (Fig. *12.15*).

This is encouraging and clears the decks for more such rounds with various other channel-deficient cancer cell types which presently are being smoked out of the closet. This phase shouldn't take long. Even as I write these lines, the word is out that a human cancer cell—a cell from a breast tumor—was tamed by a connexin-gene transplant and has passed the tumor test in mice, with flying colors.

So, one may now seriously think of a medical sortie—of scaling up the test-tube runs to target tumor cells inside the human body. Our doctors have maintained a long and spirited battle with that relentless predator on the human race. They have scored their victories with scalpel, gamma rays, drugs, and other good weapons. Though all too often these were Pyrrhic ones and in the end there was the epitaph—*hic iacet sepultus*, "here lies buried the predator with its prey."

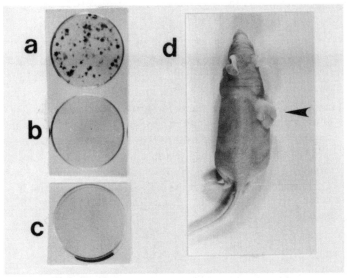

Figure 12.15   An experimental cancer cure. *Left*, photographs of culture dishes, illustrating the growth of the mouse cancer cells *(a)* before and *(b)* after they received the channel gene transplant. These cells were seeded sparsely on top of a confluent layer of normal mouse cells (10T ½) and allowed to grow for three weeks. The original cancer cells proliferated, forming many colonies (the dark spots); whereas the transplant recipients did not grow at all (the culture dishes are transparent). The transplant recipients behaved just like the normal cells seeded on top of the normal cell layer, shown in *c*. The single cell layer is transparent; the dark stain is seen only where cells pile up. *Right*, a sample of a test of tumor growth. This (immunosuppressed) mouse had been injected with 1,000,000 transplant-recipient cancer cells into the left flank, and with the same number of cancer cells that had received a mock gene transplant (which lacked the channel-protein sequence) into the right flank. The photograph shows the animal eleven weeks after the inoculation; the arrow points to the tumor produced by the cancer cells with the mock gene transplants. The cell recipients of the channel-gene transplants did not produce tumors. [From Mehta, Hotz-Wagenblatt, Rose, Shalloway, and Loewenstein, *J. Membr. Biol.* 124: 207 (1991), and Rose, Mehta, and Loewenstein, *Carcinogenesis* 14:1073 (1992), by permission of Springer Verlag and of Oxford University Press.]

But now, perhaps, as we are finally coming to grips with the information side of growth control, things may change. Progress in medicine is closely tied to progress in physiology—two thousand years of the Hippocratic pursuit attest to that. It will not be easy to bring the gene-transplant approach to the hospital bedside. But now that the fog is starting to lift on the physiological scene, there is a ray of hope.

To the question of the relationship between brain and mind the answer given by a physiologist 60 years ago was Ignorabimus. But to-day less than yesterday do we think the definite limits of exploration yet attained. The problem I have too grossly touched to-day has one virtue at least, it will long offer to those who pursue it the comfort that to journey is better than to arrive; but that comfort assumes arrival. Some of us, perhaps because we are too old—or is it, too young—think there may be arrival at last. And when, and if, that arrival comes, there may be still regret that the pursuit is over.

—Sir Charles Sherrington, Rede Lecture, 1933

Neuronal Communication, Neuronal Computation, and Conscious Thought

Neuronal Communication and Computation

This concludes the story of how the information in the genetic library—the pre-coded information in each cell—gets disseminated through the organismic mass. But before we close the books on intercellular communication, I shall briefly deal with a special brand in the nervous system. This intercellular operation started rather late in evolution, well after the Grand Climacteric. It filled the niche for fast, long-distance communication, a niche that came with multicellularity; but eventually it did much more than that. Nerve cells have long extensions—in our own organism they get to be meter long. These extensions or *axons* carry information with speeds ranging from about 1 to 100 meters per second, several orders of magnitude higher than those of the four basic communication modes we have seen. The nerve cells and their extensions form highly organized, richly structured communication networks which do not depend solely on the vagaries of molecular diffusion as the other communication modes do—the neurons have weaned themselves of the ancient seek and find game, we might say.

The first neuronal networks were of modest size, consisting of just a few hundred cells. Networks of this sort can still be found at the lower phylogenetic rungs today. But as the organismic cell mass grew, so did the networks, which eventually counted billions of neurons at the higher rungs—1000 billion (10^{12}) in humans. That number is truly astronomical; indeed it is close to the count of stars in the Milky Way. And such plenitude opened vast computational possibilities, as we shall see.

Among the various cell types in an organism, the nerve cells are the most specialized ones. As adults, they no longer engage in all of the four basic communication modes, though many types still use the hormonal mode, and a few still use cell-to-cell channels for electrical intercellular communication. Neurons tend to stand aloof from the cellular crowd and are different from other cells in many ways—they have different shapes, highly developed cytoskeletons, special cytoplasmic transport mechanisms, and they no longer divide or do so rarely—but what stands out in boldest relief is their capacity of generating electrical signals. This capacity is owed to just a few phrases in the cells' genetic library—phrases for voltage-sensitive membrane protein of the likes we already have come across before in connection with the G-system (Figs. *10.13* and *10.14*). The electric signals are monotone—at least, those carried by the axons are—they are brief pulses (a few thousandths of a second in duration) of invariant strength all along an axon. Thus, if we wiretap an axon, we either get an electric pulse there or not; what varies are the intervals between the pulses, not the pulses themselves. Neuronal networks resemble digital computers in this regard. The functional units of a computer, the logic gates made of silicon, are controlled by invariant electric pulses—there either is an electric pulse down the wire or there is not. And when such units are adequately connected, the ensemble can compute any logic or arithmetic function. How much it can compute depends on the number of units—and in our brain there is certainly no shortfall of that.

Neuronal networks are the products of a natural selection for fast information-conducting and -processing devices. The first devices were quite simple, probably not more than a handful of neuronal units connected in an information circle. Such devices satisfied the needs of small multicellular beings; by virtue of their rapidly conducted electric pulses, they allowed such organisms to respond to environmental changes more quickly than their masses of cells could on their own. But as those beings grew, the neuronal networks expanded and got adapted to ever more demanding muscular and secretory tasks: feeding, mating, hunting, . . . and eventually the ultimate in biological communication: speech.

The biggest expansion occured sometime between 6 and 10 million years ago—only yesterday on our cosmic time scale (Fig. *3.3*)—when our hominid ancestors broke loose from the primate crowd. The number of neurons then approached the 1,000 billion mark. Anatomically, this is a record number—and this has long been recognized. But more to the point, it is an information record, for the total organismic information capacity here reached unprecedented heights. Consider the computational capacity of those 10^{12} connected neurons in combination with the precoded genetic information in each organismic cell, the neurons included. We have seen in Chapter 1 how immense that precoded information is. But here on top of that immensity now comes another one inherent in the organization of the neuronal ensembles, the computational capacity of the brain network, which raises the total organismic information capacity to dizzying heights. Those heights are certainly unprecedented on Earth, and we may rightly consider this point on our chart the pinnacle of biological evolution. But keeping the cosmic perspective, let's bear in mind that the unprecedented has a penchant for becoming the ordinary in this three-and-a-half-billion-year-old work-in-progress.

Now, turning from capacity to speed of computation, a neuronal network may at first sight not seem much. Neuronal signal rates rarely are much higher than 800 per second—not much against the 120,000,000-per-second rate of an Intel silicon chip. But they are adequate for our organismic needs; the neuronal signal rates are well-matched to the speeds of our muscle cells—they were naturally selected for that. They get us through our daily muscular tasks and with capacity to spare for little extras, like the tremolos of an Isaac Stern or the Grand Slams of a Babe Ruth. And then consider this: if only one-hundredth of those 10^{12} neurons were operative at a given time and firing at 300 per second, we could do on the order of 10^{12} computations per second. This still tops the best of Silicon Valley—though, I concede, at the clip things are going there, that advantage may not last long.

But I doubt that this will be so regarding overall performance. Even in what concerns only the number of connections, machines would have a long, long way to catch up. In today's electronic computing machines, the connections generally are 1–2 per computing unit—in the cream of the crop, 4; whereas in our brain, numbers of the order of 1,000 connections (*synapses*) per neuron are not uncommon, not to speak of the cerebellum where those numbers get up to the order of 100,000 per neuron. And totaled up over the whole network, the connections become countless as the sands—one thousand trillion connections in our cerebral cortex alone.

Informational convergence and divergence are the hallmarks of all forms of biological communications, as we have seen, but nowhere is the arborization as expansive and the connectivity as rich as in neuronal networks; here, apart from the electric signals conducted by all the neuronal branchwork, hormonal signals get blended in. This profuse connectivity gives the cell masses in our brain an exquisitely fine control over the digital activity of the neuronal network. And all this well-modulated activity goes on day in and day out, encoding or decoding the huge amounts of digital information that come in from our sense organs or go out to our muscles and glands. So, everywhere in our brain, the circuits are pulsing, flickering, beating with the business of moving, feeling, remembering, and figuring things out.

Consciousness

Now, all this—and it's a big "all"—so far concerns but the digital operations of the nervous system, the operation which has been probed with everything in our arsenal, ever since the physiologist Emil Dubois-Reymond recorded the first electric pulse from a nerve (1848). These indagations brought us an understanding of the nature of the neuronal signal and of the mechanisms of its generation and transmission, and some glimpses (reductionist slices) of the neuronal network. All this is what we might call the classical aspects of the nervous system. But what of the less tangible aspects—awareness, thinking, mind? Here by and large we are still groping in the dark and, for all the search and advances, Sherrington's words of sixty years ago, which are heading this chapter, might well have been spoken today.

However, there is now a fresh ferment. A new generation of neuroscientists with computer know-how is opening, through analysis and modeling of neuronal circuitry, a window to the mechanisms of memory, attention, and learning. This has rekindled the interest in the brain/mind problem and raised hope that a solution may at last be near. The age-old question of what is consciousness is now moving to center stage and there are two major currents of thought. There are those who hold that the states of consciousness result from the digital-computer operations of neuronal networks and those who hold that they are noncomputational or that they transcend reductionist approach altogether.

The first camp believes that consciousness is an emergent property of the operation of digital-computational systems of neurons—"emergent" in the same sense as, say, fluidity or solidity emerges from systems of molecules. Implicit in this tenet is the assumption that computers can, in principle, exhibit states of consciousness—that if the number of connections in a digital electronic computer could be made high enough and the circuitry adequately organized, a consciousness would follow. The other camp is wedded to the opposite, that consciousness is sui generis of brains—that electronic computers, however complex, are inherently incapable of consciousness. The skirmishes between the two camps are being fought with a wide range of arguments, and the major battle lines are being drawn in quite disparate terrains—neurology, philosophy, formal logic, computer science, even quantum theory of gravity. In short, a hot dispute.

Well, that should tip us off. When there is scientific controversy over two mutually exclusive alternatives, particularly when the debate gets overheated, one can usually bet that an essential piece of knowledge is missing—the polemic trying to pass off for the missing piece, vigor fronting for rigor. The history of science is full of examples of this sort. We have seen one in Chapter 3 pertaining to our own time, an example where a century-old polemic went poof as soon as the missing piece turned up. The piece then came from the physics of chaos, seemingly out of the blue. But that's how things often go in science; the crucial clues can come from the most unexpected quarters.

So, heeding this lesson, we will keep an open mind and pursue some off-the-beaten-track leads concerning the states of consciousness. This will take us into the elementary-particle sphere, the quantum world. I have intentionally held that strange world at arm's length so far. In dealing with the cellular and intercellular information flow—a flow carried by large masses of molecules—there was no need to go below the familiar sphere of large-scale things. But here at the brain/mind frontier, I feel, I can no longer skirt the quantum world, lest we risk missing a crucial piece of knowledge.

But first, a few words about what we mean by consciousnesss. True consciousness involves the awareness of an endless number of hues of things—the shades of green of a meadow, the smells of honeysuckle, the sounds of rushing wind. It involves the remembrance of things—a beloved face, the "touch of a vanish'd hand and the sound of a voice that is still." It involves the awareness of the passing of time, our joys and worries, our wonderings and will—the whole polychrome of the brain's I.

What mechanism could possibly give rise to that?! However much we may yearn for a physicochemical explanation, we cannot cut ourselves loose from the subjective here. And we should not. Indeed, there is much to be learned from introspection, and it may be useful, at least for starters, to look for common ground between the results of introspection and objective experiment. Then, a feature comes to the surface: the globality or oneness of the conscious state. Through introspection, we experience this as an all-excluding oneness (philosophers have called it "the oneness of thought") and through experimental probings, as a pervasive electrical brain activity encompassing vast numbers of cells. Consider, for example, what happens when an image, say, a green square, is flashed before our eyes on a TV screen. There is then, within a fraction of a second, a flurry of electric signals in our retina, in the lateral geniculate nucleus (a way station in the brain), and higher up in various regions and layers of the occipital cortex—a myriad of neurons in far-flung places become detectably active in that fraction of a second when we *see* the square. Or consider what happens when we consciously will ourselves to move a hand. There, too, is then a burst of detectable electric signal activity spreading far and wide through the brain.

All this leads us to suspect that whatever gives rise to consciousness must operate at a rather large cellular scale. And if we consider the conscious states to be information states in the sense of equation *1*, then the corresponding information transforms—transforms from the TV image and from previous sensory experience stored in the brain (memory)—must spread over a large cellular space, an immense space in molecular terms.

But what could possibly give rise to such a fast spread of information in so large a space? We can dismiss offhand the ordinary molecular seek and find game: mechanisms based solely on molecular diffusion (and this includes communication via cell-to-cell channels) would be orders of magnitude too slow. However, a mechanism combining molecular diffusion with electrical signal transmission perhaps may be fast enough. It takes the brain network about 5–10 thousandths of a second to transmit an electrical signal down an axon and across a synapse (this includes the time of signal integration at dendrites and soma), and about 2–3 tenths of a second for us to become aware of that TV image with all its qualities of squareness, greenness, and brightness. This sets the time constraints for any theory of consciousness. These are severe for digital computation theories, as they leave only some 20–60 neuronal steps for the digital operations to effect a unified conscious state, a visual perception.

The stringency of it all sinks in if we consider what those operations must entail. Apart from the digitized information from the retina image, which, through divergent pathways, gets scattered widely through the brain, intrinsic brain information, that is, stored information from previous experience, gets mixed in. This is the memory component of consciousness. Thus, apart from the transmission and processing of the digitized image information in the neuronal network, there must be integration of this information with that in the network's memory—and all this within the aforementioned time frame (and if anything, my upper-limit estimate of neuronal processing steps is generous).

However, those constraints immediately get relaxed if we are willing to look beneath the molecular surface. There, new and vast possibilities for transmitting and processing information open up. Indeed, when elementary particles enter a condition known as *quantum coherence*, the informational operations get cast in a very different mold from that in the molecular sphere. Quantum coherence is a state of matter where large numbers of elementary particles behave in unison. In this state, information can be transmitted virtually instantaneously through a system and, given the appropriate conditions, a practically limitless number of parallel computations could be performed in the available 2–3 tenths of a second.

Quantum Coherence

Let's toy with this idea a little and pursue it for the case of electrons, the particles likely to be of interest here. In the quantum-coherent condition large numbers of electrons in a population occupy the same quantum state. Ordinarily this doesn't happen; electrons occupy only a particular quantum state—fermions are ruled by Pauli's exclusion principle (box, p. 100). But in certain conditions multitudes of electrons (typically on the order of 10^{23}) can be forced into the same state; they then tend to form pairs ("Cooper pairs")—in effect, dielectronic molecules with such a large radius that there are immense numbers of other electrons between the partners, each forming their own pairs. Such quantum coherencies can operate over vast molecular spaces.

Cooper pairing is a pairing in an abstract sense. It is a collective paired state—each electron here tends to be near another electron but not, as it were, any particular one. It is not easy to find similies for the abstract, often counterintuitive, quantum world (though with electrons it's not as bad as with bosons). But as a crude analogy, one may envisage the transition to the coherent quantum behavior this way: to start with, the particles are milling around randomly (their usual behavior); but all of a sudden many of them begin to move, like well-drilled soldiers, in unison. And if the paired electrons were to light up, we would see the quantum-coherent matter scintillating far and wide.

Although such coherency is not precisely garden variety, there is nothing outlandish here in energy terms; it is the configuration of the particles that costs the least energy in this case—just one more example of nature seeking out the state of lowest energy. In general, one speaks of coherence in physics when oscillations occurring at different places beat time with one another. Here, the term refers to oscillations of elementary particles, their wave-like behavior. That such particles have a wave associated with themselves (a *wavefunction*) is one of the basic concepts of quantum physics. The notion took its start from observations that such particles can undergo diffraction. Just like a beam of light (photons), a stream of electrons, protons or neutrons is associated with a wave field which can interfere with the field of another particle, reinforcing or cancelling each other out. The wave length depends on the particle's mass and velocity. [1] Such quantum waves, like water waves, can be added together or superposed, and when two or more superposed waves behave like one, they are called *coherent*. In the case of the Cooper pairing, we have a collective wavefunction, one that refers to a large

---

1.   The wavelength λ associated with a particle quantum state is given by the de Broglie relation

$$\lambda = h/mv$$

where h is Planck's constant and m and v the mass and velocity of the particle (this neglects relativistic effects, a valid approximation in a low-temperature context). The λ dependence on particle velocity translates to a temperature maximum at which a wave-like behavior is observed; if the system is in contact with an environment in thermal equilibrium at temperature T, then, by equating the kinetic energy $mv^2/2$ to the mean thermal energy $3 kT/2$, that temperature maximum is roughly

$$T_o \lesssim h^2/3 \, mka^2$$

where k is Boltzmann's constant and a, the mean spacing between particles (the equivalent of the aperture width in a classical Young's slits experiment)—typically of the order of 2 or 3 angstroms for liquids or solids.

The energy associated with the electron quantum states does not vary continuously with decreasing wavelength. Instead, there are energy bands with gaps in between—the *energy gaps*. In a good insulator, the lowest energy bands are completely filled with electrons up to such a gap, and the electrons, by virtue of Pauli's exclusion principle, cannot adjust to an electric field.

assembly of particles, a superposition of many quantum states. It is as though the entire assembly behaved as the quantum state of a single particle, but everything is greatly scaled up.

This is the *Bose-Einstein condensation* in quantum physics, a state of nonlocality, where one part of a system is able to affect other parts instantaneously. This sort of situation has an eerie feel to it—it defies our intuition—Einstein once called it "the spooky action at a distance". But it is as legitimate as it is interesting from the information point of view. And as for its realness, that is attested by a variety of spectacular phenomena, like superconductivity—a state of matter where all electric resistance disappears—or the "Meissner effect"—where the magnetic flux is excluded from the superconducting material, turned out of doors, so to speak—an effect which, at a critical distance, gives rise to levitation! Such quantum coherent states are not restricted to electrons. They occur with other fermions as well, and also with bosons, as attested by the phenomenon of superfluidity—a state where friction between particles disappears.

Not long ago such phenomena would hardly have been worth a note in a book on biology—they had been found at temperatures far too low to be of biological interest. But in recent years such quantum-coherence manifestations have turned up also quite a few notches higher on the temperature scale, and nowadays, with a better understanding of the physics involved, they keep cropping up by the dozen. Indeed, temperature is not a fundamental condition for such coherence. It was natural for physicists to look for it at low temperatures because even the unruly quantum world tends to become more orderly in the deep cooler. What is basic for a Bose-Einstein condensate is the presence of an energy gap, an insulation between the particle system and its environment. Without such an insulation, the subtle quantum-coherence phenomenon could never come off the ground—all those simultaneously superposed quantum actions would be drowned out by the environmental noise. And here is precisely where the temperature comes in. If the environmental temperature is too high (the limit T_o in footnote 1), the energy of its particles will be high enough to breach the gap and wipe out the coherence. That is why superconductivity and superfluidity had originally been found only way down on the temperature scale, just a few degrees above absolute zero. However, Bose-Einstein condensates turned up in some rather ordinary solids and liquids (that satisfied the energy-gap condition) closer to the comfort index—and the past fifteen years have seen these temperatures getting milder and milder.

Now, to come back to our question of the brain, it is not the first time that the possibility of quantum coherence has been considered for a biological tissue. Already in 1968 Herbert Fröhlich, a leading figure in the physics of superconductivity, proposed, against a background of general skepticism, that the conditions for such quantum coherences might be fulfilled at ambient temperatures in living matter. What Fröhlich had in mind was electrically well-insulated living matter, dielectric material, like that found in cell surface membranes. Indeed, one could hardly fail to be impressed by the dielectric quality of the membrane—its molecules hold up under voltage gradients of tens of thousands of volts per centimeter. But dieletctic materials are not restricted to cell surface membranes. There are quite a few inside the cells in the form of lipid and (hydrophobic) protein molecules. In particular, the latter may be of more than passing interest here to us; the electric charges (dipoles) of those protein molecules can get rapidly redistributed in a changing electric field, triggering changes in protein conformation on a time scale of nanoseconds (which one can follow with the aid of infrared-spectroscopy, laser scattering, and nuclear-magnetic-resonance techniques).

It was knowledge of this and another sort pertaining to quantum polymer chemistry that led Fröhlich to propose his far-reaching hypothesis: the dielectric proteins would form coupled, coherently moving dipole systems. For a good while there was not much to show for this idea. Quantum-coherent oscillations are difficult to demonstrate on living systems—heat effects, due to resonance in thermal frequency modes, all too readily intrude, jumbling the experimental picture. I can speak with feeling here—I tried my hand at this myself in the 1960s, without success. Meanwhile, however, some positive findings have been reported in probings of various types of cells with low-intensity millimeter waves. Grundler and Keilmann, for example, found a resonance response in cellular growth of yeast—a sharp nonthermal resonance at 5×10^{10} Hz (1 Hz = 1 oscillation per second). The result is within expectation of the Fröhlich hypothesis—the coupled-dipole idea implied that, at a critical threshold of energy input, the system would oscillate coherently in a single vibrational mode in the range of 10^9–10^{11} Hz (all other modes being in thermal equilibrium).

This gives some corporeal ground to the idea of quantum coherence in living systems. However, regarding consciousness, there is not much here to take to the bank. The aforementioned experiments were not done on brain, or even on nerve cells. Moreover—and fundamentally more important—they don't tell us whether the quantum-coherent space found went beyond the boundaries of the cells (the resonance observed may have taken place within the confines of the individual cell units), whereas consciousness is by all indications a mulicellular state. But the experiments offer some encouragement and preliminary guideposts for the next step, a search of a quantum coherence that transcends the cell membrane, an *inter*cellular quantum coherence.

Quantum Computations?

Let us restate, then, our question in intercellular terms: Could a collective quantum state over a space transcending the cell boundaries—a single state with a large number of quantum activities occurring simultaneously in many neurons—be the stuff of our conscious thoughts? If at first this comes on as odd, this feeling subsides as soon as we consider our conscious states as information states in the sense of equation *1*. Information, as we have seen, takes up residence in all possible realms of the universe—it has been doing so from the beginning. So, it is not stranger to think of such states residing in quantum coherent matter of the neuronal network than residing in digital patterns of that network. The quantum possibility attracts one's interest because it has such extraordinary prospects for parallel computation. But before we put our shoulder to the wheel, we turn to an argument of a different sort, a line of reasoning concerning the computational basis of conscious thought.

This line was spearheaded by the mathematician and theoretical physicist Roger Penrose. Going to the roots of logic, Penrose argues that our conscious thinking is not reducible to ordinary computational rules, the mathematical logic of ordinary computers. I deal with this line of reasoning here at the start because, if its argument succeeds, it would cut to the chase—it would make quantum coherence a Hobson's choice.

To appreciate the full depth and consequence of this point, let us set Penrose's view against the extreme opposite, which some of the "Artificial Intelligence" (AI) advocates in the digital computer camp have espoused. For them, mental qualities are associated with the logical functions of computers. Any computing device, be that one with chips and wires or one with cogs and wheels, would exhibit such qualities. Conscious mental states would be simply inherent in the carrying out of some systematic sequence of calculational operations (algorithms). To think would mean to activate an algorithm.

Penrose argues against that point of view. His argument develops from Gödel's theorem. In this theorem, a cornerstone of formal logic, Gödel proves that for any consistent formal system that can do arithmetic, say a computer program specified by an algorithm, there will be a true sentence (the system's *Gödel sentence*) which the system cannot prove—if the sentence is true, then it cannot be proved, and if it is false, the system is inconsistent. Now, since we humans, Penrose reasons, understand the meaning of Gödel's theorem—we do see that the Gödel sentence is true—we must possess a capacity that is lacking in a purely algorithmic system; hence, there would be facets of consciousness—those involving understanding—which have ingredients that cannot be emulated by mere algorithmic operations, let alone be embodied by them. He makes this conclusion extensive to *any* form of computation based on the logical principles underlying electronic or mechanical computing devices—regardless of whether they use algorithmic, loosely programmed ("bottom-up"), or random calculational operations. It is in this profound sense that Penrose considers our mental qualities noncomputational.

A number of objections have been raised to Penrose's argument. The issues are too complex to be dealt with cogently in this brief chapter (I discuss them at length in a forthcoming book devoted to the informational aspects of the brain/mind question). However, as an abrégé of the arguments and counterarguments, I think it is fair to say that the case for a noncomputational mental world of consciousness is yet to be made. It would be nice, indeed, if one could

walk the high road of mathematical logic here, but the Gödelian path can succeed only if one can show that we humans are always able to see whether a formal system is consistent, and there is no evidence that even the best of minds can
actually do so in the case of complex systems. But it also is fair to say that the
case for an AI consciousness has yet to be made.

So, let's clamber out of this impasse and see what the alternative of a quantum coherence has to offer in its own right. I already mentioned one attraction,
large-scale parallel computation—*quantum computation*. Quantum computation is a concept put forward in the 1980s by Richard Feynman, David Deutsch
and Paul Benioff. The concept extends the notion of an algorithmic computer to
the quantum level; the discrete character of quantum mechanics matches that of
digital information processing—the quanta of information, the bits, are register-
able as quanta of quantum mechanics. A quantum computer makes use here of
the simultaneous linear superpositions (the simultaneous wavefunctions) in the
quantum sphere—it extracts information from the superposed states. The principle is inherent already in the classical experiments which gave birth to quantum
theory. Consider, for instance, the experiment where a photon is being simultaneously transmitted through a half-silvered mirror and reflected by it. These two
superposed things represent two different computations, and information can be
drawn from the superposed pair. This illustrates the basic operation of a quantum computer and its potential usefulness. And just as such an elementary device
could perform two simultaneous computations, a large-scale one operating with
many superposed quantum states could do many more. Indeed, with a big
enough hunk of matter obeying the laws of quantum physics, only the sky would
be the limit—in principle, such a computer could perform *all* possible computations at once. A computer like that has lots of winning features, but one alone is
enough to put a bee in one's bonnet here: it could run these computations while
the individual results may not be of interest to us, but certain combinations of all
the results may be.[2]

---

2.   Such a computer, as Peter Shor has shown, could factor large numbers in very short times
(something a digital computer, even a supercomputer, has difficulties with). It could also generate information by chance—true random bits (a no-go for a digital computer—such a computer can only generate pseudo-random numbers). Another appealing feature, which was
brought to light earlier by Charles Bennett and Rolf Landauer, is the extremely low energy
expenditure in quantum computations; the aforesaid kTln2 minimum of the linear case does
not apply. Such a computer, as Wojciech Zurek has shown, is not only microscopically but also
operationally reversible. It can be arbitrarily fast with arbitrarily small dissipation (though measurement and resetting require the minimal entropy increment), and the dissipation can be used
(along with redundancy) to prevent errors. As for practical realization of such computers, which
until quite recently seemed forbiddenly difficult, one is now beginning to see the light at the
end of the tunnel. In the past four years a broad class of physical systems with inherent quan-
tum-logic gates, which might serve as computers, has been identified by Seth Lloyd, and the
first constructions of prototypes are now underway in several labortories.

Now, a computer of this sort—a device subject to the laws of quantum physics—is what Penrose postulates to be operating in the brain. This view does not disregard the classical digital brain activity. The two computational operations would exist side-by-side in our brain, though at different levels of energy and matter. We will deal with the two-level aspect further on, but first this question: What may be the substrate for such computations? This may seem an adventurous question to ask at this early stage, but it strikes at the heart of the matter. So, let us at least write down the a priori requisites for that substrate. This will be useful in sieving through the proposals that are already on the table.

The requisites are easily specified: (1) the substrate must be insulated from the cellular-sap, or else, as we have seen, the coherent quantum state gets drowned in the gross; and (2) it must provide intercellular continuity in order to allow a multicellular quantum-coherent state. As to the identity of the substrate, all we can usefully do at this time is to weigh some general possibilities in this a priori light.

We will start with a possibility which Penrose has raised, that the microtubules—the 140-angstrom-wide tubes found in virtually all sorts of cells crisscrossing their interior—might be the substrate. This seems unlikely because they do not fulfill condition 2. While they have attractive dielectric features that might conceivably enable them to serve as *intra*cellular conduits of quantum-coherent waves[3], it seems unlikely that they could do so as *inter*cellular conduits; all the available evidence indicates that they do not traverse the cell membrane.

However, this does not exhaust the possibilities. There are other dielectric candidates in cells that show more promise. They form part of the cytoplasmic ground substance, a net stretching through the interior of cells that is made of finer filaments than the microtubules. That net, or at least a major component of it, fulfills requisite 2 and bids fair for requisite 1. Indeed, this is no ordinary net—its mesh holds organized water! Or perhaps I should say the mesh organizes the water, because the protein constituents make hydrogen bondings with the surrounding water, ordering it. But it's not the information in the molecular order that I am harping on, but that in the quantum order—the information

---

3.   The hypothesis of microtubules as quantum-coherent conduits and computation devices was originally advanced by Stuart Hameroff to explain concerted movement of flagellates, single-celled organisms, like *Paramecium*, which have no nervous system. The microtubule subunits would switch between dipole states at the rate of 5×10^{10} Hz (Fröhlich's resonance range), and information would be transmitted and processed down the length of the tubules as coherent waves of different polarization states. Indeed, calculations by Emilio Del Guidice and colleagues, based on quantum-field theory, show that the 140-angstrom diameter of the tubules would be compatible with the waveguide idea; it is in the right ballpark to focus electromagnetic energy that exceeds a certain threshold, and to conduct massless particle waves. However, while such a microtubular quantum coherence conceivably might be involved in *intra*cellular controls, such as those pertaining to ciliary and flagellar movement, there is nothing to indicate that microtubules could serve as *inter*cellular information conduits.

inherent in the quantum states of the hydrogen atoms. This information changes in an oscillating electric field. Hydrogen atoms flip between quantum states in such a field. They can act as logic gates—bits of information can be registered by the quantized spins of the electron and proton in a hydrogen atom. Thus, the hydrogen atoms here offer natural quantum gates, and a string of such atoms may serve as a quantum-computer circuit that could operate over and over. On something like this, I'd take a gamble. The beauty is that such quantum gates are able to function in linear and nonlinear mode—they permit full- and half-way flipping, or, to use the language of the trade, the gates can register *bits* and *qubits*. And a device that can do both things is able to perform any arithmetic operation.

Thus, the view on the quantum side of the neuronal network is not without prospects. And there may be some more if we consider that, apart from the oscillating electric fields in individual brain cells, there are such fields where masses of brain cells beat in unison. But I will put a stop to my musings here, lest they coddle a panglossian view. Anyway, it's time to put the whole idea to the test. And the time seems ripe for that—things have advanced a good deal since my original foray into the question of biological quantum coherence. Especially the past four years have seen the physics of quantum computation (formerly purely theoretical) going experimental. And just as I write these lines, the first computer prototypes are being tested (see Footnote 2). These experiments with relatively simple (nonbiological) systems are providing us with clues about the question of how quantum gates could be linked. Such clues are badly needed—that question is a puzzler. The coupling of quantum gates is a fiendishly difficult technical proposition. Where for the linking of digital silicon gates just a wire will do, for the linking of quantum gates more subtle stuff is needed, something that provides a coupling that is immune to the outside peturbations destroying quantum coherence. The first laboratory hardware of this sort has just been achieved by isolating qubits on their way from gate to gate in an ion trap or by means of a nonlinear photon interaction. But ingenious as these devices are, it wouldn't surprise me if the lady with the billion-year-long lead has come up with something better. Considering all the marvels she wrought from protein under the unending information pressures, this would be just one more, to top her enterprise.

The notion that that glittering jewel of human intellect, conscious thought, could be reducible to protein and water, may be a blow to one's self-esteem. But really no more than any other notion of evolution—Darwinian schemes always step on the peacock tail. And if one comes to think of it in electron detail, a dielectric protein-water association is stuff with qualities that are found nowhere else in earthly matter.

Two Worlds Unbridged

So, the pursuit begun by Sherrington seventy years ago is now entering a new phase. There is a whole range of questions now facing us, and one looms large:

how might a quantum-coherent state be coupled to the digital operation in the brain? This question follows in the wake of answers obtained by the Sherrington school, answers concerning the digital operation on the input side as well as the output side of the brain. Take the input side, for instance. The information our brain receives about the world outside gets digitally encoded well before it reaches the brain. Our sense organs already convert that information into digital electric form, the nerve impulses, and in that form it flows along the sensory nerve fibers to the brain. This is the solid legacy which the physiologist Lord Edgar Adrian and two generations of investigators of the nervous system after him have left us. Thus, if quantum computation is to serve consciousness, it must somehow be linked to the digital operation that precedes it.

The strong electric field in neurons would seem a natural for such a linkage. The changes in this fields associated with a nerve impulse—a toroidal field—are felt by cytoplasmic matter on the underside of the membrane. Such changes may flip quantum gates, as exemplified by the hydrogen-string model we considered above. But the coupling between this field and the gates must be subtle, just as the coupling in between the gates must be, or else the macroscopic environment would destroy the quantum-coherent states. And this is the reason why the coupling is of primary concern.

The physicist here may have a sense of déjà vu. Ever since the formulation of quantum theory, there loomed the issue of the interface between the quantum sphere and the more familiar sphere of large-scale things. Indeed, this is the central question of quantum theory: How are the two worlds bridged? How do things jump from the dual state of superposed quantum levels to the state where, in the formulism of quantum theory, large things can exist simultaneously in two totally different states? This paradoxical situation is highlighted by Schrödinger's famous cat. In this gedankenexperiment, Erwin Schrödinger, one of the pioneers of quantum physics, envisaged a cat in a box containing a flask of cyanide gas rigged to a quantum trigger mechanism, say, a mechanism activated by a photon coming from a half-silvered mirror (Fig. *13.1*). When the photon gets reflected onto the trigger mechanism, the cat is killed; when it goes through the mirror, the cat is saved. So, there is a 50:50 probability of the transmitted part of the photon's wavefunction killing the cat, and in quantum terms the system is described as a superposition corresponding to that probability. Thus, if we treat cat and box as a single quantum system, then the linear superpositions between alternatives must be maintained right up to the scale of the cat; hence, the cat, too, must be in a superposition of the two states—it would be dead and alive.

This surreal fate befalls the cat as we carry the logic of quantum physics through to the realm of large things. We may not be able to define where precisely on the dimensional scale that realm begins; but a cat, a conglomerate of some 10^{27} particles, certainly qualifies as a large thing. Something must be missing here from our description of the physical world; somehow, in going from the quantum to the large-thing sphere, the dual ghost-like state vanishes in thin air.

Figure 13.1   Schrödinger's cat. The release of cyanide gas from the bottle is triggered by a photon. If the photon, represented by the wave, is reflected by the half-silvered mirror, it hits the mechanism that triggers the hammer which smashes the cyanide bottle, releasing the lethal gas. If the photon goes through the mirror, it merely hits the wall of the box. Does quantum physics imply that the cat is dead *and* alive?

The quantum physicist calls this the *collapse of the wavefunction*, meaning that the number of state vectors is reduced. However, a complete mathematical description of these events, a description encompassing both spheres, still escapes us—Schrödinger's deep paradox remains unsolved.

The problem has deep roots. It lies in the way the evolution of a system of particles is described. We have here two completely different mathematical ways in which the state vector is described as changing with time. While in the ghost world of particles the description of the time-development of a wave packet is indeterministic (the *Schrödinger equation*), in the "real" world of large objects the description of the development of the system is deterministic. This uncomfortable dichotomy has led Niels Bohr to bring forward the concept of *Complementarity*, which has profound philosophic implications to which I will turn later on. Here, without explicit reference or mathematics, I just sum up Bohr's point of view as a complementarity of information—two aspects of a description that are mutually exclusive yet both necessary for a full understanding of what is to be described. However, if one takes the unitary view that all things in nature can be comprehended by a single mathematical formulation—something scientists have yearned and striven for over two thousand years, ever since Democritus put that pea under their mattress—this merely means that our *present* descriptions are incomplete. It is easy to draw a dimensional scale, as I did in Figure *2.1*, and extend it down to 10^{-20}cm, or what have you, in the particle world, or even play mathematics at those dimensions. But to tell the truth, we simply do not yet understand the nature of space at those unfathomable dimensions. In conclusion, the dichotomy between the small- and the large-scale world—Schrödinger's paradox—still is what it was from the start: the central mystery of quantum theory.

And so mystery meets mystery. We get here to a sphere where the journeys of the biologist and the physicist meet, but neither has yet come to an end. However, if the foregoing notions about consciousness are correct, or even half-way correct, we may dare hope that the key may be waiting at the brain-mind interface where Evolution had found the solution long ago.

Part Four

The Science of the Peculiar

How Can You Explain So Wise an Old Bird in a Few Words?

Here ends our excursion along the biological information stream. This is as far as I dared to go; farther "down" it gets too labyrinthine for safe travel. But no journey would be complete without a little philosophy, especially not a journey so full of wonders. It has been said by Epicurus—and he was probably right—that all philosophy takes its origin from wondering. So here we'll ponder over the peculiar character of biology as a science and the prospects of a unifying theory. I will draw a parallel with physics to improve the sibyllic odds.

Though we kept mainly to the sheltered waters of the mainstream, we caught sight of what lay beyond: a basin of endless circles leavening the stream, loops-within-loops engendering a vast variety of bioforms. That variety leaves one rapt in wonder, but the most staggering thing from the scientific point of view is its singular dependence on time. Each organism—indeed, each biomolecular entity—is peculiar in its own space and time and asks to be dealt with on its own terms; but there are seemingly no "absolutes," no phenomena of universal timeless validity.

A physicist coming round to biology can't help feeling at a loss in such a void. What is one to make of a science where everything is so uniquely time-bound? The crucial dimension here, by all appearances, is on a calendar, not on a map. The discipline has more elements of history than of scientific prediction. Where are the reassuring unifying laws? It is certainly not coherency that is wanting. There is palpably a continuous thread going through those kaleidoscopic forms, a thread extending through an immensity of time; but there are no comprehensive explanations, no mathematical formulas accounting for the whole or major parts. The physicist Max Delbrück, who spent a lifetime in biology, once wittily summed the situation up like this: "You cannot expect to explain so wise an old bird in a few simple words."

An Occam's Razor for Biology

How, then, may we hope to come to grips with that strange bird? Is there a biological theory in the offing, a rule book with power of prediction? Game theory, as we have seen, holds out promise here, but we are still a long way off from formulating precognizant laws; the specifying of the pertinent mathematical saddlepoints will keep us busy for a good while. However, there is now something to guide us in that task: the Principle of Information Economy. Which brings us back to the evolutionary canon that has been the underlying theme of much of this book (p. 91). This principle also has its epistemological side. It offers us a handhold to help guide our choices among alternate hypotheses in the day-by-day theoretical chores. It will be our cutting tool—a sort of Occam's Razor—and it promises to be every bit as sharp as the one in physics.

Occam's Razor was the brainchild of the fourteenth-century theologian William of Ockham (which later became Occam). The Razor wasn't actually meant for physics—it dates back several hundred years before the modern era of science—it started as a rule for making abstractions in scholastic philosophy. But, more than philosophy, this rule, which demands that unnecessarily complex entities be cut out from our abstractions, was to profoundly influence physics. It became the means for cutting out unnecessarily complex hypotheses in the waxing science; indeed, it is hard to imagine a physics without it.

Now, in biology, our information economy principle may be put to just such a use: alternate hypotheses may be ranked according to their information cost and the pricey ones are cut out. The ranking is measured in information units; and so perhaps, step by analytical step, one may arrive at a saddlepoint strategy.

Thus, imagine the future practitioner of this art: he assesses the information value of each biological event in space-time and, with Sherlockian sparkle in his eyes, strops Occam's Razor and cuts along the most information-economical path. With an adequate reductionist undergirding, that way of sleuthing may go a long way.

The Mistress We Can Live Without

One now also may bid farewell to teleology—its last afterglow. In physics, what is left of that Aristotelian doctrine are but faint memories. Not so in biology. Here, the creed of immanent goals (*teleological* means "goal-oriented"; from the Greek, *telos*, end or goal) gained a stronger hold. And for good reasons: from whichever angle the anatomical components of living beings were looked at, they always seemed exactly right for their physiological task—an aptness that, from Aristotle to Teilhard de Chardin, has mesmerized philosophers into believing in an intrinsic goal or purpose of things.

There must be some soft spot in the human mind for such a credenda, and it isn't altogether confined to biological notions. Nothing could show this better than what happened in the eighteenth century, at the very height of the "Age of Reason." The mathematician and physicist Joseph Lagrange just then had come up with a most elegant shortcut to Newton's laws. Instead of the usual instant-by-instant trajectory computations from the forces at play, he used a quantity called *action*, a quantity based on the mass and velocity of moving objects and the space they traversed. This brought out something with a wondrous quality to it: no matter what forces were at work, the moving object—a planet, a missile, a ball—always took the path of smallest action. To many a dabbler in philosophy, this was teleological music to one's ear—a planet's ellipse or a ball's parabola seemed to hold an intrinsic goal or purpose—and pundits, literati, and philosophasters returned to the Aristotelian fold in droves.

Well, Aristotle once said himself:

"There's many a slip 'twixt cup and lip."[1]

Fortunately, those slips don't last very long, and this one has long sunk into merciful oblivion. Actually, it was more like a slip-back; teleology already had lost its appeal in physics well before. It had fallen apart toward the beginning of the seventeenth century under Bacon's and Descartes' positivist fire; and later in that century, Newton's laws, accounting for the movement of all macroscopic things, dispatched what was left of the creed of intrinsic purpose. In biology, though, it lingered on. The theory of evolution sapped some of teleology's strength by offering in natural selection a rationale for the "aptness." But it provided no legality, and so the Darwinian stick did not beat as hard as the Newtonian one.

Now, however, we have a stronger one: thermodynamics and information theory—the two close ranks. And as we use them to make inquisition with, we see at once what's wrong with teleology: it violates equation *3*—information and entropy don't balance. The idea of an immanent goal or purpose implies that the biological system knows beforehand which among the many possible paths is the most information-economical one. Or, put in terms of an evolving system, it implies that, at each evolutionary bifurcation, the system possesses the information for the most information-economical path. This foreknowledge would amount to giving a runner in our labyrinth metaphor (Fig. *6.2*) the information for the shortest route—a foul by all the rules of the game! (No less a foul than a ball or a planet knowing beforehand at each instant of its parabola or ellipse the next most *action*-economical path.)

Teleology loses some of its nonscientific flavor when it is stripped of the canon of immanent purpose, and the idea of goal-orientedness is used in the cybernetic sense—in the same sense as automata are goal-oriented but without

1. For the doubting Thomas, I cite Aristotle's own words: "πολλὰ μεταξὺ πέλει κὺλικος καὶ χείλιος ἄκρου."

the goal being the cause of the machine's operation. In this format (often referred to as *teleonomy*) it has actually been quite useful to biologists, especially for making physiological hypotheses. Teleology then officiates in the argumentation for choosing among possible mechanisms. The *modi ponentes* and *modi tollentes* here, to be sure, would hardly meet Bacon's standards—they usually start with "why." Nevertheless, they have been useful to the biological scientist; they have lent him wings to rise above the void of absolutes—and many an advance in physiology has come about this teleological way.

Consider, for example, the argument Wilhelm Roux once used for his momentous hypothesis concerning the unit makeup of chromosomes. All he then had to go by were some observations made under a low-resolution microscope of chromatin figures—a curious sequence of spatial arrangements of the cellular chromatin during cell division. Why this elaborate process, argued Roux, why does nature not simply split the nucleus in half?

> We wish to find out why there is this elaborate deployment of forms ("Formenspiel"), what use it has for the objective, the division of the single nucleus in two halves . . . If the objective . . . were just a simple halving of the nuclear mass and the spatial separation of the two halves, the process of the indirect nuclear division would constitute an enormous detour . . .

wrote Roux in his classic paper (1883). And pondering various possibilities, he advanced the hypothesis that *the chromatin consisted of many qualitatively distinct particles* that needed to be lined up for their splitting in halves—a teleological thunderbolt whose afterclaps can still be heard in every molecular genetics paper today.

Such teleological arguments are the lifeblood of physiological thought. Any biologist who has delved into mechanisms has, at one time or another, turned to them—though rarely are they stated as clearly as Roux's, if they are at all admitted. The physiologist John Haldane once quipped that "teleology is like a mistress one is ashamed to be seen with in public, but one cannot live without."

But now, at last, we may hope that that lady will relax her hold and be replaced with an avocation more becoming to mature age. There is now an Occam's Razor to cut through the biological jungle and chart a course to unifying laws.

The Arduous Minuet

The thought of such laws may strike one as odd. Biology is so dominated by serendipity that it may take awhile for one to warm to the notion of precognizant laws—one may get easily deluded that biology progresses differently from physics. But all intellectual endeavors worthy of the name of science gather way in like manner: through an interplay between theoretical and experimental effort. The ascendancy of serendipity in biology (or of empiricism in general) merely reflects the state of a younger science. Only when a science cuts its wisdom teeth does theory take the commanding lead. Physics has made it there; some of its optimistic practitioners even think they are close to the last *eureka*, a single all-embracing law.

Things have not always been so. Only thirty years ago the two efforts were rather close—the experimenter was then still breathing down the theoretician's neck—and at the beginning of this century one could often hardly tell the two apart, let alone see who led whom. Rather than such a mathematical fanfare, physics then was more like a strenuous minuet between theoretician and empiricist, where the participants were often not in step and occasionally changed hats. This way of doing things may have been racked with wasteful effort, but it ensured that speculation was disciplined by a massive body of facts and not by logic alone.

Well, biology is in that minuet phase now. Not much precognizance here— not yet. So far, there have been four major intellectual thrusts: Darwin's theory of natural selection, providing a rationale for how organisms evolved(1859); Schleiden's cell theory, giving a unit for natural selection and organismic evolution (1838); Mendel's theory of heredity, offering a unit for the organismic continuity of biological information (1865); and Crick's and Watson's DNA model and "Central Dogma," giving the molecular basis for this information and its transmission (1953). The next thrust, the one that may bring us closer to the Archimedian ideal of science, we may expect to come from information theory.

That passage won't be short; the beast is so immensely more complex than that in physics. However, the complexity by itself is no insurmountable barrier; there are no inherent limits to the analysis of the complex. Fortunately, individual natural processes can be isolated from their context in order to be explained mathematically. To understand the workings (dynamics) of a system we don't need to know its detailed structure, but only that part of it that is crucial to abstraction—the laws governing macroscopic matter were deduced long before there was a knowledge of molecules; the principles of molecular interaction were inferred before there was an atomic theory; and the behavior of atoms was perceived before coming to grips with subatomic structure—this has been the epistemological legacy for us of three hundred years of physics.

The Processing of Information Called *Science*

We don't know, of course, what form the impending biological theory will take. It is in the very nature of scientific inquiry not to know the end from the beginning. We cannot tell even now what part of our current empirical capital will be crucial to our abstractions and generalizations. And such noodling, after all, is what science is about. A science is not a stamp collection, but a search for commonalties and regularities—an ordering of recurrences in the collected observational data that leads to the formulation of laws.

The overarching operation here is a processing of information: one extracts information from the pile of observational data, arranges that information in strings, and then condenses it into a shorthand formula or algorithm—a long series of transformations. The final information transform, the algorithm—and this gets us to the critical point—must end up containing the same amount of information as the string, or almost the same (and the perennial minuet sees to it that it does).

In principle, the informational condensation could be carried out in any language. But there is no human language that could do it even remotely as well and as compactly as mathematics does. This is what I meant above by "Archimedian ideal of science"—a science expressed in mathematical form—a trend begun by the mathematician Archimedes two thousand years ago. Physics offers us example after example of the condensing power of that language. Newton squeezed all the information then collected about movement—"the motion of the bodies in the heavens and earth"—in three simple algorithms plus a law of gravitation; Lagrange condensed the information further to an overriding principle, the principle of stationary action, which Newton's laws can be seen to be derived from; and the great coups of this century—Einstein's general relativity theory, Weinberg's and Salam's unified electromagnetic-and-weak-forces theory, and the more recent unified three-forces theory (Chapter 2)—delivered ever more encompassing compressions. The unassailable logic of such abstractions and their explicative power of the past, present, and future give us confidence in the scientific process. And there are no reasons to expect any inherent limitations of the application of this process to biology.

The obvious place to start is the one-dimensional information script, the DNA or RNA. There, the commonalties and recurrences are in plain sight. That script offers us example after example—in Evolution's own handwriting, as it were—of thematic recurrences, seemingly endless uses and reuses of information over the aeons. It is as if Evolution could not help repeating herself, and we have seen why: in her realm everything is pledged to the principle of information economy—her using the same old linear pieces again and again was but obeying the rules of the game. This is why in her sphere there is no such thing as flogging a dead horse; there, when the horse is no longer flogged, it is dead.

Thus, the DNA seems a logical place to center our initial reductionist efforts. Its unit-sequence contains a collectanea of raw data for scientific processing—a richer mine than physicists ever had. What we presently see of the (unprocessed) three-dimensional somatic realm, it is true, seems extravagantly elaborate, tangled, and labyrinthine. But it wouldn't be the first time that what to the naked eye appears as complex and muddled, our mind's eye eventually resolves as simple and full of regularity. Thus, we have all the reason to hope that the right reductionist approach will bare the legal essence.

In the two-thousand-year pursuit in physics, that essence has invariably proven to be of simple elegance and beauty. Lagrange's equations, Planck's $E = h\nu$, Einstein's $E = mc^2$, and the unifying formulations thereafter all bear testimony to that. Looked at from our present perch in biology, those monumental achievements seem hard to match. But consider, it took a full two hundred years to wangle those things out. It wasn't so long ago when phenomena in nature often didn't fit the existing physics molds; those with sanguine expectations from theory then weren't precisely flushed with joy. Someone—a past master of the fine art of mixing chutzpah with gallows humor—captured the spirit of the time with this piece of advice: "Don't let an ugly fact spoil a beautiful hypothesis."

This may touch a sympathetic cord in biologists. But we really needn't fear on that account. There is an all-pervading beauty to that lively twig of the cosmic tree. And as for its impending laws, they are bound to be no less elegant than those for the rest of the tree; the principle of information economy guarantees that.

The Limits of Scientific Knowledge

But lest we get too uppity, there is a limitation, an epistemological limit, from which neither biology nor physics is exempt. We have grazed that limit earlier on in this book, and this is the second bequeathal from physics to us—from quantum physics. The practitioners of this branch have had to cope with a world that is not amenable to rigorously deterministic mathematical description. One cannot determine simultaneously both the position and the velocity of an elementary particle with any prescribed degree of accuracy; and this fundamental uncertainty is behind the two-world dilemma we came upon in the preceding chapter. We skirted that strange quantum world as much as we could by keeping to the large-scale molecular in the book. However, if a comprehensive theory is our goal, the time will come when we will have to look under that coarse-grained skin. And it is good to know in advance that the legality awaiting us then will no longer have the deterministic bedrock we are used to from macroscopic systems.

There is, strange as that may seem, an irreducible limit of knowability, because in the elementary-particle universe it is impossible to account for the physical reality with any single description; two complementary views seem to be needed to fully account for any particular physical event—two views that are (simultaneously) mutually exclusive, yet both necessary for a full understanding of what is to be described. Take, for example, the carbon atom, one of the building-stones of our body. This object can be described as a planetary system with a nucleus around which electrons revolve, or as a nucleus surrounded by a system of stationary waves. Each of these descriptions is legitimate in its own right and valid for a particular set of empirical conditions. But they are not complete by themselves. Only together, through complementation, do they provide a full picture. This is Bohr's concept of complementarity, which I have touched on before.

To read Bohr's papers in *Physical Review* (1935) and in an Einstein festschrift (1949) which contains an illuminating thought experiment, is to understand one of the fundamental epistemological questions of our time. Yet, outside physics his complementarity concept is widely misunderstood, especially among professional philosophers. Perhaps, this situation may be helped if we reduce Bohr's argument to information terms. In the transparent clarity of Heisenberg's mathematics, the quantities describing the momentum (p) and position (q) of a particle are revealed as unmeasurable conjointly; when one tries to measure either quantity, information about the other is necessarily lost (the product of the uncertainties Δp and Δq can never be less than Planck's constant). *There is, thus, an **intrinsic** shortfall of information in our descriptions of objects in the elementary-particle sphere; any single description contains less information than the object to be described, but two descriptions can complement one another*—they make up for the shortfall, as it were.

The upshot is a strangely blurred world where particles have no definite trajectories (their motions are probability functions)—a world where our predictions about spatial particle distributions of energy and matter are statements of likelihood. So, one makes the best of it with complementary descriptions.

In biology, Bohr's complementarity has yet had few repercussions. It is easy to see why: till now, quantitative biologists have stayed with large-scale macroscopic systems. The elemental information deficit doesn't surface there, and so whatever mathematical synopses have been achieved contain about the same information as the physical or chemical events they attempt to account for; or, to lapse into the vernacular, those events are of such a high probability that to all intents and purposes we may consider them deterministic.

But at a deeper level the deterministic mask must come off. Even staying close to the cellular information "mainstream," as we did, we continuously bumped up against phenomena that are kindled by processes of individual atoms. And we didn't even go into processes like the triggering of electric signals in our eyes by photons—a triggering of a macroscopic signal by a single elementary particle—where quantum mechanical effects must obviously play a role. Thus, the probabilistic underpinning cannot be ignored; a comprehensive biological theory will have to cover it, and the descriptions may well be of Bohr's complementary sort . . . at least, for awhile.

I do not imply here a complementarity between the laws of biology and physics, but a complementarity of laws *within* physics. Though the former may once, when the chasm between the two sciences seemed so profound, have been a reasonable thought,[2] that time is well past. Ever since physiologists in the past century began to take the lid off the organismic box, the belief took hold that biological phenomena can be explained in terms of pure physics and chemistry; and, in this century, our confidence has steadily grown as biochemistry and molecular biology crossed part of the chasm. And when eventually in 1953 the DNA structure was bared, even the mystery of mysteries, how like begets like, seemed no longer quite so stark.

Thus, it is but in the spirit of the age if we now try to press the analysis of life to the limit, the quantum realm. It is here where some of the frontiers of contemporary biology may lie. One instinctively shies away from that weird, counterintuitive particle world, in the hope that all the answers may lie in the macroscopic realm. But they don't. We had a little taste of that in Chapter 13, and if the perspective of a quantum-coherent cellular ground substance I offered there stands the test, then it is at the quantum level of protein-water organization in the brain where we must seek the physical basis of mind. And, if I may, cathedra ignorantiae, enlarge on that, it is also at this level where we may learn the solution of that stubborn two-world riddle of physics—a solution Evolution may have hit upon long ago. Well, it wouldn't be the first time that she takes the wind out of our inflated sails.

It is not easy to make friends with that Alice-in-Wonderland world where things—the elementary stationary states of matter—turn into one another and are not legislated in every detail by an incontestable logic of cause and effect. But it is in this wonderland-at-the-bottom-of-everything where biology and physics become one. It is that that one may hope one day to find the stationary states which are giving the slip to the present unifying physics theories, the states which may allow us to understand whatever it is that underlies the very nature of matter—all matter, including the biological one. Alas, it may take a while until that unifying sermon will be preached on the streets of Gotham.

2.   Bohr himself once espoused such a view in the 1930s and 1940s, as a deeper scholium of the old notion of a special biological force. In "Light and Life" (1933), he expressed the opinion that biological phenomena might not be reducible to physics—that some fundamental laws pertaining to life remained still to be discovered—and raised the possibility of a complementarity relation between the physiological and physical facets of life. But after the explosive progress in molecular biology in the 1950s, he no longer saw any "inherent limitation of the application of elementary physical and chemical concepts to the analysis of biological phenomena" (1959). In "Light and Life Revisited" (1962), he dropped all reference to complementarity.

Epilogue

"Grant me eternal life," asked the RNA of the genie. "That's not in my power to give," answered the genie. "Grant me then at least a wish," the wily RNA said. "One wish?" laughed the genie. "Yes," said the RNA, "only one." And so the wily RNA asked its wish: "Make me thrifty." To this the genie cheerily gave the nod . . . and the RNA lived forever after.

appendix

A Brief Chronicle of "Information"

The Form-Giving Principle

Information, as a quantitative concept, as we have seen in Chapter 1, comes from physics, but the roots are in biology. Aristotle used the concept two thousand years ago exactly in the same sense as we use it in biology today. In his "Generation of Animals" he expounds that the information for the developing organism resides in what he calls *eidos*, the form-giving principle (in the semen). He defines this principle as something that shapes the embryo, without itself being changed in the process, and likens it to a building plan:

It contributes nothing to the material body of the embryo but only communicates its program of development. . . . It does not become part of the embryo, just as no part of the carpenter enters into the wood he works . . . but the form is imparted by him to the material by means of the changes he effects." (De generatione animalium I, 21, 22)

It is hard to imagine a better definition for DNA at that time and era or even only fifty years ago. Yet, in the modern era of science starting in the seventeenth century, this great concept of a form-giving principle was given the cold shoulder. The idea that something could shape things without itself being changed or, as Max Delbrück aptly put it, move things without itself being moved, clashed head on with the Newtonian precept, "action equals reaction." So, for three centuries the Aristotelian biological information concept was assigned to the trash heap. It was dug out by Delbrück, and back on the socle it belongs in a delightful essay he entitled "Aristotle-totle-totle" (1971).

In Latin, *eidos* eventually became *informatio*, from *informare*: to "give form," to "shape," to "guide." This meaning persisted through medieval times in the Indo-European languages, as in Middle French, *enfourmer*, and Middle English, *enfourmen*. That connotation was lost in modern times, until it was resuscitated for use in physics and eventually found its way back to biology.

The Tribulations of the Two Great Laws

The idea of information in physics begins with the notion of *entropy*. That word was made up by the German physicist Rudolf Clausius in the 1850s from two Greek stems meaning "turning into." Clausius had the turning of heat into degraded form in mind (what he called *Verwandlungsinhalt*), but the name came out adroitly in more than one way. With the insights of statistical mechanics and information theory gained later, the term "entropy" became truly chameleonic—which is only right for such a cosmic entity.

The notion of entropy was closely entwined with the first two laws of thermodynamics. Clausius wrote them in two lines: "The energy of the universe is constant. The entropy of the universe tends to a maximum."

The discoverer of the second law was the French engineer Sadi Carnot, who published his "Réflexions sur la puissance motrice du feu et sur machines" in 1824. Although this memoir was in easy language and sold for only three francs, it had few takers and was soon forgotten. Twenty-five years later, Rudolf Clausius in Germany and Lord Kelvin in England reformulated the law in general terms.

The first law fared scarcely better. The German surgeon Julius Mayer postulated it in 1842: "A force once in existence cannot be annihilated" (he used "force" where we would use "energy" today). Mayer could not get his paper published in the reputable German physics journal of his time and had it privately printed in a pamphlet on nutrition. The German physicist and physician Hermann von Helmholtz was on the same quest and independently propounded the law in greater detail in 1847. He too found his paper rejected by that journal and also had it privately printed. There followed a period when the accolades for the discovery of the law even went to others, and it was not until the 1860s that the fair-minded English physicist John Tyndall straightened out the record, and Mayer's achievement was recognized.

Two of the most momentous laws in science needed the passing of almost one generation to gain acceptance!

The Ripening of an Afterthought

The distilling of the concept of information to its abstract essence began as an afterthought on entropy. By the 1880s, the Austrian physicist Ludwig Boltzmann had developed a statistical molecular basis for entropy—one of the great theoretical achievements of all times (equation *2* in Chapter 1 gives Boltzmann's original formulation, only in slightly different form). In a later work, in 1894, and, almost in passing, Boltzmann remarked that the statistical entropy could be interpreted as a measure of missing information.

It took another thirty years for this "little" postscript to come into its own. In 1928, the American engineer R. Hartley, pondering the problem of transmitting information by telegraph, concluded that the information in a message is the logarithm of the number of possible messages. And in 1929, the Hungarian physicist Leo Szilard, then in Germany, took up Maxwell's half-century-old paradox and showed that what the demon "knew" amounted to k log 2 in entropy, each time he selected a molecule. As it happens often in science, the encounter of what appeared to be a deep paradox led to a higher level of understanding. Both men had no difficulties in getting their papers accepted by the right journals—Hartley's appeared in the *Bell System Technical Journal* and Szilard's in the *Zeitschrift für Physik*—nevertheless, their discoveries rode at anchor for twenty years.

Information theory finally gathered momentum in the 1940s at the hands of the American mathematicians Claude Shannon, Norbert Wiener, John von Neumann, and others. Shannon elegantly related statistical mechanics to information transmission. His theorems—they, too, appeared in the *Bell System Technical Journal*—showed that the information conveyed from a source to a receiver, by selecting a subset of messages out of a set, is equal to the difference between the entropies of the two sets (equation *1*). Shannon, as the story goes, was first inclined to use a more neutral word than entropy for the measure of information. He used it, though, upon the prankish advise of von Neumann that "in a debate it would always give him an advantage, since anyway no one knows what entropy is."

Shannon's generalizations had immediate practical repercussions: recovery of garbled messages, transmission of television pictures, early-warning radar, communication by space satellites, and so on. These successes alone would have given information theory a secure home in today's science. But there was more.

At about the same time as Shannon worked out the principles of information transmission, there grew another branch of information theory, the theory of feedback automata. Norbert Wiener, one of its foremost exponents, named it *cybernetics*, after Kybernes, in Greek, the steersman. This is the science of controls: the control of systems by information that is fed back into them to maintain a determinate state.

This branch of information theory already has revolutionized a good deal of our technology in automatic machinery and robots, as has information theory as a whole through its most visible products, computers, and it is far from being through with the revolutionizing, at this point. It is, however, too powerful a theory to stay limited in its applications to engineering. Rather, it may be expected to shed light on fundamental problems of control, particularly biological control. Everything we know about the turnings of organismic wheels tells us that they are controlled by intricate, multilevel feedbacks. It seems likely, therefore, that the strongest impact of this branch of information theory is yet to come in the understanding of the controls within ourselves. In forecasting the answers that cybernetics might hold in store for biological problems, one usually thinks of the riddle of the brain and mind. This may eventually come true, but probably closer at hand are answers to questions of basic organismic control, somatic and genetic. Three hundred years ago, the mathematician and philosopher Gottfried Wilhelm Leibniz (inventor of the differential calculus) offered an ingenious solution to the problem of causality in living beings: automata within automata, wheels within wheels, in infinite regression. I suspect we are about to learn in the next ten or twenty years what these automata are in molecular terms and what the "infinite" regression is in information terms.

Recommended Reading

Physics

Atkins, P.W. 1995. *Physical Chemistry*. Freeman Press, New York.

Davies, P. 1996. *The New Physics*. Cambridge University Press, New York.

Feynman, R., Leighton, R.V. and Sands, M.L. 1989. The Feynman Lectures on Physics. Commemorative issue. Addison-Wesley, Reading, Mass.

Feynman, Richard. 1982. *The Character of Physical Law*. MIT Press, Cambridge, Mass.

Gell-Mann, M. 1994. *The Quark and the Jaguar: Adventures in the Simple and the Complex*. Freeman Press, New York.

Polkinghorne, J.C. 1984. *The Quantum World*. Longman, London.

Rae, A. 1986. *Quantum Physics: Illusion or Reality?* Cambridge University Press, Cambridge.

Information Theory and Statistical Mechanics

Chomsky, N. 1957. *Syntactic Structures*. Mouton and Co., The Hague.

Elsasser, W.M. 1966. *Atom and Organism. A New Approach to Theoretical Biology*. Princeton University Press, Princeton, N.J.

Fano, R.M. 1961. *Transmission of Information*. MIT Press, Cambridge, Mass.

Penrose, R. 1970. *Foundations of Statistical Mechanics: A Deductive Treatment*. Pergamon, Oxford.

Raisbeck, G. 1964. *Information Theory: An Introduction for Scientists and Engineers*. MIT Press, Cambridge, Mass.

Ruelle, D. 1969. *Statistical Mechanics*. Benjamin Inc., New York.

Wiener, N. 1948. *Cybernetics or Control and Communication in the Animal and the Machine*. MIT Press and John Wiley and Sons, New York.

Chemistry, Biochemistry, and Bioenergetics

Cantor, C.R. and Shimmel, P.R. 1980. *Biophysical Chemistry.* Freeman Press, San Francisco.

Dickerson, R.E., Gray, H.B. and Haight, G.P. 1979. *Chemical Principles.* 3rd Edition. Benjamin/Cummings, Menlo Park, Calif.

Fersht, A. 1983. *Enzyme Structure and Mechanism.* 2nd Edition. Freeman Press, San Francisco.

Harold, F.M. 1986. *The Vital Force: A Study of Bioenergetics.* Freeman Press, San Francisco.

Lehninger, A.L. 1973. *Bioenergetic: The Molecular Basis of Biological Energy Transformation.* Benjamin/Cummings, Menlo Park, Calif.

Lehninger, A.L. 1994. *Biochemistry.* 2nd Edition. Worth, New York.

Nomura, M. 1973. *Assembly of Bacterial Ribosomes.* Science 179:864–873.

Schulz, G.E. and Schirmer, R.H. 1979. *Principles of Protein Structure.* Springer-Verlag, New York.

Stryer, L. 1995. *Biochemistry.* 4th Edition. W.H. Freeman, San Francisco.

Cell and Molecular Biology

Alberts, B., Bray, D., Lewis, J., Raff, M., Roberts, K. and Watson, J.D. 1994. *Molecular Biology of the Cell.* 3rd Edition. Garland Publishing, New York and London.

Branden, C. and Tooze, J. 1991. *Introduction to Protein Structure.* Garland Publishing, New York.

Darnell, J., Lodish, H. and Baltimore, D. 1995. *Molecular Cell Biology.* 3rd Edition. Scientific American Books, W.H. Freeman, New York.

Dawkin, R. 1990. *The Selfish Gene.* Oxford University Press, Oxford.

Lewin, B. 1990. *Genes IV.* Oxford University Press, New York.

Stent, G. 1968. *That Was the Molecular Biology That Was.* Science 160:390–395.

Watson, J.D., Hopkins, N.H., Roberts, J.W., Argetsinger, F., Steitz, J. and Weiner, A.M. 1987. *Molecular Biology of the Gene.* 4th Edition. Benjamin/Cummings, Menlo Park, Calif.

Wolpert, L. 1991. *The Triumph of the Embryo.* Oxford University Press, New York.

Neurosciences

Barlow, H.B., Blakemore, C., Watson-Smith, M. 1990. *Images and Understanding.* Cambridge Universtiy Press, Cambridge.

Churchland, P.S. and Sejnowski, T. J. 1992. *The Computational Brain.* MIT Press, Cambridge, Mass.

Crick, F. 1995. *The Astonishing Hypothesis.* Simon and Schuster, New York.

Eccles, J.C. 1994. *How the Self Controls Its Brain.* Springer-Verlag, Berlin.

Hall, Z.W. ed. 1992. *An Introduction to Molecular Neurobiology.* Sinauer Publishers, Sunderland, Mass.

Hille, B. 1992. *Ionic Channels of Excitable Membranes.* Sinauer Publishers, Sunderland, Mass.

Hubel, D.H. 1988. *Eye, Brain and Vision.* Freeman Press, San Francisco. Scientific American Library Series.

Kandel, E.R., Schwartz, D.H. and Jessell, T.M. 1991. *Principles of Neural Science.* Elsevier, New York.

Levitan, I.B. and Kaczmarek, L.K. 1991. *The Neuron: Cell and Molecular Biology.* Oxford University Press, New York.

Nicholls, J.G., Martin, A.R. and Wallace, R.G. 1992. *From Neuron To Brain.* Sinauer Publishers, Sunderland, Mass.

Searle, J.R. 1996. *The Rediscovery of the Mind.* MIT Press, Cambridge, Mass.

Shepherd, G.M. *Neurobiolgy* 1988. Oxford University Press, New York.

Self-Organization and the Origins of Life

Crick, F. 1981. *Life Itself: Its Origin and Nature.* Simon and Schuster, New York.

Eigen, M. and Oswatitsch, R.W. 1992. *Steps toward Life: A Perspective on Evolution.* Oxford University Press, New York.

Gedulin, B. and Arrhenius, G. 1994. Sources and geochemical evolution of RNA precursor molecules: the role of phosphate. *In: Early Life on Earth.* Nobel Symposium 84. Columbia University Press, New York. 91–110.

Gesteland, R.F. and Atkus, J.F. eds. 1993. *The RNA World.* Cold Spring Harbor Laboratory Press, New York.

Holland, J. 1995. *Adaptation in Natural and Artificial Systems.* MIT Press, Cambridge, Mass.

Kauffmann, S.A. 1993. *The Origins of Order: Self-Organization and Selection in Evolution.* Oxford University Press, New York.

Langton, C.G., Taylor, C., Farmer, J.D. and Rasmussen, S. eds. 1992. *Artificial Life II.* Addison-Wesley, Reading, Mass.

Miller, S.L. and Orgel, L.E. 1974. *The Origins of Life on the Earth.* Prentice Hall, Englewood Cliffs, N.J.

Nicolis, G. and Prigogine, I. 1985. *Exploring Complexity.* Freeman, New York.

Orgel, L.E. 1973. *The Origin of Life: Molecular and Natural Selection.* John Wiley & Sons, New York.

References

INTRODUCTION

Goodstein, D.L., Neugebauer, G. 1989. *In*: The Feynman Lectures on Physics. Commemorative issue. Addison-Wesley Publishing Company. Reading, Mass. pp. xiv, volume II.

1. INFORMATION AND ORGANISMS

Maxwell's Demon

Maxwell, J.C. 1880. *Theory of Heat*. 6th Edition, D. Appleton Co., New York.
Szilard, L. 1929. Z. Phys. 53:840. *On the Decrease of Entropy in a Thermodynamic System by the Intervention of Intelligent Beings*. Reprinted in Quantum Theory and Measurement, J. A. Wheeler and W. H. Zurek, eds. Princeton University Press, Princeton, N.J., 1983.

What Is Information? / The Information-Entropy Trade-Off

Bridgman, P.W. 1969. *The Nature of Thermodynamics*. Peter Smith, Gloucester, Mass.
Brillouin, L. 1962. *Science and Information Theory*. Academic Press, New York.
Fano, R.M. 1961. *Transmission of Information*. MIT Press, Cambridge, Mass.
Rothstein, J. 1951. *Information, measurement and quantum mechanics*. Science 114:171–175.
Schrödinger, E. 1967. *What is Life? The Physical Aspects of the Living Cell*. Cambridge University Press, Cambridge.

Shannon, C.E. and Weaver, W. 1959. *The Mathematical Theory of Communication*. University of Illinois Press, Urbana, Ill.
Withrow, G.J. 1972. *The Nature of Time*. Holt, Rinehart & Winston, New York.

The Act of a Molecular Demon / How the Molecular Demons in Organisms Gather Information

Bennett, C.H. 1982. *The thermodynamics of computation*. Internat. J. Theoret. Phys. 21:905–940.
Bennett, C. H. 1987. *Demons, engines and the second law*. Sci. Amer. 257:108–116.
Brillouin, L. 1962. *Science and Information Theory*. Academic Press, New York.
Landauer, R. 1989. Matrix Computations and Signal processing *In*: Selected Topics in Signal Processing. S. Haykin, ed., pp. 18–47. Prentice-Hall, Englewood Cliffs, N.J.
Landauer, R. 1993. *Statistical Physics of Machinery: Forgotten Middle-Ground*. Physica A 194:551.

2. THE COSMIC ORIGINS OF INFORMATION

The Bigger Picture of Organizations → Debut of the Cosmic Quartet

Barrow, J.D. and Silk, J. 1983. *The Left Hand of Creation: The Origin of Evolution of the Expanding Universe*. Basic Books, New York.

Davies, P. 1985. *Superforce.* Simon & Schuster, New York.

Feinberg, G. 1977. *What is the World Made Of? The Achievements of Twentieth Century Physics.* Anchor Press/Doubleday, Garden City, N.Y.

Feynman, R. 1982. *The Character of Physical Law.* MIT Press, Cambridge, Mass.

Sato, H. 1981. *The early universe and clustering of the relic neutrinos.* Symposium on Relativistic Astrophysics. Ann. NY Acad. Sci. 375:44.

Origins

Hawking, S.W. 1988. *A Brief History of Time: From The Big Bang to Black Holes.* Bantam Books, New York.

Hawking, S.W. and Ellis, G.F.R. 1977. *The Large Scale Structures of Space Time.* Cambridge University Press, New York.

Hodge, P.W. 1981. *The extra-galactic distance scale.* Ann. Rev. Astron. Astrophys. 19:357.

Rozental, I.L. 1988. *Big Bang. Big Bounce. How Particles and Fields Drive Cosmic Evolution.* Springer-Verlag, Berlin.

Smolin, L. 1997. *The Life of the Cosmos.* Oxford University Press, New York.

Sunyaev, R.A. and Zeldovich, Y.B. 1980. *Microwave background-radiation as a probe of the contemporary structure and history of the universe.* Ann. Rev. Astron. Astrophys. 18:537.

Weinberg, S. 1977. *The First Three Minutes. A Modern View of the Origin of the Universe.* Basic Books, New York.

Weinberg, S. 1985. *Origins.* Science 230:15–18.

3. THE GROWTH OF THE COSMIC INHERITANCE ON EARTH

The Randomness that Contains Information

Prigogine, I. 1955. *Introduction to Thermodynamics of Irreversible Processes.* 3rd Edition. Interscience Publishers, New York.

Prigogine, I. 1980. *From Being to Becoming: Time and Complexity in the Physical Sciences.* Freeman & Co., San Francisco.

Molecular Organization from Chaos

Balesca, R. 1979. *Equilibrium and Non-Equilibrium Statistical Mechanics.* John Wiley & Sons, New York.

Crutchfield, J.P. 1984. *Space-time dynamics in video feedback.* Physica. 10D:225–245.

DeKepper, P., Castets, V., Dulos, E. and Boissonade, J. 1991. *Turing-type chemical patterns in the chlorite-iodide-malonic acid reaction.* Physica D. 49:161–169.

Ouyang, Q. and Scrinney, H.L. 1991. *Transition from a uniform state to hexagonal and striped Turing patterns.* Nature 352:610–612.

Prigogine, I. and Stengers, I. 1984. *Order Out of Chaos.* Bantam, Toronto.

Turing, A.M. 1952. *The chemical basis of morphogenesis.* Phil. Trans. R. Soc. London B 327:37.

Winfree, A.T. 1974. *Rotating chemical reactions.* Sci. Amer. 230: 82–94.

Winfree, A.T. 1980. *The Geometry of Biological Time.* Springer-Verlag, New York.

The New Alchemy and the First Stirrings in the Twiglet of the Cosmic Tree

Dobbs, D.J. 1975. *The Formulations of Newton's Alchemy.* Cambridge University Press, Cambridge.

Monod, J. 1971. *Chance and Necessity.* Vintage Books, New York.

A Question of Time

Hoyle, F. and Wickramasinghe, N.C. 1981. *Evolution from Space.* Simon and Schuster, New York.

Kauffman, S. 1995. *At Home in the Universe.* Oxford University Press, New York.

Levin, R. 1992. *Complexity: Life on the Edge of Chaos.* Macmillan, New York.

Schopt, J.W. 1983. *Earth's Earliest Biosphere: Its Origins and Evolution.* Princeton University Press, Princeton, N.J.

Shapiro, R. 1986. *Origins: A Skeptic's Guide to the Creation of Life on Earth.* Heinnemann, London.

4. FROM ENERGY TO BONDS TO MOLECULAR INFORMATION: THE INFORMATION/ENTROPY BALANCING ACT

The Energy Flux Nursing Biomolecular Organization

Edsall, J.T. and Gutfreund, H. 1983. *Biothermodynamics: The Study of Biochemical Processes at Equilibrium.* John Wiley & Sons, New York.

Harold, F.M. 1986. *The Vital Force: A Study of Bioenergetics.* W.H. Freeman, New York.

Morowitz, H.J. 1968. *Energy Flow in Biology: Biological Organization as a Problem in Thermal Physics.* Academic Press, New York.

Light into Bonds into Information

Mayer, J.R. 1845. *Die Organische Bewegung in ihrem Zusammenhang mit dem Stoffwechsel.* Heilbronn.

5. THE THREE-DIMENSIONAL CARRIERS OF BIOLOGICAL INFORMATION

The Molecular Jigs → Transmission of Information by Weak Forces Hinges on Goodness of Molecular Fit

Davidson, N. 1967. *Weak Interactions and the Structure of Biological Macromolecules. In:* The Neurosciences. G.C. Quarton, T. Melnechuk, and F.O. Schmitt, eds., pp. 46–56. Rockefeller University Press, New York.
Fischer, E.H. 1894. *Einfluss der Configuration auf die Wirkung der Enzyme.* Ber. Deutsche Chem. Ges. 27:2985.
Jencks, W.P. 1973. *Approximation, chelation and enzymic catalysis.* P.A.A.B.S. Rev. 2:235–319.

Molecular Relays: The Foolproof Genies / The Molecular Machines for Coherent Transfer of Information

Karplus, M. and McCammon, J.A. 1983. *Dynamics of proteins: elements and function.* Annu. Rev. Biochem. 52:263–300.
Koshland, D.E. 1970. *The Modular Basis of Enzyme Regulation. In:* The Enzymes. P.D. Boyer, ed., pp. 341–396. 3rd Edition. Vol. 1. Academic Press, New York.
Koshland, D.E. 1973. *Protein shape and biological control.* Sci. Amer. 229:52–64.
Monod, J., Changeux, J-P., and Jacob, F. 1963. *Allosteric proteins and cellular control systems.* J. Mol. Biol. 6:306–329.
Ringe, D., Petsko, G.A. 1985. *Mapping protein dynamics by X-ray diffraction.* Progr. Biophys. Biol. 45:197–235.

The Proteus Stuff

Rossmann, M.G., and Argos, P. 1981. *Protein folding.* Annu. Rev. Biochem. 50:457–532.

Transmission of Information by Weak Forces Hinges on Goodness of Molecular Fit

Davidson, N. 1967. *Weak Interactions and the Structure of Biological Macromolecules. In:* The Neurosciences. G.C. Quarton, T. Mel-
nechuk, and F.O. Schmitt, eds., pp. 46–56. Rockefeller University Press, New York.
Dickerson, R.E. and Geis, I. 1969. *The Structure and Action of Proteins.* Benjamin/Cummings, Menlo Park, Calif.
Dickerson, R.E. and Geis, I. 1979. *Chemistry Matter and the Universe.* Benjamin/Cummings, Menlo Park, Calif.
Phillips, D.C. 1966. *The three-dimensional structure of an enzyme molecule.* Sci. Amer. 215:78–90.
Stroud, R.M. 1974. *A family of protein-cutting proteins.* Sci. Amer. 231:74–88.

6. THE GROWTH OF BIOLOGICAL INFORMATION

How Sloppy Demons Save Information

Jencks, W.P. 1982. *Rules and Economics of Energy Balance in Coupled Vectorial Processess. In:* Membranes and Transport, A.N. Martinosi, ed., Vol. 1. Plenum Press, New York.
Jencks, W.P. 1983. *On the Economics of Binding Energies.* Solvay Conference (Nov. 1983).

Keeping Things in the Family of Adam and Eve Has Its Advantages

Brenner, S. 1988. Forbes Lecture, Marine Biological Laboratory.

The Vestiges of Ancient Information

Brazas, R. and Ganem, D. 1996. *A cellular homolog of hepatitis delta antigen: implications for viral replication and evolution.* Science 274:90–94.
Chambon, P. 1981. *Split genes.* Sci. Am. 244(5):60–71.
Crick, F. 1979. *Split genes and RNA splicing.* Science 204:264–271.
Darnell, J.E., Jr. 1978. *Implications of RNA-RNA splicing in evolution of eukaryotic cells.* Science 202:1257–1260.
Darnell, J.E., Jr. 1982. *Variety in the level of gene control in eukaryotic cells.* Nature 297:365–371.
Darwin, C. 1871. *The Descent of Man* (quotation from Chapter 21). D. Appleton, New York.
Doolittle, W.F. 1978. *Genes in pieces: were they ever together?* Nature 272:581–582.
Gilbert, W. 1978. *Why genes in pieces?* Nature 271:501.
Robertson, H.D. 1992. *Replication and evolution of viroid-like pathogens.* Curr. Top. Microbiol. Immunol. 176:213–219.

The Basic Loop

Simon, H. A. 1988. *The Sciences of the Artificial.* 2nd edition. MIT Press, Cambridge, Mass.

Why Mutations Don't Rub Off Folded Molecular Structures

Kekez, D., Ljubici, A. and Logan, B.A. 1990. *An upper limit to violations of the Pauli Exclusion Principle.* Nature 348:224.

Ramberg, E. and Snow, G.A. 1990. *Experimental limit on a small violation of the Pauli Principle.* Phys. Lett. B. 238:438–441.

Sudbery, T. 1990. *Exclusion Principle still intact.* Nature 348:193–194.

Natural Selection Is Not a Gentle Game

Darwin, C. 1890. *The Origin of the Species.* Chp. IV; p. 34, Humbolt, New York.

The Garden of Eden

Bernal, J.D. 1967. *The Origin of Life.* Weidenfeld & Nicholson, London.

Böhler, C., Nielsen, P.E. and Orgel, L.E. 1995. *Template switching between DNA and RNA oligonucleotides.* Nature 376:578–581.

Bruick, R.K., Dawson, P.E., Kent, S.B.H., Usman, N. and Joyce, G.F. 1996. *Template-directed ligation of peptides to oligonucleotides.* Chemistry and Biology 3:49–56.

Cairns-Smith, A.G. 1982. *Genetic Takeover and the Mineral Origins of Life.* Cambridge University Press, Cambridge.

Calvin M. 1969. *Chemical Evolution: Molecular Evolution Towards the Origin of Living Systems on Earth and Elsewhere.* Oxford University Press, London.

Cech, T.R. and Bass, B.L. 1986. *Biological catalysis by RNA.* Annu. Rev. Biochem. 55:599–629.

Crick, F.H.C. 1968. *The origin of the genetic code.* J. Mol. Biol. 38:367–379.

Crick, F. 1982. *Life Itself: Its Origin and Nature.* Simon and Schuster, New York.

Crick, F.H.C., Brenner, S., Klug, A. and Pieczenik, G. 1976. *A speculation on the origin of protein synthesis.* Origins of Life. 7:389–397.

Crick, F.H.C. and Orgel, L.E. 1973. *Directed panspermia.* Icarus 49:341–348.

Dabbs, E.R. 1991. *Mutants lacking individual ribosomal proteins as tools to investigate ribosomal properties.* Biochemie 73:639–645.

Eigen, M. 1971. *Self-organization of matter and the evolution of biological macromolecules.* Naturwiss. 58:465–523.

Eigen, M. and Schuster, P. 1977. *The Hypercycle: a principle of natural self-organization.* Part A. Emergence of the Hypercycle. Naturwiss. 64:541–565.

Eigen, M. and Schuster, P. 1978. *The Realistic Hypercycle.* Part C. Emergence of the Hypercycle. Naturwiss. 65:341–369.

Ferris, J.P., Hill, A.R., Liu, R. and Orgel, L.E. 1996. *Synthesis of long prebiotic oligomers on mineral surfaces.* Nature 381:59–61.

Fox, S.W. 1973. *Origin of the cell: Experiments and premises.* Naturwiss. 60:359–369.

Gedulin, B. and Arrhenius, G. 1994. *Sources and geochemical evolution of RNA precourser molecules: the role of phosphate. In*: Early Life on Earth, Nobel Symposium 84 (Columbia University Press, New York), pp. 91–110.

Gesteland, R.F. and Atkus, J.F., eds. 1993. *The RNA World.* Cold Spring Harbor Laboratory Press, New York.

Joyce, G.F. and Orgel, L.E. 1993. *Prospects for Understanding the Origin of the RNA World. In*: The RNA World. R.F. Gesteland and J.F. Atkus, eds. Cold Spring Harbor Laboratory Press, New York.

Miller, S.L. and Orgel, L.E. 1974. *The Origin of Life on Earth.* Prentice-Hall, Engelwood Cliffs, N.J.

Moore, P.B. 1993. *Ribosomes and the RNA World. In*: The RNA World, R.F. Gesteland and J.F. Atkus, eds. Cold Spring Harbor Laboratory Press, New York.

Noller, H.F., Hoffarth, V. and Zimniak, L. 1992. *Unusual resistance of peptidyl transferase to protein extraction procedures.* Science 256:1416–1419.

Oparin, A.I. 1964. *Life: Its Origin, Nature and Development.* Academic Press, New York.

Orgel, L.E. 1968. *Evolution of the genetic apparatus.* J. Mol. Biol. 38:381–393.

Orgel, L.E. 1973. *The Origin of Life: Molecules and Natural Selection.* John Wiley & Sons, New York.

Robertson, M.P. and Miller, S.L. 1995. *An efficient prebiotic synthesis of cytosine and uracil.* Nature 375:772–774.

Schopf, J.W., ed. 1983. *Earth's Earliest Biosphere: Its Origins and Evolution.* Princeton University Press, Princeton, N.J.

Woese, C. 1967. *The Evolution of the Genetic Code. In*: The Genetic Code, pp. 179–195. Harper and Row, New York.

7. THE TWO INFORMATION REALMS

The Quintessence of Conservation of Information: DNA

Pauling, L. 1940. *A theory of the structure and process of formation of antibodies.* J. Am. Chem. Soc. 62:2643–2653.

Pauling, L. 1948. *Molecular Architecture and the Processes of Life*. The 21st Sir Jesse Boot Lecture. Nottingham, Eng. Sir Jesse Boot Foundation.

Watson, J.D. and Crick, F.H.C. 1953. *Genetical implications of the structure of deoxyribonucleic acid*. Nature 171:964–967.

The Information Subsidy for DNA Replication and Error Correction

Alberts, B.M. 1985. *Protein machines mediate the basic genetic processes*. Trends Genet. 1:26–30.

Kornberg, A. 1980. *DNA Replication*. W.H. Freeman Press, New York; and 1982 supplement to this book.

Kornberg, A. 1984. *DNA replication*. Trends Biochem. Sci. 9:122–124.

The Spooling of the Golden Thread

Felsenfeld, G. and McGhee, J.D. 1986. *Structure of the 30 NM chromatin fiber cell*. Cell 44: 375–377.

Kornberg, R.D. 1977. *Structure of chromatin*. Annu. Rev. Biochem. 46:931–954.

Lewin, B. 1980. *Gene Expression*. Vol 2. *Eukaryotic Chromosomes*. 2nd Edition. John Wiley & Sons, New York.

Losa, R. and Brown, D.D. 1987. *A bacteriophage RNA polymerase transcribes in vitro through a nucleosome core without displacing it*. Cell 50:801–808.

Pederson, D.S., Thoma, F., and Simpson, R.T. 1986. *Core particle, fiber, and transcriptionally active chromatin structure*. Annu. Rev. Cell Biol. 2:117–147.

Richmond, T.J., Finch, J.T., Rushton, B., Rhodes, D. and Klug, A. 1984. *Structure of the nucleosome core particle at 7 Å resolution*. Nature 311:532–537.

The Genetic Code

Brenner, S. 1957. *On the impossibility of all overlapping triplet codes in information transfer from nucleic acid to proteins*. PNAS 43: 687–694.

Crick, F.H.C. 1958. *On protein synthesis*. Symp. Soc. Exptl. Biol. 12:138–163.

Crothers, D.M. 1982. *Nucleic acid aggregation geometry and the possible evolutionary origin of ribosomes and the genetic code*. J. Mol. Biol. 162:379–391.

Gamow, G. 1954. *Possible relation between deoxyribonucleic acid and protein structures*. Nature 173:318.

Gamow, G. and Iça, M. 1955. *Statistical correlation of protein and ribonucleic acid composition*. PNAS 41:1019–1101.

Khorana, G. 1968. *Nucleic acid synthesis in the study of the genetic code*. In: Nobel Lectures: Physiology and Medicine. 1963–1970, pp. 341–369. American Elsevier, New York. (1973).

Nirenberg, M. 1968. *The genetic code*. In: Nobel Lectures: Physiology and Medicine. 1963–1970, pp. 372–395. American Elsevier, New York. (1973).

Nirenberg, M.W. and Leder, P. 1964. *RNA codewords and protein synthesis*. Science 145: 1399–1407.

Nirenberg, M.W. and Mattaei, J.H. 1961. *The dependence of cell-free protein synthesis in E.Coli upon naturally occurring or synthetic polyribonucleotides*. Proc. Natl. Acad. Sci. USA 47:1588–1602.

Stent, G. 1971. *Molecular Genetics*. Freeman, San Francisco.

Transcription

Darnell, J.E. 1985. *RNA*. Sci. Am. 253:68–78.

Dynan, W.S. and Tjian, R. 1985. *Control of eukaryotic messenger RNA synthesis by sequence-specific DNA-binding proteins*. Nature 316:774–778.

Platt, T. 1986. *Transcription termination and the regulation of gene expression*. Annu. Rev. Biochem. 55:339–372.

Splicing

Aebi & Weissmann. 1987. *Precision and orderliness in splicing*. Trends Genet. 3:102–107.

Darnell, J.E. 1978. *Implications of RNA-RNA splicing in evolution of eukaryotic cells*. Science 202:1257–1260.

Darnell, J.E. 1983. *Processing of RNA*. Sci. Am. 249:90–100.

Sharp, P.A. 1987. *Splicing of messenger RNA precursors*. Science 235:766–771.

The Genie that Dances to Its Own Tune

Bass, B.L. and Cech, T.R. 1985. *Specific interaction between the self-splicing RNA of Tetrahymena and its guanosine substrate: implications for biological catalysis by RNA*. Nature 308: 820–826.

Cech, T.R. 1983. *RNA splicing: three themes with variations*. Cell 34:713–716.

Cech, T.R. 1986. *RNA as an enzyme*. Sci. Am. 255:64–75.

Crick, F. 1979. *Split genes and RNA splicing*. Science 204:264–271.

Darnell, J.E. 1978. *Implications of RNA-RNA splicing in evolution of eukaryotic cells*. Science 202:1257–1260.

Doolittle, W.F. 1978. *Genes in pieces: were they ever together?* Nature 272:581–582.

Gilbert, W. 1978. *Why genes in pieces?* Nature 271:501.

Gilbert, W. 1985. *Genes in pieces revisited.* Science 228:823–824.

How to Bridge the Information Gap

Kirkwood, T.B.L., Rosenberg, R.F., and Galas, D.J. 1986. *Accuracy in Molecular Processes:. Its Control and Relevance to Living Systems.* Chapman & Hall, Philadelphia, Pa.

Rich, A., and Kim, S.H. 1978. *The three-dimensional structure of transfer RNA.* Sci. Amer. 238:52–62.

Schimmel, P.R., Söll, D., and Abelson, J.N. eds. 1980. *Transfer RNA: Structure Properties and Recognition.* 2 vols. Cold Spring Harbor Laboratory Press, New York.

Sussman, J.L., and Kim, S.H. 1976. *Three-dimensioned structure of a transfer RNA in two crystal forms.* Science 192:853–858.

8. THE GAME OF SEEK AND FIND

Seek and Find

Whitman, Walt. 1892. *Leaves of Grass.* "Songs of Parting." David McKay Publisher, Philadelphia.

The Paradox of the Progressive Reactionary

Darwin, C. 1859. *The Origin of Species.* Chapter V: *Laws of Variation.* Humboldt, New York. 1890.

Dover, G. 1982. *Molecular drive: a cohesive mode of species evolution.* Nature 299:111–117.

Eigen, M. and Schuster, P. 1977. *The Hypercycle: a principle of natural self-organization.* Part A. Emergence of the Hypercycle. Naturwiss. 64:541–565.

Shannon, C.E. 1948. *A mathematical theory of information.* Bell System Tech. J. 27: 279–423; 27:623–656.

Smith, J.M., ed. 1982. *Evolution Now. A Century After Darwin.* W. H. Freeman, San Francisco and Oxford.

What if the Cellular Information Stream Ran Backward?

Cairns, J., Overbaugh, J. and Miller, S. 1988. *The origin of mutants.* Nature 335:142–145.

Lewin, R. 1983. *How mammalian RNA returns to its genome.* Science 219:1052–1054.

Reanney, D. 1984. *Genetic noise in evolution?* Nature 307:318–319.

Sharp, P.A. 1983. *Conversion of RNA to DNA in mammals: Alu-like elements and pseudogenes.* Nature 301:471–472.

Steele, E.J. 1979. *Somatic Selection and Adaptive Evolution.* Williams Wallace. International and Croom Helm, Toronto and London.

Can Organisms Promote Useful Mutations?

Cairns, J., Overbaugh, J. and Miller, S. 1988. *The origin of mutants.* Nature 335:142–145.

Cairns, J. and Foster, P.L. 1991. *Adaptive reversion of frameshift mutations in the lacz gene of Escherichia Coli.* Genetics 128:695–701.

Cavalli-Sforza, L.L. and Lederberg, J. 1956. *Isolation of pre-adaptive mutants in bacteria by sib selection.* J. Genetics 41:367–381.

Hall, B.G. 1990. *Spontaneous point mutations that occur more often when advantageous than when neutral.* Genetics 126:5–16.

Jagadeeswaran, P., Forget, B.G. and Weissman, S.M. 1981. *Short interspersed repetitive DNA elements in eukaryotes: transposable DNA elements generated by reverse transcription of RNA POL III transcripts?* Cell 26:141–142.

Le Carré, J. 1989. *Call for the Dead.* Alfred Knopf, New York. (The quotation in the text is a question spymaster Smiley asks of Mrs. Fenman in one of the scenes of the novel.)

Lederberg, J. and Lederberg, E. 1952. *Replica plating and indirect selection of bacterial mutants.* J. Bact. 63:399–406.

Luria, S.E. and Delbrück, M. 1943. *Mutations of bacteria from virus sensitivity to virus resistance.* Genetics 28:491–511.

Stahl, F.W. 1988. *A unicorn in the garden.* Nature 335:112–113.

The Solid-State Apparatus at the Headgates of the Mainstream

Johnson, P.F. and McKnight, S.L. 1989. *Eukaryotic transcriptional proteins.* Annu. Rev. Biochem. 58:799–839.

Ptashne, M. 1988. *How eukaryotic transcriptional activators work.* Nature 335:683–699.

Yamamoto, K. R. 1985. *Steroid receptor regulated transcription of specific genes and gene networks.* Annu. Rev. Genet. 19:209–252.

Three Protein Designs for Tapping Positional Information from DNA

Berg, J.M. 1986. *Potential metal-binding domains in nucleic acid binding proteins.* Science 232: 485–487.

Klug, A. and Rhodes, D. 1987. *"Zinc fingers": a novel protein motif for nucleic acid recognition.* Trends Biochem. Sci. 12:464–469.

McNight, S.L. 1991. *Molecular zippers in gene regulation.* Sci. Amer. 264:54–64.

Pabo, C.O. and Sauer, R.T. 1984. *Protein-DNA recognition.* Annu. Rev. Biochem. 53:293–321.

What Do Proteins See Where They Plug into the DNA?

Calladine, C.R. 1982. *Mechanics of sequence-dependent stacking of bases in B-DNA.* J. Mol. Biol. 161:343–352.

Dickerson, R.E. 1983. *Base sequence and helical structure variation in B and A DNA.* J. Mol. Biol. 166:419–441.

9. RULES OF THE INTERCELLULAR COMMUNICATION GAME

The End of the Cell Era: The Merger of Circuses

Buffon, G-L.L., Count de. Histoire naturelle, génerale et particulière, avec la description du Cabinet du Roi. 44 vols. Paris, 1749–1804.

The Innards of the Cell Membrane

Blaurock, A.E., and Stoeckenius, W. 1971. *Structure of purple membrane.* Nature 233:152–155.

Blobel, G. 1980. *Intracellular protein topogenesis.* Proc. Natl. Acad. Sci. USA 77: 1496–1500.

Blobel, G., and Dobberstein, B. 1975. *Transfer of proteins across membranes.* J. Cell Biol. 67:835–862.

Danielli, J.F. and Davson, H. 1935. *A contribution to the theory of permeability of thin films.* J. Cell Comp. Physiol. 5:495.

Danielli, J.F. 1975. *The bilayer hypothesis of membrane structure in cell membranes.* In: Biochemistry, Cell Biology and Pathology. G. Weisemann and R. Claiborne, eds. H.P. Publishing, New York.

Deisenhofer, J., Epp, O., Miki, K., Huber, R., and Michel, H. 1985. *Structure of the protein subunits in the photosynthetic reaction centre of Rhodopseudonomonas viridis at 3 Å resolution.* Nature 318:618–624.

Engelman, D.M., Steitz, T.A., and Goldman, A. 1986. *Identifying nonpolar transbilayer helices in amino acid sequences of membrane proteins.* Annu. Rev. Biophys. Chem. 15:321–353.

Kennedy, S.J. 1978. *The structure of membrane proteins.* J. Membr. Biol. 42:265–279.

Lenard, J., and Singer, S.J. 1966. *Protein conformation in cell membranes as studied by optical rotation, dispersion and circular dichroism.* Proc. Natl. Acad. Sci. USA 56:1828–1835.

Pinto da Silva, P., and Branton, D. 1970. *Membrane splitting in freeze-etching.* J. Cell Biol. 45:598–605.

Singer, S.J. 1992. *The structure and function of membranes—a personal memoir.* J. Membr. Biol. 129:3–12.

Singer, S.J. and Nicolson, G.L. 1972. *The fluid mosaic model of the structure of cell membranes.* Science 175:720–731.

Wallach, D.F.H., and Zahler, P.H. 1966. *Protein conformations in cellular membranes.* Proc. Natl. Acad. Sci. USA 56:1552–1559.

Weiss, M.S., Abele, V., Weckesser, J., Welte, W., Schietz, E., and Schultz, G.E. 1991. *The molecular architecture and electrostatic properties of bacterial porin.* Science 254:1627–1630.

Wickner, W.T. and Lodish, H.F. 1985. *Multiple mechanisms of protein insertion into and across membranes.* Science 230:400–406.

The Intercellular Communication Channels / The Imperturbable Biological Signal Stuff

Fano, R.M. 1961. *Transmission of Information.* MIT Press, Cambridge, Mass.

Gallager, R. 1968. *Information Theory and Reliable Communication.* John Wiley & Sons, New York.

Marko, H. 1973. *The bidirectional information theory—a generalization of information theory.* IEEE Comm. 21: 1345–1351.

Raisbeck, G. 1964. *Information Theory: An Introduction for Scientists and Engineers.* MIT Press, Cambridge, Mass.

Shannon, C.E. 1948. *A mathematical theory of communication.* Bell Syst. Tech. J. 27: 379–423.

Shannon, C.E. and Weaver, W. 1959. *The Mathematical Theory of Communication.* University of Illinois Press, Urbana, Ill.

"Nature's Sole Mistake"

Gilbert, W.S. and Sullivan, A.S. Opera *Princess Ida*.

How to Get a Message Through a Noisy Channel: The Second Theorem

Baker, R. 1992. *"Systems Going Down."* The New York Times. June 27, p. 23.

Shannon, C.E. 1949. *Communication in the presence of noise.* Proc. J.R.E. 37:10.

Shannon, C.E. 1956. *Zero-error capacity of noisy channels.* I.R.E. Trans. I.T. 2:8.

The Question of the Origin of Natural Codes

Crick, F., Brenner, S., Klug, A., and Pieczenik, G. 1976. *A speculation on the origin of protein synthesis.* Origins of Life. 7:389-397.

Miller, S.L., and Orgel, L. 1973. *The Origins of Life.* Prentice Hall. Englewood Cliffs, N.J.

Popper, K.R. 1972. *Objective Knowledge.* Clarendon Press, Oxford, p. 241.

Popper, K.R. 1978. *Natural Selection and the Emergence of Mind.* Darwin Lecture at Cambridge University. *Dialectica.* 22:39–355.

Waddington, C.H. 1960. *Evolutionary adaptation. In:* Evolution After Darwin, Vol. I, The Evolution of Life. University of Chicago Press, Chicago, pp. 381–402.

Waddington, C.H. 1975. *The Evolution of an Evolutionist.* Cornell University Press, Ithaca, p. 41.

The Why and Wherefore of Encoding

Abramson, N. 1963. *Information Theory and Coding.* McGraw Hill, New York.

Gatlin, L. 1972. *Information Theory and the Living System.* Columbia University Press, New York.

Khinchin, A.I. 1957. *Mathematical Foundations of Information Theory.* Dover, New York.

What Can We Learn from Game Theory?

Von Neumann, J. 1928. *Zur Theorie der Gesellschaftspiele.* Mathematische Annalen 100: 295–320.

Von Neumann, J., and Morgenstern, O. 1947. *Theory of Games and Economic Behavior.* Princeton University Press, Princeton, N.J.

Telecontrol

Hofstadter, D. 1980. *Gödel, Escher, Bach: An Eternal Golden Braid.* Vintage Books, New York.

Gödel, K. 1931. *Monatshefte f. Math u. Physik.* 38:179.

10. THE BROADCASTING OF CELLULAR INFORMATION

The "Emanations from Organs"

Bernard, C. 1855. *Léçons de Physiologie Experimentale Appliqué à la Medicine.* Vol. I. Paris.

Brown-Séquard, C.F., D'Arsonval, A. 1891. Compt. rend. Soc. Biol., Paris 9(iii):248.

Brown-Séquard, C.F. and D'Arsonval, A. 1891. *De l'injection des estraits liquides provenant des glandes et des tissus de l'organisme comme méthode thérapeutique.* Compt. Rendu. Hebdomadaires des Séances et Mémoires de la Société de Biologie, Paris, Vol 43, series 9, tome 3, 248–250.

De Bordeu, Théophile. 1775. *Recherches sur les maladies chroniques.* Oeuvres complètes, tome ii, p. 942.

Elliot, T.R. 1904. J. Physiol. Lond. xxxi, Proc. Physiol. Soc. xx.

Loewi, O. 1921. *Über humorale Übertragbarkeit der Herznervenwirkung.* Pflügers Arch. 189: 239–242.

Loewi, O. 1935. *Problems Connected with the Principle of Humoral Transmission of Nervous Impulses. The Ferrier Lecture.* Proc. R. Soc. Lond. Biol. Sci. 118:299–316.

Needham, J. 1983. *Science and Civilization in China.* Cambridge University Press, Cambridge.

Rolleston, H.D. 1936. *The Endocrine Organs in Health and Disease, with an Historical Review.* Oxford University Press, London.

Starling, E.H. 1905. *The Croonian Lectures on the Chemical Correlation of the Functions of the Body.* Lancet 2:339–341.

The Pervasive Messengers → The Transfer of Steroid Hormone Information

Block, K. 1983. *Sterol Structure and Membrane Function.* Crit. Rev. Biochem. 14:47–92.

Evans, R.M. 1988. *The Steroid and Thyroid Hormone Receptor Superfamily.* Science 240:889–895.

Gehring, U. 1976. *Steroid Hormone Receptors: Biochemistry, Genetics and Molecular Biology.* Trends Biochem. Sci. 12:399–402.

O'Malley, B.W. and Schrader, W.T. 1976. *The Receptors of Steroid Hormones.* Sci. Amer. 234:32–43.

Wilson, A.C., Carlson, S.S., White, T.J. 1977. *Biochemical Evolution.* Annu. Rev. Biochem. 46:573–639.

Yamamoto, K.R. 1985. *Steroid Receptor Regulated Transcription of Specific Genes and Gene Networks.* Annu. Rev. Genet. 19:209–252.

The DNA Multiplex and Remote Gene Control

Attardi, B., and Ohno, S. 1978. *Physical Properties of Androgen Receptors in Brain Cytosol from Normal and Testicular Feminized Mice.* Endocrinology 103: 760–770.

Griffin, J.E., and Wilson, J.D. 1980. *The Syndromes of Androgen Resistance.* N. Engl. J. Med. 302: 198–209.

The Rise of a Demon

Tomkins, G.M. 1975. *The Metabolic Code.* Science 189:760–763.

Yamamoto, K.R. 1985. *Steroid Receptor Regulated Transcription of Specific Genes and Gene Networks.* Annu. Rev. Genet. 19:209–252.

Demons with an Esprit de Corps

Agard, D.A., and Sedat, J.W. 1983. *Three-dimensional Architecture of a Polytene Nucleus.* Nature 302: 676–681.

Ashburner, M., Chihara, C., Meltzer, P., and Richards, G. 1974. *Temporal Control of Puffing Activity in Polytene Chromosomes.* Cold Spring Harbor Symp. Quant. Biol. 38:655–662.

Beermann, W. 1972. *Chromosomes and Genes. In*: Developmental Studies of Giant Chromosomes. W. Beerman, ed., pp. 1–33. Springer-Verlag, New York.

Clever, U. and Karlson, P. 1960. *Induktion von Puff-Veränderungen in den Speicheldrüsenchromosomen von Chironomus Tentans durch Ecdyson.* Exp. Cell Res. 20:623–626.

Lamb, M.M., and Daneholt, B. 1979. *Characterization of Active Transcription Units in Balbiani Rings of Chironomus Tetans.* Cell 17: 835–848.

A Smoke Signal

Bohme, G.A., Bon, C., Stutzmann, J.-M., Doble, A. and Blanchard, J.-C. 1991. *Possible Involvement of Nitric Oxide in Long-Term Potentiation.* Eur. J. Pharmacol. 199:379–381.

Bredt, D.S., and Snyder, S.H. 1992. *Nitric oxide, a Novel Neuronal Messenger.* Neuron 8:3–11.

Furchgott, R.F. and Zawadzki, J.V. 1980. *The Obligatory Role of Endothelial Cells in the Relaxation of Arterial Smooth Muscle by Acetylcholine.* Nature 288:373–376.

Ignarro, L.J. 1991. *Signal Transduction Mechanisms Involving Nitric Oxide.* Biochem. Pharm. 41:485–490.

Cyclic AMP

Pastan, I. 1972. *Cyclic AMP.* Sci. Am. 227: 97–105.

Pastan, I., and Perlman, R. 1970. *Cyclic Adenosine Monophosphate in Bacteria.* Science 169: 339–344.

Sutherland, E.W. 1972. *Studies on the Mechanisms of Hormone Action.* Science 177:401–408.

The Information Relays in the Cell Membrane

Dohlman, H.G., Thorner, J., Caron, M.G., and Lefkowitz, R.J. 1991. *Model Systems for the Study of Seven-Transmembrane Segment Receptors.* Annu. Rev. Biochem. 60: 653–688.

Gilman, A.G. 1987. *G Proteins: Transducers of Receptor-Generated Signals.* Annu. Rev. Biochem. 56: 615–649.

Rodbell, M. 1980. *The Role of Hormone Receptors and GTP-Regulatory Proteins in Membrane Transduction.* Nature 284:17–22.

Stryer, L., and Bourne, H.R. 1986. *G Proteins: A Family of Signal Transducers.* Annu. Rev. Cell Biol. 2:391–419.

The Alpha Demon / The Alpha Merry-Go-Round / The G-Show Libretto

Bourne, H.R., Sanders, P.A., and McCormick, F. 1991. *The GTPase Superfamily: Conserved Structure and Molecular Mechanisms.* Nature 349:117–127.

Jurnack, F., Heffron, S., and Bergmann, E. 1990. *Conformational Changes Involved in the Activation of ras p21: Implications for Related Proteins.* Cell 60:525–528.

Lipmann, F. 1941. *Metabolic Generation and Mobilization of Phosphate Bond Energy.* Advances in Enzymology 1:99–162.

Milton, J. *L'Allegro.* Line 31.

Pai, E.F., Kabsch, W., Krengel, U., Holmes, K.C., John, J., and Wittinghofer, A. 1989. *Structure of the Guanine-Nucleotide-Binding Domain of the Ha-ras Oncogene Product p21 in the Triphosphate Conformation.* Nature 341: 209–214.

Racker, E. and Stoeckenius, W. 1974. *Reconstitution of Membrane Vesicles Catalyze Light-Driven Proton Update and ATP Formation.* J. Biol. Chem. 249:662–663.

Ion Channels as Maxwell Demons

Catterall, W. A. 1992. *Cellular and Molecular Biology of Voltage-Gated Sodium Channels.* Physiol. Rev. 72 (Suppl.): S15-S48.

Hille, B. 1992. *Ionic Channels of Excitable Membranes.* Sinauer Associates, Suderland, Mass.

Hodgkin, A.L. and Huxley, R.F. 1952. The Dual Effect of Membrane Potential on Sodium Conductance in the Giant Axon of *Loligo.* J. Physiol. 116:497–506.

Miller, C. 1992. *Ion Channel Structure and Function.* Science 258:240–241.

Pongs, O. 1993. *Structure-Function Studies on the Pore of Potassium Channels.* J. Membr. Biol. 136:1–8.

Unwin, N. 1989. *The Structure of Ion Channels in Membrane of Excitable Cells.* Neuron 3:665–676.

Unwin, N. 1993. *Neurotransmitter Action: Opening of Ligand-Gated Ion Channels.* Cell 72/Neuron 10 (Suppl.): 31–41.

How to Make Signals from Informational Chicken Feed

Brown, A. 1993. *Membrane-Delimited Cell Signalling Complexes: Direct Ion Channel Regulating G Proteins.* J. Membr. Biol. 131:93–104.

Hille, B. 1992. *Ionic Channels of Excitable Membranes.* Sinauer Associates, Sunderland, Mass.

Hodgkin, A.L. and Huxley, R.F. 1952. *A Quantitative Description of Membrane Current and Its Application to Conduction and Excitation in Nerve.* J. Physiol. 117: 500–544

Katz, B. 1966. *Nerve, Muscle and Synapse.* McGraw Hill, New York.

Levitan, I.B. and Kaczmarek, L.K. 1991. *The Neuron: Cell and Molecular Biology.* Oxford University Press, New York.

Nicholl, R.A., Malenka, R.C. and Kauer, J.A. 1990. *Functional Comparison of Neurotransmitter Receptor Subtypes in Mammalian Central Nervous System.* Physiol. Rev. 70:513–565.

Sakmann, B. and Neher, E., eds. 1983. *Single Channel Recording.* Plenum Press, New York.

A Small Ion with Cachet. The Calcium Signal

Baker, P.F., Hodgkin, A.L., Ridgway, E.B. 1971. *Depolarization and Calcium Entry in Squid Giant Axons.* J. Physiol. (Lond.) 218:709.

Baker, P.F. 1976. *The Regulation of Intracellular Calcium. In:* Calcium in Biological Systems. Symp. Soc. Exptl. Biol. 30:55.

Kretsinger, R. 1979. *The Informational Role of Calcium in the Cytosol. In:* Advanced Cyclic Nucleotide Research Vol. 11. P. Greengard and G.A. Robison, eds. Raven Press, New York.

Weber, A. 1976. *Synopsis of the Presentations. In:* Calcium in Biological Systems. Symp. Soc. Exptl. Biol. 30:445–454.

Signal Mopping: An Economical Cure for Cross Talk

Loewenstein, W.R., and Rose, B. 1978. *Calcium in (Junctional) Intercellular Communication and a Thought on Its Behavior in Intracellular Communication.* Annu. N.Y. Acad. Sci. 307:285–307.

Rose, B., and Loewenstein, W.R. 1975. *Calcium Ion Distribution in Cytoplasm Visualized by Aequorin: Diffusion in the Cytosol Is restricted due to Energized Sequestering.* Science 190: 1204–1206.

The Shoo-in Called Calcium

Del Castillo, J., and Stark, L. 1952. *The Effect of Calcium Ions on the Motor End-Plate Potentials.* J. Physiol. (Lond.) 116:507–515.

Douglas, W.W. and Rubin, R.P. 1961. *The Role of Calcium in the Secretory Response of the Adrenal Medulla to Acetylcholine.* J. Physiol. (Lond.) 159:40–57.

Katz, B. 1962. *The Croonian Lecture: The Transmission of Impulses from Nerve to Muscle, and the Subcellular Unit of Synaptic Action.* Proc. Roy. Soc. (Lond.) Ser. B 155:455–477.

Katz, B., Miledi, R. 1965. *The Effect of Calcium on Acetylcholine Release from Motor Terminals.* Proc. Roy. Soc. (Lond.) Ser. B 161:496–503.

Snapshots of Two T Demons

Babu, Y.S., Sack, J.S., Greenbough, T.J., Bugg, C.E., Means, A.R., and Cook, W.J. 1985. *The Three-Dimensional Structure of Calmodulin.* Nature 315:37–40.

Cheung, W.Y. 1980. *Calmodulin Plays a Pivotal Role in Cellular Regulation.* Science 207:19–27.

Krebs, E.G. and Beavo, J.A. 1979. *Phosphorylation and Dephosphorylation of enzymes.* Annu. Rev. Biochem. 48:923–959.

Kretsinger, R.H. 1976. *Calcium-Binding Proteins.* Annu. Rev. Biochem. 45:239–266.

Kretsinger, R.H. 1979. *The Informational Role of Calcium in the Cytosol. In:* Adv. Cyclic Nucleotide Res. Vol. 11. P. Greengard and G.A. Robison, eds. Raven Press, New York.

Kretsinger, R.H. 1992. *The Linker of Calmodulin—To Helix or Not to Helix.* Cell Calcium 13:376–388.

Levine, B.A., Dalgarno, D.C., Esnouf, M.P., Klevit, R.E., and Williams, R.I.P. 1982. Ciba Foundation Symp. 83:81.

11. INTERCELLULAR COMMUNICATION CHEEK-BY-JOWL

The Case for Syncretic Communication

Anklesaria, P., Teixidó, J., Laiho, M., Pierce, J.H., Greenberger, J.S., and Massagué., J. 1990. *Cell-Cell Adhesion Mediated by Binding of Membrane-Anchored Transforming*

Growth Factor a to Epidermal Growth Factor Receptors Promotes Cell Proliferation. Proc. Natl. Acad. Sci. 87:3289–3293.

Brachmann, R., Lindquist, P.B., Nagashima, M., Kohr, W., Lipari, T., Napier, M., and Derynck, R. 1989. *Transmembrane TGFα Precursors Activate EGF/TGFα Receptors.* Cell 56:691–700.

Massagué, J. 1990. *Transforming Growth Factor α: A Model for Membrane Anchored Growth Factors.* J. Biol. Chem. 265:21393–21396.

Pandiela, A., and Massagué, J. 1990. *Cleavage of the Membrane Precursor for Transforming Growth Factor α is a Regulated Process.* Proc. Natl. Acad. Sci. 88:1726–1730.

How Syncretic Information Flows

Hernández-Sotomayor, S.M.T. and Carpenter G. 1992. *Epidermal Growth Factor Receptor: Elements of Intracellular Communication.* J. Membr. Biol. 128:81–89.

Mroczkowski, B., Reich, M., Chen, K., Bell, G.I., and Cohen, S. 1989. *Recombinant Human Epidermal Growth Factor Precursor is a Glycosylated Membrane Protein with Biological Activity.* Mol. Cell Biol. 9:2771–2778.

The Information-Crossover Gambit

Margolis, B., Li, N., Koch, A., Mohammadi, M., Hurwitz, D.R., Zilberstein, A., Ullrich, A., Pawson, T., and Schlessinger, J. 1990. *The Tyrosine Phosphorylated Carboxy Terminus of the EGF Receptor is a Binding Site for GAP and PLC.* Embo. J. 9:4375–4380.

Schlessinger, J. 1986. *Allosteric Regulation of the Epidermal Growth Factor Receptor Kinase.* J. Cell Biol. 103:2067–2072.

Ulrich, A., and Schlessinger, J. 1990. *Signal Transduction by Receptors with Tyrosine Kinase Activity.* Cell 61:203–212.

Banerjee, U., and Zipursky, S.L. 1990. *The Role of Cell-Cell Interaction in the Development of the Drosophila Visual System.* Neuron 4:177–187.

Horvitz, H.R., and Sternberg, P.W. 1991. *Multiple Intercellular Signalling Systems Control the Development of the Caenorhabditis Elegans Vulva.* Nature 351:535–541.

Singer, S.J. 1992. *Intercellular Communication and Cell Adhesion.* Science 255:1671–1677.

12. INTERCELLULAR COMMUNICATION VIA CHANNELS IN THE CELL MEMBRANES

The Cell-to-Cell Channels → How They Form → How Cells Are Excommunicated and When

Brown, D.A. and London, E. 1998. *Structure and Function of Ordered Lipid Domains in Biological Membranes.* J. Membr. Biol. 164:103–114.

Epstein, M.L., Sheridan, J.D. and Johnson, R.G. 1977. *Formation of low-resistance junctions in vitro in the absence of protein synthesis and ATP production.* Exp. Cell Res. 104:25–30.

Ito, S., Loewenstein, W.R. 1969. *Ionic Communication between Early Embryonic Cells.* Dev. Biol. 19: 228–243.

Loewenstein, W.R. 1981. *Junctional intercellular communication. The cell-to-cell membrane channel.* Physiol. Rev. 61:829–913.

Loewenstein, W.R. 1984. *From cell theory to cell connectivity: Experiments in cell-to-cell communication. In:* Membrane Transport in Physiology: People and Ideas. 100th Anniversary of the American Physiological Society. Chp. 13, pp. 303–335. Bethesda, Md.

Loewenstein, W.R., and Kanno, Y. 1964. *Studies on an epithelial (gland) cell junction. I. Modifications of surface membrane permeability.* J. Cell Biol. 22: 565–586.

Loewenstein, W.R., Kanno, Y., and Socolar, S.J. 1978. *Quantum jumps of conductance during formation of membrane channels at cell-cell junction.* Nature 274: 133–136.

Loewenstein, W.R., Socolar, S.J., Higashino, S., Kanno Y. and Davidson, N. 1965. *Intercellular communication: renal, urinary bladder, sensory, and salivary gland cells.* Science 149:295–298.

Rose, B., and Loewenstein, W.R. 1975. *Permeability of cell junction depends on local cytoplasmic calcium activity.* Nature 254:250–252.

Socolar, S.J., and Loewenstein, W.R. 1979. *Methods for studying transmission through permeable cell-to-cell junctions. In:* Methods in Membrane Biology. Vol. 10, pp. 123–179. Plenum Press, New York.

The Channel Shutter

Loewenstein, W.R. 1967. *Cell surface membranes in close contact. Role of calcium and magnesium ions.* J. Colloid. Interface Sci. 25:34–46.

Loewenstein, W.R., Nakas, M., and Socolar, S.J. 1967. *Junctional membrane uncoupling. Permeability transformations at a cell membrane junction.* J. Gen. Physiol. 50:1865–1891.

Makowski, L., Caspar, D.L.D., Phillips, W.C., and Goodenough, D.A., 1977. *Gap junction structure. II. Analysis of the X-ray diffraction data.* J. Cell Biol. 74:629–645.

Milks, L.C., Kumar, N.M., Houghton, R. Unwin, P.N.T., and Gilula, N.B. 1988. *Topology of the 32-kd liver gap junction protein.* EMBO J. 7:2967–2975.

Oliveira-Castro, G.M. and Loewenstein, W.R. 1971. *Junctional membrane permeability: Effects of divalent cations.* J. Membr. Biol. 5:51–77.

Rose, B. and Loewenstein, W.R. 1975. *Permeability of cell junction depends on local cytoplasmic calcium activity.* Nature 254:250–252.

Stauffer, K.A., and Unwin, N. 1992. *Structure of gap junction channels.* Sem. Cell Biol. 3:17–21.

Tibbits, T.T., Caspar, D.L.D., Phillips, W.C., and Goodenough, D.A. 1990. *Diffraction diagnosis of protein folding in gap junction connexons.* Biophys. J. 57:1025–1036.

Unwin, P.N.T., and Zampighi, G. 1980. *Structure of the junction between communicating cells.* Nature 283:545–549.

Unwin, P.N.T., and Ennis, P.D. 1984. *Two configurations of a channel-forming membrane protein.* Nature 307:609–613.

A Channel Yin-Yang

Beyer, E.C., Paul, D.L., and Goodenough, D.A. 1990. *Connexin family of gap junction proteins.* J. Membr. Biol. 116:187–194.

Kumar, N.M., and Gilula, N.B. 1992. *Molecular biology and genetics of gap junctions.* Sem. Cell Biol. 3:3–17.

Unwin, P.N.T. 1989. *The structure of ion channels in membranes of excitable cells.* Neuron 3: 665–676.

What Goes Through the Cell-Cell Channels

Flagg-Newton, J.L., Simpson, I., and Loewenstein, W.R. 1979. *Permeability of the cell-to-cell membrane channels in mammalian cell junction.* Science 205:404–407.

Rose, B., Simpson, I., and Loewenstein, W.R. 1977. *Calcium ion produces graded changes in permeability of membrane channels in cell junction.* Nature 267:625–627.

Schwartzmann, G.O.H., Wiegandt, H., Rose, B., Zimmerman, A., Ben-Haim, D., and Loewenstein, W.R. 1981. *The diameter of the cell-to-cell junctional membrane channels, as probed with neutral molecules.* Science 213: 551–553.

How Cell Communities Stay on an Even Keel

Loewenstein, W.R. 1981. *Junctional intercellular communication: The cell-to-cell membrane channel.* Physiol. Rev. 61:829–913.

Loewenstein, W.R. 1982. *Cell-to-cell communication: One decade and the next. In:* Functional Regulation at the Cellular and Molecular Levels. Robert A. Carradino, ed., pp. 217–229. Elsevier, Amsterdam.

E Pluribus Unum / How to Ferret Out the Functions of a Channel

Bernard, C. 1878. *Leçons sur les phénomnes de la vie communes aux animaux et aux végetaux.* Bailliere, Paris.

Bürk, R.R., Pitts, J.D., and Subak-Sharpe, J.H. 1968. *Exchange between hamster cells in culture.* Exp. Cell Res. 53:297.

Furshpan, E.J., and Potter, D.D. 1968. *Low resistance junctions between cells in embryos and tissue culture.* Curr. Top. Dev. Biol. 3:95–127.

Subak-Sharpe, H., Bürk, R.R., and Pitts, J.D. 1966. *Metabolic cooperation by cell-to-cell transfer between genetically different mammalian cells in tissue culture.* Heredity 21:342.

The Regulation of Cellular Growth

Goethe, W. *Dichtung und Wahrheit* ("Es ist dafür gesorgt, dass die Bäume nicht in den Himmel Wachsen." / "It is seen to that the trees don't grow into the sky."

A Model of Growth Control for Small Cell Populations

Loewenstein, W.R. 1979. *Junctional intercellular communication and the control of growth.* Biochim. Biophys. Acta Cancer Rev. 560:1–65.

A Model for Large Cell Populations

Nonner, W.F., and Loewenstein, W.R. 1989. *Appendix: A growth control model with discrete regulatory centers.* J. Cell Biol. 108:1063–1065.

λ Determines Cellular Growth

Azarnia, R., Reddy, S., Kmiecik, T.E., Shalloway, D., and Loewenstein, W.R. 1988. *The cellular src gene product regulates junctional cell-to-cell communication.* Science 239:398–401.

Mehta, P.P., Bertram, J., and Loewenstein, W.R. 1989. *The actions of retinoids on cellular growth correlate with their actions on gap junctional communication.* J. Cell Biol. 108:1053–1065.

What Is Cancer?

Bishop, J.M. 1983. *Cellular oncogenes and retroviruses.* Annu. Rev. Biochem. 52:301–354.

Bishop, J.M. 1987. *The molecular genetics of cancer.* Science. 235:305–311.

Hunter, T., and Cooper, J.A. 1985. *Protein-tyrosine kinases.* Annu. Rev. Biochem. 54:897–930.

Rous, P. 1966. *The Challenge to Man of the Neoplastic Cell.* Nobel Lectures. Elsevier, Amsterdam, 1972, pp. 220–231.

Stehelin, D., Varmus, H.E., Bishop, J.M., and Vogt, P.K. 1976. *DNA related to the transforming gene(s) of avian sarcoma viruses is present in normal avian DNA.* Nature 260:170–173.

Stoker, M.G.P. 1967. *Transfer of Growth Inhibition between normal and virus-transformed Cells.* J. Cell. Sci. 2: 293–304.

Takeya, T., and Hanafusa, H. 1983. *Structure and sequence of the cellular gene homologous to the RSV src gene and the mechanism for generating the transforming virus.* Cell 32:881–890.

λ *Is Reduced in Cancer*

Azarnia, R., Reddy, S., Kmiecik, T.E., Shalloway, D., and Loewenstein, W.R. 1988. *The cellular src gene product regulates junctional cell-to-cell communication.* Science 239:398–401.

Azarnia, R., Mitcho, M., Shalloway, D., and Loewenstein, W.R. 1989. *Junctional intercellular communication is cooperatively inhibited by oncogenes in transformation.* Oncogene 4:1161–1168.

Fentiman, I.S., Taylor-Papadimitriov, J., and Stoke, M. 1976. *Selective contact-dependent cell communication.* Nature 264:760–762.

Land, H., Parada, L.F., and Weinberg, R.A. 1983. *Tumorigenic conversion of primary embryo fibroblasts requires at least two cooperating oncogenes* Nature 304:596–597.

Loewenstein, W.R. 1985. *Regulation of cell-to-cell communication by phosphorylation.* Biochem. Soc. Symp. Lond. 50:43–58.

Loewenstein, W.R., and Rose, B. 1992. *The cell-cell channel in the control of growth.* Seminars in Cell Biol. 3(1):59–79.

Swenson, K.I., Piwnica-Worms, H., McNamee, H. and Paul, D.L. 1990. *Tyrosine phosphorylation of the gap junction protein connexin43 is required for the pp60v-src-induced inhibition of communication.* Cell Regul. 1:989–1002.

A Cancer Cure in the Test Tube

Loewenstein, W.R., and Rose, B. 1992. *The cell-cell channel in the control of growth.* Seminars in Cell Biol. 3(1):59–79.

Mehta, P.P, Hotz-Wagenblatt, A., Rose, B., Shalloway, D., and Loewenstein, W.R. 1991. *Incorporation of the gene for the cell-cell channel protein into transformed cells leads to normalization of growth.* J. Membr. Biol. 124:207–225.

Our First Line of Defense

Hossain, M.Z., Wilkens, L.R., Mehta, P.P., Loewenstein, W.R., and Bertram, J.S. 1989. *Enhancement of gap junctional communication by retinoids correlates with their ability to inhibit neoplastic transformation.* Carcinogenesis 10:1743–1748.

Loewenstein, W.R. 1979. *Junctional intercellular communication and the control of growth.* Biochim. Biophys. Acta Cancer Rev. 560:1–65.

Nonner, W.F. and Loewenstein, W.R. 1989. *Appendix: a growth control model with discrete regulatory centers.* J. Cell Biol. 108:1063–1065.

A Ray of Hope

Rose, B., Mehta, P.P., and Loewenstein, W.R. 1992. *Gap-junction protein gene suppresses tumorigenicity.* Carcinogenesis 14:1073–1075.

13. NEURONAL COMMUNICATION, NEURONAL COMPUTATION, AND CONSCIOUS THOUGHT

Neuronal Communication and Computation

Barlow, H.B. 1972. *Single units and sensation: a neuronal doctrine for perceptual psychology?* Perception 1:371–394.

Barlow, H.B., Blakemore, C., and Watson-Smith, M. 1990. *Images and Understanding.* Cambridge University Press, Cambridge.

Chalmers, D.D. 1996. *The Conscious Mind.* Oxford University Press, New York.

Changeux, J.-P. 1985. *Neuronal Man: The Biology of Mind.* Pantheon, New York.

Churchland, P.J. and Sejnowski, T.J. 1992. *The Computational Brain: Models and Methods on the Frontiers of Neuroscience.* MIT Press, Cambridge, Mass.

Dowling J.E. 1987. *The Retina: An Approachable Part of the Brain.* Harvard University Press, Cambridge, Mass.

Edelman, G.M. 1987. *Neural Darwinism.* Basic Books, New York.

Furshpan, E.J. and Potter, D.D. 1959. *Transmission at the Giant Motor Synapses of the Crayfish.* J.Physiol. 145:289–325.

Hebb, D.D. 1964. *Organization of Behavior.* John Wiley & Sons, New York.

Hofstädter, D.R. and Dennett, D.C., eds. 1981. *The Mind's I.* Penguin Books, Harmondsworth, England.

Horgan, J. 1994. *Can Science Explain Consciousness?* Sci. Am. 271:88–95. [A report on a meeting held at the University of Arizona, Tucson, April 1994, presenting the multifaceted problem of consciousness, the current trends in the field and some of the personalities behind them.]

Hopfield, J.J. 1982. *Neural networks and physical systems with emergent collective computational abilities.* Proc. Natl. Acad. Sci. USA 79:2554–2558.

Hubel, D.H. and Wiesel, T. N. 1968. *Receptive fields and functional architecture of monkey striate cortex.* J. Physiol. 195:215–243.

Lehky, S.R. and Sejnowski, T.J. 1990. *Neural network model of visual cortex for determining surface curvature from images of shaded surfaces.* Proc. Roy. Soc. Lond. B 240:251–278

McCulloch, W.S. and Pitts, W.H. 1943. *A logical calculus of the idea immanent in neuron activity.* Bull. Math. Biophys. 5:115–133.

Mesulam, M.-M. 1990. *Large-scale neurocognitive networks and distributed processing for attention, language and memory.* Annu. Neurol. 528:597–613.

Moravec, H. 1988. *Mind Children: The Future of Robot and Human Intelligence.* Harvard University Press, Cambridge.

Nicholls, J.G., Martin, R. and Wallace, B.G. 1992. *From Neuron to Brain.* Sinauer Publishers, Sunderland, Mass.

Putnam, H. 1960. *Minds and machines. In:* Dimensions of Mind. S. Hook, ed. Reprinted in Minds and Machines, A.R. Anderson, ed., pp. 43–59, Prentice-Hall, 1964.

Ramón y Cajal S. 1909–11. *Histologie du système nerveux de l'homme et des vértebrés.* Maloine, Paris. (Reprinted, Consejo Superior de Investigaciones Científicas, Instituto Ramón y Cajal, Madrid, 1972.)

Rosenblatt, F. 1962. *Principles of Neurodynamics.* Spartan Books. New York.

Rumelhart, D.E. and McClelland, J.L. 1986. *Parallel Distributed Processing.* MIT Press, Cambridge, Mass.

Sejnowski, T.J. and Rosenberg, C.R. 1987. *Parallel Networks that Pronounce English Text. In:* Complex Systems 1:145–168.

Sherrington, Sir Charles. 1941. *Man on his Nature. In:* The Gifford Lectures, Edinburgh, 1937–1938. Macmillan, New York, and Cambridge University Press, New York.

Searle, J.R. 1992. *The Rediscovery of Mind.* MIT Press, Cambridge, Mass.

Thach, W.T., Goodkin, H.G. and Keating, J.G. 1992. *The cerebellum and the adaptive coordination of movement.* Ann. Rev. Neurosci. 15:403–412.

Turing, A.M. 1950. *Computing machinery and intelligence.* Mind 59, No. 236. Reprinted in *The Mind's I* (ed. D.R. Hofstäder, D.C. Dennet, eds. Penguin Books, Harmondswork, England, 1981).

Consciousness

Beck, F. and Eccles, J.C. 1992. *Quantum aspects of consciousness.* Proc. Nat. Acad. Sci. 89:11357–11361.

Crick, F. and Koch, C. 1990. *Towards a neurobiological theory of consciousness.* Seminars Neurosc. 2:263–275.

Dubois-Reymond, E. 1848–84. *Untersuchungen über tierische Elektrizität.* Reimer, Berlin.

Eccles, J.C. 1992. *Evolution of consciousness.* Proc. Nat. Acad. Sci. 89:7320–7324.

Edelman, G.M. 1992. *Bright Air, Brilliant Fire: On the Matter of The Mind.* Penguin Press, London.

Hark, E. 1982. *Windows on the Mind: Reflections on the Physical Basis of Consciousness.* Harvester, Brighton, Sussex.

Horgan, J. 1994. *Can Science Explain Consciousness?* Sci. Am. 271:88–95. (A report on a meeting held at the University of Arizona, Tucson, April 1994, presenting the multifaceted problem of consciousness, the current trends in the field, and some of the personalities behind them.)

Hubel, D.H. 1988. *Eye, Brain And Vision.* Freeman Press, San Francisco. Scientific American Library Series.

Llinás, R.R. and Ribary V. 1993. *Coherent 40 Hz oscillation characterizes dream state in humans.* Proc. Nat. Acad. Sci. 90:2978-2081.

Pribram, K.H. 1991. *Towards a holonomic theory of perception. In:* Gestalttheorie in der modernen Psychologie. S.E. Ertel, ed., pp. 161–184. Wegenroth, Köln.

Tennyson, Lord Alfred. 1842. *Break, Break, Break.*

Quantum Coherence

Del Guidice, E., Doglia, S. and Milani, M. 1983. *Self-focusing and ponderomotive forces of coherent electric waves—a mechanism for cytoskeleton formation and dynamics. In:* Coherent excitations in biological systems. H. Fröhlich and F. Kramer, eds., Springer-Verlag, Berlin.

DeWitt, B.S. 1970. *Quantum Mechanics and Reality.* Physics Today (September).

Fröhlich, H. 1968. *Long-range coherence and energy storage in biological systems.* Int. Quantum Chemistry II:641–649.

Fröhlich, H. 1984. *General theory of coherent excitations in biological systems.* In: Nonlinear Electrodynamics in Biological Systems. W.R Adey and A.F. Lawrence, eds. Plenum Press, New York.

Grundler, B. and Keilmann, F. 1983. *Sharp resonances in yeast growth proved by non-thermal sensitive to microwaves.* Phys. Rev. Letts. 51:1214–1216.

Hameroff, S.R. 1987. *Ultimate Computing: Biomolecular Consciousness and Nano-Technology.* Elsevier, North Holland, Amsterdam.

Hameroff, S.R., Rasmussen, S. and Manson, B. 1988. *Molecular Automata in microtubules: basic computational logic of the living state?* In: Artificial Life, Santa Fe Institute Studies in the Sciences of Complexity. C. Langton, ed. Addisson-Wesley, Reading Mass.

Hirokawa, N., Shiomura, Y. and Okabe, S. 1988. *Tau proteins: the molecular structure and mode of binding on microtubules.* J. Cell Biol. 107:1449–1461.

Leggett, A. 1996. *Low temperature physics, superconductivity and superfluidity. In:* The New Physics. P. Davies, ed. Cambridge University Press, Cambridge.

Lockwood, M. 1989. *Mind, Brain and Quantum: The Compound "I".* Blackwell Publ., Oxford.

Webb, S.J. and Booth, A.D. 1969. *Absorption of microwaves by microorganisms.* Nature 222: 1199–1200.

Quantum Computations?

Albert, D.Z. 1983. *On quantum mechanical automata.* Phys. Lett. 98A:249–252.

Benioff, P. 1982. *Quantum mechanical Hamiltonian models of Turing machines.* J. Stat. Phys. 29:515–546.

Bennett, C.H. 1988. *Notes on the history of reversible computation.* IBM J. Res. Dev. 32:1–6.

Bennett, C.H., Brassard, G., Breidbard, S. and Weisner, S. 1983. *Quantum cryptography, or unforgettable subway tokens.* In: Advances in Cryptography. Plenum, New York.

Cirac, J.I. and Zoller P. 1995. *Quantum computations with cold trapped ions.* Phys. Rev. Lett. 74:4091–4094

Deutsch, D. 1985. *Quantum theory, the Church-Turing Principle and the universal quantum computer.* Proc. Roy. Soc. (Lond.) A400: 97–117.

Deutsch, D. 1992. *Quantum computation.* Phys. World 5:57–61.

Feynman, R.P. 1982. *Simulating physics with computers.* Int. J. Theor. Phys. 21:467–488.

Feynman, R.P. 1986. *Quantum Mechanical Computers. In:* Foundations of Physics.

Landauer, R. 1996. *Minimal Energy Requirements in Communication.* Science 272: 1914–1918.

Lloyd, S. 1993. *A potentially realizable quantum computer.* Science 261:1569–1571.

Lucas, J.R. 1961. *Minds, Machines, and Gödel* Philosophy 36:120–124.

Penrose, R. 1993. *An emperor still without mind.* Behavioral and Brain Sciences 16:616–622.

Penrose, R. 1994. *Shadows of the Mind. A Search for the Missing Science of Consciousness.* Oxford University Press, New York.

Shimony, A. 1996. *Conceptual foundations of quantum mechanics. In:* The New Physics. P. Davies, ed. Cambridge University Press, Cambridge.

Shor P.W. 1994. *Algorithms for quantum computation: discrete logarithms and factoring.* In: 35th Ann Symp. On Foundations of Computer Science, S. Goldwasser, ed. IEEE Computer Soc Press 1994.

Zurek, W.H. 1984. *Reversibility and Stability of Information Processing Systems.* Phys. Rev. Lett. 53:391–394.

Two Worlds Unbridged

Adrian, Lord Edgar. 1946. *The Physical Background of Perception.* Clarendon Press, Oxford.

Home, D. and Nair, R. 1994. *Wave function collapse as a nonlocal quantum effect.* Phys. Lett. A187:224–226.

Schrödinger, E. 1935. *Die gengenwärtige Situation in der Quantenmechanik.* Naturwiss. 23:807–812; 844–849.

Sherrington, C.S. 1933. *The brain and its mechanisms. In:* The Rede Lecture. Macmillan, New York and Cambridge University Press, New York.

Wigner, E.P. 1961. *Remarks on the mind-body question. In:* The Scientist Speculates. I.J. Good, ed. Heinemann, London.

14. HOW CAN YOU EXPLAIN SO WISE AN OLD BIRD IN A FEW WORDS?

Delbrück, M. 1949. *A physicist looks at biology.* Trans. Conn. Acad. Arts & Sci. 38:173–190.

The Arduous Minuet

Darwin, C. 1859. *The Origin of the Species,* Chp. IV; p. 34, Humbolt, New York.

Mendel, J.G. 1866. *Versuche über Pflanzenhybriden.* Hafner Pub. Co., New York.

Schleiden, M.J. 1838. *Beiträge zur Phytogenesis.* Müller's Arch. Anat. Physiol. Wiss. Medic. 1838:137–176.

Watson, J.D. and Crick, F.H.C. 1953. *Genetical implications of the structure of deoxyribonucleic acid.* Nature 171:964–967.

The Limits of Scientific Knowledge

Baylor, D.A. 1987. *Photoreceptor signals and vision. In*: Proctor Lecture. Invest. of Ophthamalmol. Vis. Sci. 28:34–49.

Bohr, N. 1928. *The quantum postulate and the recent development of atomic theory.* Nature (Suppl.): 580–590.

Bohr, N. 1933. *Light and life.* Nature 131: 421–423, 457–459.

Bohr, N. 1935. *Can quantum-mechanical description of physical reality be considered complete?* Physical Rev. 48:696–702.

Bohr, N. 1970. *in Albert Einstein: Philosopher-Scientist.* P.A. Schilpp, ed. p. 199. Tudor, New York.

Bohr, N. 1959. *Quantum Physics and Biology.* Symposium on Models in Biology, Bristol. Reproduced in Light and Life, W.D. McElroy and B. Glass, eds. John Hopkins University Press, 1961.

Bohr, N. 1963. *Light and life revisited. In*: Essays 1958–1962 in Atomic Physics and Human Knowledge. Interscience, New York.

Bohr, N. 1970. *Discussion with Einstein on epistemological problems in atomic physics and Einstein's reply. In*: Albert Einstein: Philosopher-Scientist. P.A. Schilpp, ed., pp. 199–242 and 666. Cambridge University Press, Cambridge.

Feynman, R.P. Leighton, R.B. and Sands, M. 1963. *The Feynman Lectures on Physics.* Addison-Wesley, Reading, Mass.

Hecht, S., Shlaer, S. and Pirenne, M.H. 1941. *Energy, quanta and vision.* J. General Physiol. 25:891–940.

Heisenberg, W. 1948. *Prinzipielle Fragen der modernen Physik.* 8th Edition. Hirzel Verlag, Leipzig.

Heisenberg, W. 1958. *The Physicist's Conception of Nature.* Greenwood Press, Westport, Conn.

Weinberg, S. 1992. *Dreams of a Final Theory.* Pantheon, New York.

APPENDIX: A BRIEF CHRONICLE OF "INFORMATION"

The Form-Giving Principle

Aristotle. *De partibus animalium.*

Aristotle. *De generatione animalium, I.*

Aristotle. *De generatione animalium.*

Delbrück, M. 1971. *Aristotle-totle-totle. In*: Of Microbes and Life. J. Monod and E. Borek, eds., pp. 50–55. Columbia University Press, New York.

Düring, I. 1966. *Aristoteles. Darstellung und Interpretation Seines Denken.*, Carl Winter Universität, Heidelberg.

The Tribulations of the Two Great Laws

Carnot, S. 1824. *Réflexions sur la Puissance Motrice du Feu et sur les Machines Propres a Déveloper cette Puissance.* Chez Bachelier, Libraire, Paris. Republished by Dover, New York. Translated by R.H. Thurston.

Clausius, R. 1850. *On the motive power of heat, and on the laws which can be deduced from it for the theory of heat. In*: Poggendorff's Annalen der Physik, LXXIX, 368, 500. Translated by W.F. Magie.

Mayer, J.R. 1842. *Bemerkungen über die Kräfte der unbelebten Natur.* Liebig's Annalen der Chemie. May.

Mayer, J.R. 1845. *Die Organielle Bewegung in ihrem Zusammenhang mit dem Stoffwechsel.* Paper published privately at Heilbronn (Mayer's home town).

The Ripening of an Afterthought

Angrist, S.W. and Hepler, L.G. 1967. *Order and Chaos.* Basic Books, New York.

Boltzmann, L. 1909. *Wissenschaftliche Abhandlungen.* F. Hasenöhrl., ed. Leipzig. (His paper, in which the relation between entropy and probability was developed, appeared originally in 1877.)

Hartley, R.V.L. 1928. *Transmission of information.* Bell System Tech. J. 7:535–563.

Rothstein, J. 1951. *Information, measurement and quantum mechanics.* Science 114:171–175.

Shannon, C.E. 1948. *A mathematical theory of information.* Bell System Tech. J. 27: 279–423; 27:623–656.

Szilard, L. 1929. *Z. Physik.* 53:840.

Wiener, N. 1959. *Cybernetics.* John Wiley & Sons, New York.

Index

364   *Index*